Basic Electrical Installation Work

Basic Electrical Installation Work

Third Edition

TREVOR LINSLEY
Senior Lecturer
Blackpool and The Fylde College

AMSTERDAM • BOSTON • HEIDELBERG • LONDON • NEW YORK • OXFORD
PARIS • SAN DIEGO • SAN FRANCISCO • SINGAPORE • SYDNEY • TOKYO
Newnes is an imprint of Elsevier

Newnes
An imprint of Elsevier
Linacre House, Jordan Hill, Oxford OX2 8DP
30 Corporate Drive, Burlington, MA 01803

First published by Arnold 1998
Reprinted by Butterworth-Heinemann 2001, 2002, 2003 (twice), 2004 (twice)

British Library Cataloguing in Publication Data
A catalogue record for this book is available from the British Library

ISBN 0 340 705744

Printed and bound in Great Britain by Biddles Ltd, King's Lynn, Norfolk

CONTENTS

PREFACE

The third edition of *Basic Electrical Installation Work* has been written as a complete textbook for the City and Guilds 2360 Certificate in Electrical Installation Theory and Practice and the City and Guilds 823 Electrical Installation Practice Trade Studies syllabus. The book meets the combined requirements of the installation theory and practice parts of the 2360 syllabus and therefore students need purchase only this one textbook for all subjects in the Part 1 examinations.

The book will also assist students taking the SCOTVEC and BTEC Electrical and Utilization units at levels I and II and many taking engineering NVQ craft courses.

Although the text is based upon the City and Guilds syllabus, the book also provides a sound basic knowledge and comprehensive guide for other professionals in the construction industry and electrical trades.

Modern regulations place a greater responsibility upon the installing electrician for safety and the design of an installation. The latest regulations governing electrical installations are the 16th edition of the IEE Wiring Regulations (BS 7671: 1992). The third edition of this book has been revised and updated to incorporate the requirements and 1997 amendments of the 16th edition of the IEE Wiring Regulations.

The Part 1 examination in Electrical Installation Theory and Practice comprises a single two-hour multiple-choice written paper. For this reason multiple-choice questions can be found at the end of each chapter. More traditional questions are included as an aid to private study and to encourage a thorough knowledge of the subject.

I would like to acknowledge the assistance given by the following manufacturers and organizations in the preparation of this book:

Crabtree Electrical Industries Limited
Wylex Ltd
RS Components Ltd
The Institution of Electrical Engineers
The British Standards Institution
The City & Guilds of London Institute

I would also like to thank my colleagues and students at Blackpool and The Fylde College for their suggestions and assistance during the preparation of the third edition of this book.

Finally, I would like to thank Joyce, Samantha and Victoria for their support and encouragement.

Trevor Linsley,
Poulton-le-Fylde 1997

To Joyce, Samantha and Victoria

1

ELECTRICITY AND THE ENVIRONMENT

Environmental considerations

The electricity-generating companies are required by Parliament to develop and maintain an efficient and economical supply of electricity for all parts of the United Kingdom and to take account of the effects which proposals might have upon the natural beauty of the countryside and buildings or objects of special interest. Efficiency does not always go hand in hand with the preservation of natural beauty, but the generating companies do take their responsibilities very seriously.

Modern power stations are very large, both physically and electrically (2000 MW), and cannot be hidden from view. But, by careful siting, careful selection of building materials, suitable ground modelling and tree planting, acceptable results can be obtained.

CONVENTIONAL POWER STATIONS

To keep the modern coal-fired power stations operating requires about 100 million tonnes of coal each year. Therefore, new coal-fired and oil-fired stations are built close to fuel sources in order to reduce transport costs and so that heavy coal-train traffic does not impose upon passenger railway routes.

NUCLEAR POWER STATIONS

One of the advantages of nuclear power stations is that they involve no difficult fuel transport problems. One modern power station fuel element has the energy equivalent of 1800 tonnes of coal. A reactor contains about 25 000 fuel elements, and these are all changed in small batches over a period of 4 or 5 years.

Nuclear power plants are built to the highest safety standards and during the last 30 years have gained a great deal of practical operating experience and as a result maintain a very high safety record. There are 14 operational nuclear power stations in England and Wales, including the new nuclear stations recently built at Heysham, Teeside and Hartlepool.

COOLING WATER

Excluding fuel, the other essential requirement of all power stations is an adequate supply of cooling water. A 2000 MW station requires about 225 million litres of water per hour. Such quantities of water can only be obtained from the sea or major river estuaries. Inland power stations use water recirculated through cooling towers.

WATER DISCHARGE

Cooling towers emit plumes of visible water vapour through evaporation. Although unattractive, the plume does not pollute the atmosphere and quickly evaporates.

The cooling water discharged into the sea or river from a large power station may have been raised in temperature by 10°C. Extensive ecological research has confirmed that there are no adverse effects from warm water discharge and the evidence would suggest that fish can benefit from the warmed water.

CHIMNEY DISCHARGE

The burning of any kind of fuel produces pollutants, but modern power stations burn coal and oil in the least harmful way due to highly efficient boilers. Power stations are not sole polluters since industrial chimneys emit carbon dioxide (CO_2) and sulphur dioxide (SO_2) gases. The most controversial effect of the sulphur dioxide gas is the turning of rainfall over industrial areas into dilute sulphuric acid which attacks the stonework on buildings and inhibits the growth of plants. This has encouraged power stations to build tall chimneys to discharge the gases high into the atmosphere and so avoid harmful concentrations at ground level. However, unfortunately this does not provide a simple solution since the Norwegian and Swedish authorities, who are downwind of the UK, are now complaining that acid rain falling on their territory originates from UK power stations.

ASH DISPOSAL

The coal used in power station boilers is pulverized into very fine particles before burning. Consequently, the ash is a very fine powder which is removed from the flue gases by grit arrestors. The generating board's marketing officers have developed many uses for waste ash including the filling and formation of roads, motorways and airfield runways, the manufacture of building blocks and bricks and the reclamation of wasteland at reasonable costs.

RADIOACTIVE WASTE DISPOSAL

As the uranium fuel elements are 'burned' in the reactor they become less effective and are therefore replaced.

The spent fuel elements are transported in shielded containers to the reprocessing plant of the Windscale Works at Sellafield in Cumbria, where they are stripped of their cans and the radioactive products prepared for safe storage.

Radioactivity is the process by which an unstable atom emits high-energy particles until it becomes stable. The rate at which radioactive nuclei decay is characterized by their 'half-life', that is, the time taken for half the nuclei in a given sample to decay. For example, an unstable variety of xenon 138 has a half-life of 17 minutes, while plutonium 239 has a half-life of 24 000 years.

Short-lived radioactive waste is stored under water until it is safe for disposal like normal industrial waste. Some low-activity waste is encased in concrete and dumped deep in the ocean under strict international supervision. The long-term highly active waste products are stored at the Windscale Works as a concentrated liquid in stainless steel vessels surrounded by concrete shielding. A dozen such vessels, approximately equal to the volume of two average-sized houses, contain all the highly active waste produced by the UK since the 1950s, but storage as a non-reactive solid would be preferable in the long term. A process is being developed for converting the liquid waste into glass blocks which, when they have been cased in stainless steel containers, will be disposed of away from the human environment, probably by burying them deep in stable rock formations on land or under the deep ocean bed. It has been estimated that after about 1000 years the radioactivity of the glassified waste will have fallen to a safe level.

Power generation

POWER GENERATION IN ENGLAND AND WALES

The generating companies National Power, PowerGen and Nuclear Electric supply more than 200 000 000 MWh of electricity, meeting a peak maximum demand of about 45 000 MW from a total installed capacity of approximately 50 000 MW. About 85% of this energy is supplied from fossil fuel stations, about 14% from nuclear stations, and less than 1% from hydroelectric installations.

National Power and PowerGen are heavily dependent upon fossil fuels (coal, oil and gas), but we know from respectable scientific sources that the fossil fuel era is drawing to a close. Popular estimates suggest that gas and oil will reach peak production in about the year 2060, while British coal reserves will last until 2200 at the present rate of consumption. The generating companies must give consideration to other ways of generating electricity so that coal, oil and gas might be conserved in the next century.

POWER GENERATION IN SCOTLAND

The generation of electricity in Scotland is the joint responsibility of the Scottish Power and Hydroelectric companies. They supply about 22 000 000 MWh of electricity, meeting a peak maximum demand of about 1600 MW. This electricity is generated by nuclear and fossil fuel power stations, 59 hydroelectric power stations and two pumped water storage schemes, while the Western Isles and Shetland are supplied by diesel power generators. The smaller islands are connected to the mainland by submarine cable and the whole network consists of 8800 km of 132 kV and 270 kV transmission cables.

The mountainous regions and relative isolation of many consumers in Scotland have given the electricity boards an ideal opportunity to pioneer many renewable energy schemes. In addition to the very successful hydroelectric schemes, Orkney has successfully operated an experimental 250 kW and 300 kW wind generator, and a 3 MW wind aerogenerator was connected to the board's system in 1987 (see Fig. 1.1).

POWER GENERATION IN NORTHERN IRELAND

The generation of electricity in Northern Ireland is the responsibility of the Northern Ireland Electricity Service. This supplies a little over 5 000 000 MWh of electricity, meeting a peak demand of about 1300 MW from an installed capacity of 2050 MW. This electricity is generated by four fossil fuel stations and transmitted over 1250 km of 275 kV and 110 kV lines and cables to approximately 1.2 million customers.

The power stations operated by the Northern Ireland Electricity Service are heavily dependent upon oil and at a time when oil prices are uncertain there is a desire to have alternative fuel sources available for the long-term planning and stability of the industry. To this end the Kilroot oil-fired power station is being converted to burn either oil or coal, whichever is the

Fig. 1.1 A 3 MW wind generator, Burghar Hill, Orkney (by kind permission of the North of Scotland Hydroelectric Board).

cheaper fuel. This will reduce the dependence on oil by the Northern Ireland Electricity Service from 90% to 70% towards the end of this century.

HYDROELECTRIC POWER

Water power makes a useful contribution to the energy needs of Scotland, but the possibilities of building similar stations in England are very limited since there are not the high mountains or vast reserves of water that make these schemes attractive. However, hydroelectric power stations have many advantages over fossil fuel and nuclear stations. They can be brought up to full load very quickly and are non-polluting. Water is a so-called 'renewable' energy source and the low running costs may make them more attractive as the cost of other forms of energy increases.

WIND POWER

Modern wind machines will be very different from the traditional windmill of the last century which gave no more power than a car engine. Very large structures are needed to extract worthwhile amounts of energy from the wind. Modern wind generators are about as tall as electricity pylons, with a three-blade aeroplane-type propellor to catch the wind and turn the generator. They are usually sited together in groups of about 20 generators in what are known as 'wind energy farms'.

Each modern wind turbine generates about 600 kW of electricity; a wind energy farm of 20 generators will, therefore, generate 12 MW – a useful contribution to the National Grid, using a naturally occurring, renewable, non-polluting primary source of energy. The Department of Energy considers wind energy to be the most promising of the renewable energy sources.

The Countryside Commission, the government's adviser on land use, has calculated that to achieve a target of generating 10% of the total electricity supply by wind power by the year 2025 will require 40 000 generators of the present size. At the time of writing (1997) we have about 650 operational wind generators, mostly in Wales and Cornwall, but this is a promising start to achieving this target.

Wind power is an endless renewable source of energy, is safer than nuclear power and provides none of the polluting emissions associated with fossil fuel. However, wind farms are, by necessity, sited in some of the wildest and most beautiful landscapes in the UK, such as the west coast of Scotland, the north Pennines, the Lake District, Wales and Cornwall. Siting wind energy farms in these areas of outstanding natural beauty has outraged conservationists. The Ramblers' Association, the Council for National Parks and the Council for the Protection of Rural England, together with its sister bodies in Scotland and Wales, have joined forces to call for tougher government controls on the location of wind energy farms.

The Department of Energy has calculated that 10 000 wind machines could provide the energy equivalent of 8 million tonnes of coal per year. While this is a worthwhile saving of fossil fuel, opponents point out the obvious disadvantages of wind machines, among them the need to maintain the energy supply during periods of calm, which means that wind machines can only ever supplement a more conventional electricity supply.

HARNESSING THE SEA

Great Britain is a small island surrounded by water. Surely we could harness some of the energy contained in the tides or waves?

In March 1985 the British government committed £220 000 to studies of a barrage across the Severn Estuary as a source of tidal power and a new traffic route to relieve the Severn Bridge. The Severn Estuary has a tidal range of up to 15 m – the largest in Europe – and a reasonable shape for building a barrage or dam across it. This would allow the basin to fill with water as the tide rose, and then allow the impounded water to flow out through electricity-generating turbines as the tide fell.

A scheme proposed by the team that designed and built the Thames barrier suggests that a 14 km barrage could supply as much energy as four or five nuclear power stations at two-thirds the cost per kilowatt. The impounded water would also provide a water sports area suitable for boating, water-skiing, swimming and angling.

Unfortunately such a scheme would produce power geared to the lunar cycle of the tides and not to the demands of industry and commerce. The tidal barrier might have disastrous ecological consequences upon numerous wildfowl and wading bird species by submerging the mud-flats which now provide winter shelter for these birds. Therefore the value of the power produced would need to be balanced against the possible consequences.

Professor Salter of Edinburgh University has done a great deal of work on harnessing the energy contained in waves. He has designed a cam-shaped device which bobs up and down on the waves. The wave motion acting upon the device produces electricity which is fed to shore by submarine cables. The generated electrical energy is of a random frequency and therefore cannot be connected directly to existing distribution systems, but Professor Salter's research would suggest that about 1000 km of the devices, which he has called 'ducks', could meet about half the present electrical energy requirements.

Britain has no commercially produced wave-power electricity but at Bergen in Norway a wave-energy power station came into operation in November 1985, and France has successfully operated a 240 MW tidal power station at Rance in Brittany for the past 25 years.

SOLAR ENERGY

On a bright sunny day the sun's radiation represents a power input of about 1 kW per square metre. A 2000 MW power station occupies about 1.5 square miles, that is 4 million square metres. Therefore about 4000 MW of solar radiation falls on to the land occupied by the power station. If we could cover the same area with solar cells having an efficiency of 50% we could generate the same quantity of electricity without burning any fuel. Unfortunately, mass-produced solar cells have a conversion efficiency of about 12% and therefore to actually produce 2000 MW would require about 7 square miles of solar cells, which is quite unacceptable on a small island like Great Britain.

While there seems to be a possibility of using solar power directly, the efficiency of solar cells must be greatly improved before a commercial solar station becomes a possibility. There is also the problem of maintaining the supply through the night and during dull days in the winter, when demand is higher than in summer.

Electricity today

Our future energy requirements and resources are difficult to predict but we do need independence from fossil fuels which, at the present rate of consumption, are predicted to run out in the first quarter of the next millennium. It is therefore realistic to plan future systems which are flexible, have large safety margins, and give consideration to the renewable sources.

The generating companies will continue to meet our demand for electricity with fossil fuel and nuclear stations despite the worries caused by the Chernobyl power station disaster in April 1986. Research will continue into the renewable energy sources such as wind, wave, tidal and solar, but most of these alternative energy sources are in a very early stage of development and consequently the immediate future will undoubtedly see us remain dependent upon nuclear energy and fossil fuels.

Britain has large coal reserves, but these will be increasingly required for many uses other than electricity generation – to make fertilizers and chemicals, to produce oil substitutes and to manufacture town gas as natural gas becomes scarce.

Energy from fossil fuels made the industrial revolution possible and has enabled science and technology to reach a level of development at which the exploitation of other energy sources is within our capability. Electricity is now well established as a means of harnessing energy and bringing it to the service of man. Electricity will continue into the twenty-first century because of its many advantages: it is easy to make, distribute, and control, and, when properly installed, is safe for anyone to use.

Using the laws discovered by Michael Faraday in 1831 (the laws of electromagnetic induction described in Chapter 8), electricity may be generated in commercial quantities and distributed to its end use simply by connecting supply and consumer together with suitable cables. The supply may be controlled manually with a switch or automatically with a thermostat or circuit breaker. The wheels of industry are driven by electric motors, and sophisticated artificial heating and lighting installations make it possible for industry and commerce to work long hours in safety and comfort. Modern homes use microwave cookers to prepare meals quickly, freezers to store food safely and conveniently, and electric cleaners to speed household cleaning chores. In our leisure time we watch television and listen to music on stereo hi-fi music centres. While home computers are used to run simple programs and play electronic games, business and commerce use large-capacity computers to store data, word-process and 'number-crunch'. All the high

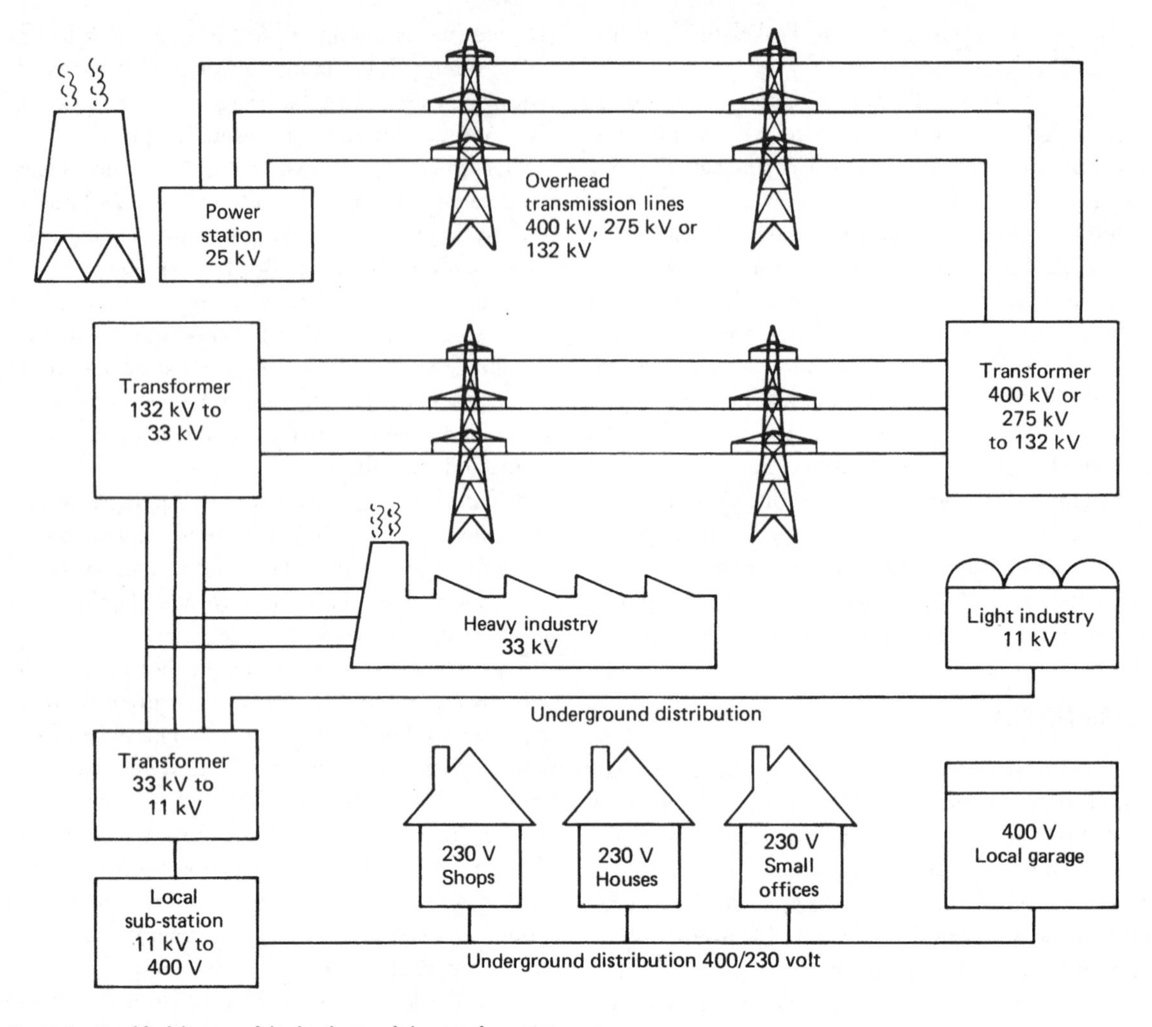

Fig. 1.2 Simplified diagram of the distribution of electricity from power station to consumer.

technology which we today take for granted is dependent upon a reliable and secure electricity supply.

Distribution of electricity

Electricity is generated in modern power stations at 25 kV and fed through transformers to the consumer over a complex network of cables known as the National Grid system. This is a network of cables, mostly at a very high voltage, suspended from transmission towers, linking together the 175 power stations and millions of consumers. There are approximately 5000 miles of high-voltage transmission lines in England and Wales, running mostly through the countryside.

Man-made structures erected in rural areas often give rise to concern, but every effort is made to route the overhead lines away from areas where they might spoil a fine view. There is full consultation with local authorities and interested parties as to the route which lines will take. Farmers are paid a small fee for having transmission line towers on their land. Over the years many different tower designs and colours have been tried, but for the conditions in the United Kingdom, galvanized steel lattice towers are considered the least conspicuous and most efficient.

For those who consider transmission towers unsightly, the obvious suggestion might be to run all cables underground. In areas of exceptional beauty this is done, but underground cables are about 16 times more expensive than the equivalent overhead lines. The cost of running the largest lines underground is about £3 million per mile compared with about £200 000 overhead. On long transmission lines the losses can be high, but by raising the operating voltage and therefore reducing the current for a given power, the I^2R losses are reduced, the cable diameter is reduced and the overall efficiency of transmission is increased. In order to standardize equipment, standard voltages are used. These are:

- 400 kV and 275 kV for the Super Grid;
- 132 kV for the original Grid;
- 66 kV and 33 kV for secondary transmission;
- 11 kV for high-voltage distribution;
- 400 V for commercial consumer supplies;
- 230 V for domestic consumer supplies.

A diagrammatic representation showing the distribution of power is given in Fig. 1.2.

All local distribution in the UK is by underground cables from substations placed close to the load centre and supplied at 11 kV. Transformers in these local substations reduce the voltage to 400 V, three-phase and neutral distributor cables connect this supply to consumers. Connecting to one-phase and neutral of a three-phase 400 V supply gives a 230 V single-phase supply suitable for domestic consumers.

When single-phase loads are supplied from a three-phase supply, as shown in Fig. 1.3, the load should be 'balanced' across the phases. That is, the load should be equally distributed across the three phases so that each phase carries approximately the same current. This prevents any one phase being overloaded.

CONSUMER'S MAINS EQUIPMENT

The consumer's mains equipment is normally fixed close to the point at which the supply cable enters the building. To meet the requirements of the IEE Regulations it must provide:

- protection against electric shock (Chapter 41);
- protection against overcurrent (Chapter 43);
- isolation and switching (Chapter 46).

Protection against electric shock is provided by insulating and placing live parts out of reach in suitable enclosures, earthing and bonding metal work and providing fuses or circuit breakers so that the supply is automatically disconnected under fault conditions.

To provide overcurrent protection it is necessary to provide a device which will disconnect the supply automatically before the overload current can cause a rise in temperature which would damage the

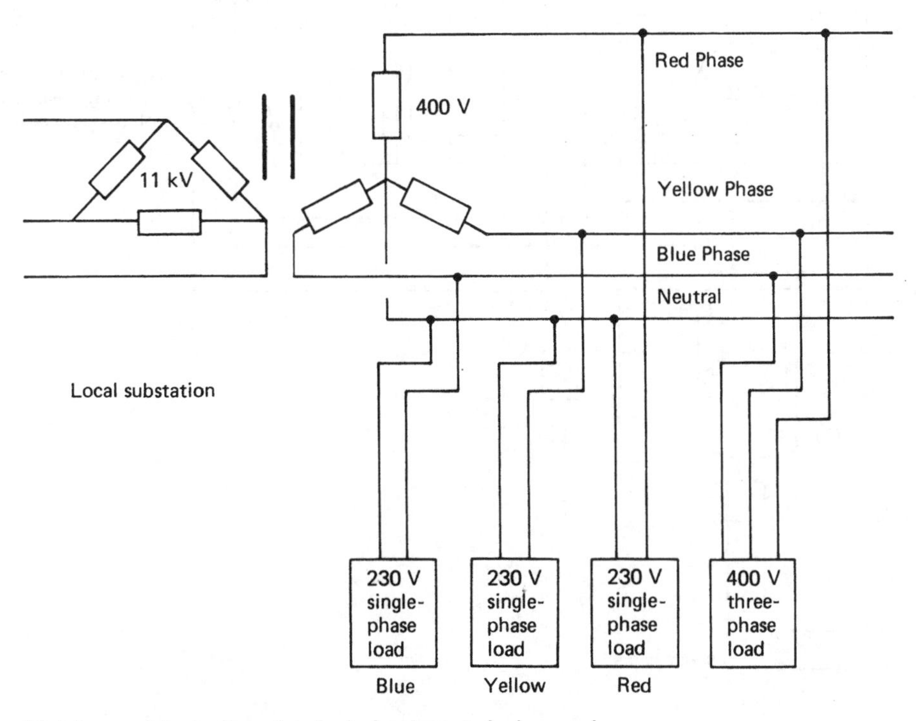

Fig. 1.3 Simplified diagram of the distribution from local substation to single-phase supply.

installation. A fuse or miniature circuit breaker (MCB) would meet this requirement.

An isolator is a mechanical device which is operated manually and is provided so that the whole of the installation, one circuit or one piece of equipment may be cut off from the live supply. In addition, a means of switching off for maintenance or emergency switching must be provided. A switch may provide the means of isolation, but an isolator differs from a switch in that it is intended to be opened when the circuit concerned is not carrying current. Its purpose is to ensure the safety of those working on the circuit by making dead those parts which are live in normal service. One device may provide both isolation and switching provided that the characteristics of the device meet the Regulations for both functions. The switching of electrically operated equipment in normal service is referred to as functional switching.

Circuits are controlled by switchgear which is assembled so that the circuit may be operated safely under normal conditions, isolated automatically under fault conditions, or isolated manually for safe maintenance. These requirements are met by good workmanship and the installation of proper materials such as switches, isolators, fuses or circuit breakers. The equipment belonging to the supply authority is sealed to prevent unauthorized entry, because if connection were made to the supply before the meter, the energy used by the consumer would not be recorded on the meter. Figures 1.4 and 1.5 show connections and equipment for different installation situations.

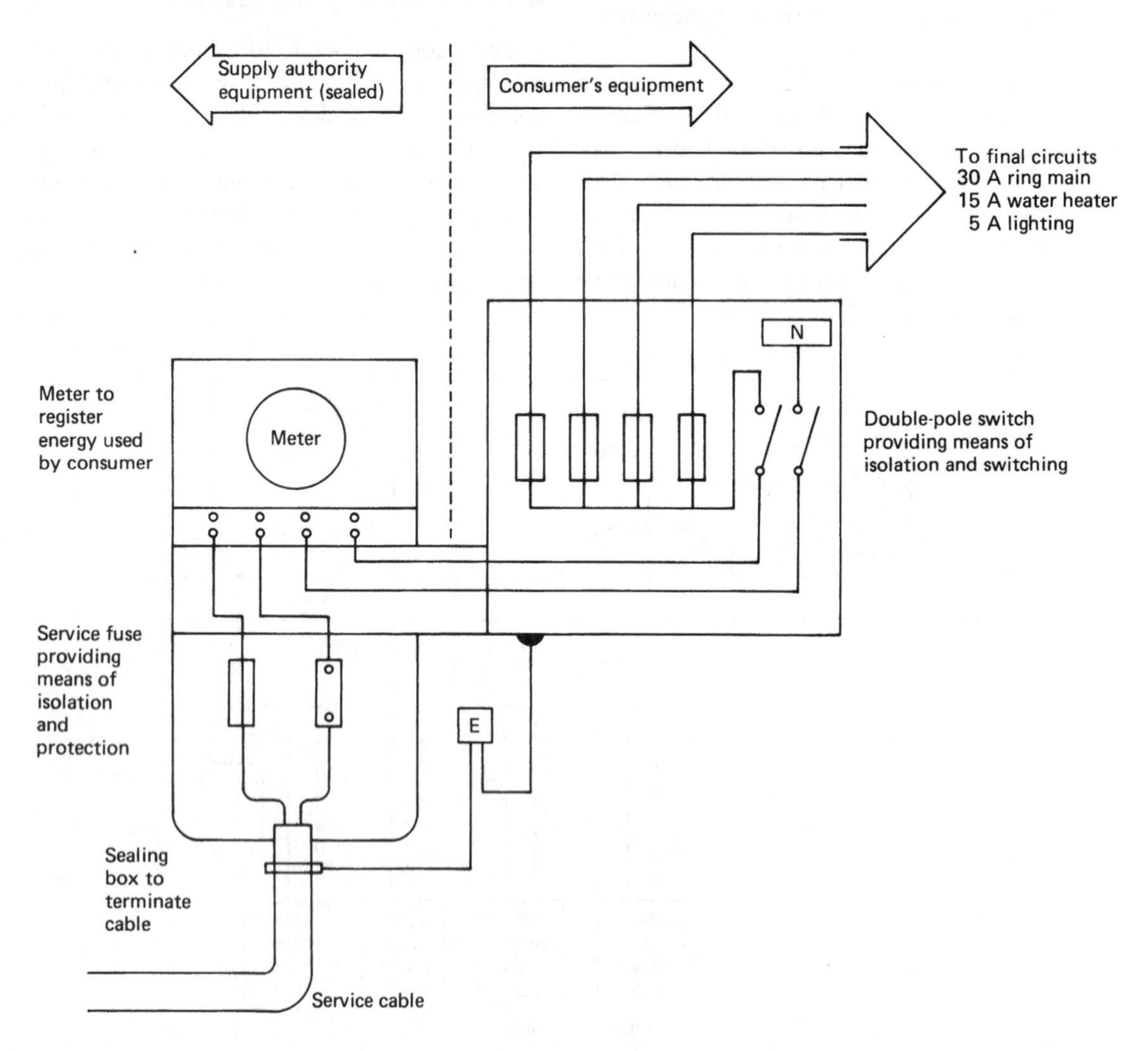

Fig. 1.4 Simplified diagram of connections and equipment at a domestic service position.

GRADED PROTECTION DEVICES FOR EFFECTIVE DISCRIMINATION

The main fuse at the consumer's service position protects the incoming cable from short circuits at the mains position and against overload and short circuits in the final circuits. However, in a properly designed installation there will be other protective devices between the main fuse and the consumer's equipment because of the subdivision of the installation, and it is these which should operate before the main fuse. The current rating of the fuses or circuit breakers should be so graded that when a fault occurs only the device nearest to the fault will operate. The other devices should not operate, so that the supply to healthy circuits is not impaired.

If a fault occurs in the appliance shown in Fig. 1.6, only the plug top fuse should blow if the circuit fuses have been correctly graded. Similarly, if a fault occurs on the cable feeding the socket circuit, only that fuse in the fuse board should blow leaving the lights, cooker and bell operating normally.

ELECTRICAL BONDING TO EARTH

The purpose of the bonding regulations is to keep all the exposed metalwork of an installation at the same earth potential as the metalwork of the electrical installation, so that no currents can flow and cause an electric shock. For a current to flow there must be a difference of potential between two points, but if the points are joined together there can be no potential difference. This bonding or linking together of the exposed metal parts of an installation is known as 'equipotential bonding'.

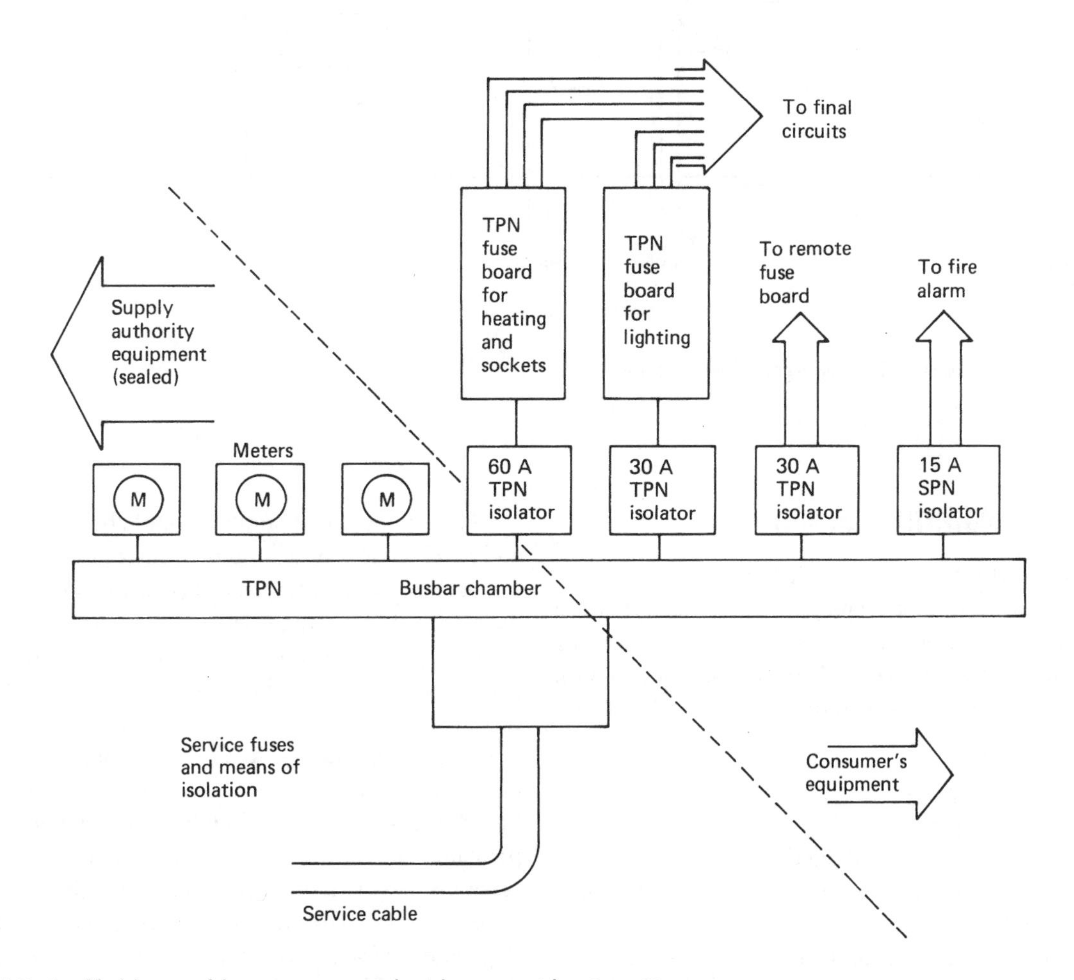

Fig. 1.5 Simplified diagram of the equipment at an industrial or commercial service position.

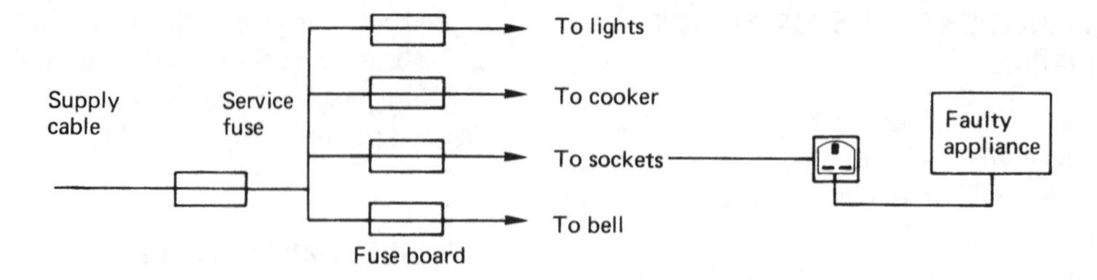

Fig. 1.6 Subdivision of circuits in a domestic position.

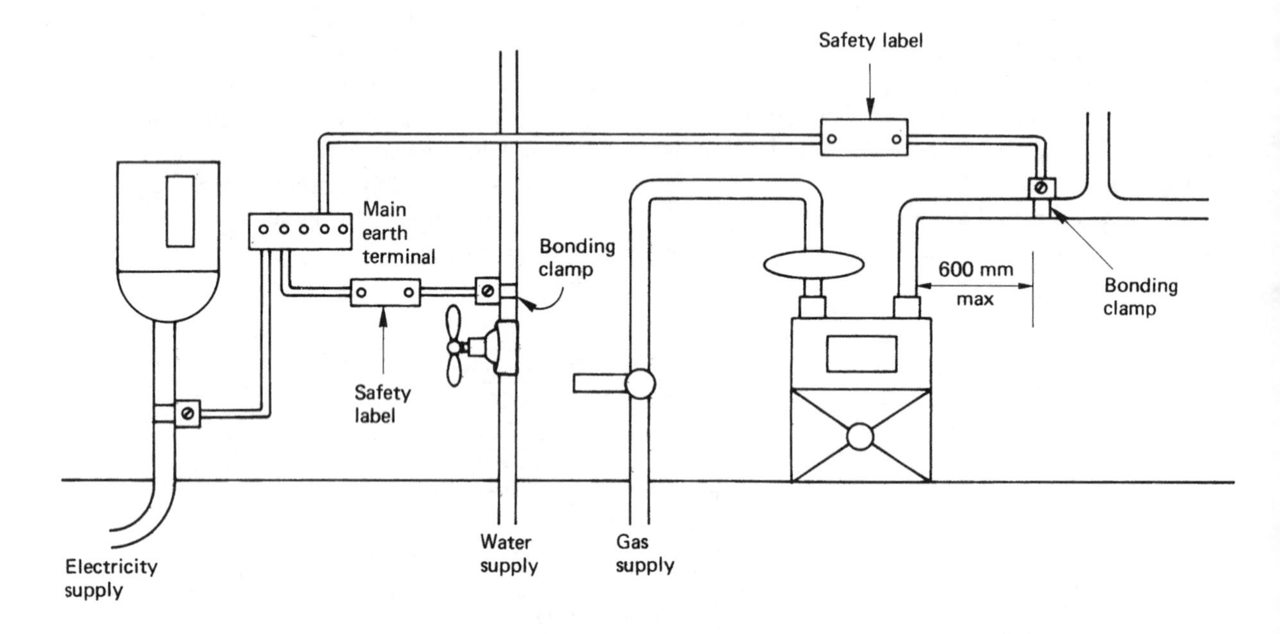

Fig. 1.7 Main equipotential bonding of gas and water supplies.

MAIN EQUIPOTENTIAL BONDING

Where earthed electrical equipment may come into contact with the metalwork of other services, they too must be effectively connected to the main earthing terminal of the installation (IEE Regulation 130–04).

Other services are described by IEE Regulation 413–02–02 as:

- main water pipes;
- main gas pipes;
- other service pipes and ducting;
- risers of central heating and air conditioning systems;
- exposed metal parts of the building structure;
- lightning protective conductors.

Main equipotential bonding should be made to gas and water services at their point of entry into the building, as shown in Fig. 1.7, using insulated bonding conductors of not less than half the cross-section of the incoming main earthing conductor. The minimum permitted size is 6 mm^2 but the cross-section need not exceed 25 mm^2 (IEE Regulation 547–02–01). The bonding clamp must be fitted on the consumer's side of the gas meter between the outlet union, before any branch pipework but within 600 mm of the meter. A permanent label must also be fixed at or near the point of connection of the bonding conductor with the words 'Safety Electrical Connection – Do Not Remove' (IEE Regulation 514–13–01). Supplementary bonding is described in Chapter 6 of this volume.

Exercises

1 All modern power stations require two of the following for their efficient operation:
(a) an adequate supply of fuel
(b) access to a main rail link
(c) a large, highly skilled labour force
(d) lots of cooling water.

2 Which of the following energy sources may be classified as 'renewable energy'?
(a) coal
(b) oil
(c) nuclear
(d) hydro.

3 'Salter's ducks' produce electrical energy directly from:
(a) wind energy
(b) fossil fuel
(c) wave motion
(d) solar energy.

4 Electricity is generated in a modern power station at:
(a) 230 V
(b) 400 V
(c) 25 kV
(d) 132 kV.

5 Electricity is distributed on the National Grid at:
(a) 230 V
(b) 400 V
(c) 25 kV
(d) 132 kV.

6 The highest transmission line voltage in Britain is:
(a) 240 kV
(b) 415 kV
(c) 400 kV
(d) 1000 kV.

7 The voltage used for transmission on the Grid is transformed to a very high voltage because:
(a) this increases the line current
(b) the p.f. of the line is improved at high values
(c) the line resistance is increased
(d) the line efficiency is increased.

8 The national transmission network in the UK is known as:
(a) the CEGB
(b) the National Grid System
(c) the National Coal Board
(d) the British Transmission System.

9 The generation of electricity in England and Wales is the responsibility of:
(a) the generating companies
(b) regional electricity companies
(c) the National Coal Board
(d) Parliament.

10 The distribution of electricity to individual consumers is the responsibility of:
(a) the generating companies
(b) regional electricity companies
(c) the National Coal Board
(d) Parliament.

11 A 230 V single-phase supply may be obtained from a 400 V three-phase supply by connecting the load between:
(a) any line and neutral
(b) any line and live
(c) any two phases
(d) neutral and earth.

12 The phase to neutral voltage of a 400 V, three-phase, four-wire supply is:
(a) 230 V
(b) 400 V
(c) 11 kV
(d) 132 kV.

13 The distribution transformer is the main source of supply to a domestic dwelling. This transformer:
(a) generates the required electrical power
(b) is star-connected to provide a single-phase supply
(c) provides a d.c. supply
(d) steps up the voltage to 230 V.

14 Discriminative operation of the excess current protection device of an installation means that when a fault occurs:
(a) the supply is cut off from the installation
(b) the main protection device opens
(c) the faulty circuit only is disconnected from the supply
(d) the distribution fuse board only is disconnected from the supply.

15 Describe what is meant by acid rainfall, and how it causes problems. Who is claimed to be responsible for creating acid rainfall, and what steps are being taken to eliminate it?

16 Describe how waste products from nuclear power

stations are transported for reprocessing. How are long-term highly active waste products stored safely?

17 Describe the main advantages of generating electricity by nuclear power stations.

18 Describe how electricity may be generated by one renewable energy source. State the advantages and disadvantages over present methods of generating commercial quantities of electricity.

19 State five advantages of using electricity as a means of harnessing energy.

20 Describe how electricity is distributed from the power station to a domestic consumer's terminals. State the voltages used at each stage and describe the advantages and disadvantages of overhead and underground cables.

2

THE ELECTRICAL CONTRACTOR

The construction industry

An electrician working for an electrical contracting company works as a part of the broader construction industry. This is a multi-million-pound industry carrying out all types of building work, from basic housing to hotels, factories, schools, shops, offices and airports. The construction industry is one of the UK's biggest employers, and carries out contracts to the value of about 10% of the UK's gross national product.

Although a major employer, the construction industry is also very fragmented. Firms vary widely in size, from the local builder employing two or three people to the big national companies employing thousands. Of the total workforce of the construction industry, 92% are employed in small firms of less than 25 people.

The yearly turnover of the construction industry is about £35 billion. Of this total sum, about 60% is spent on new building projects and the remaining 40% on maintenance, renovation or restoration of mostly housing.

In all these various construction projects the electrical contractor plays an important role, supplying essential electrical services to meet the needs of those who will use the completed building.

The building team

The construction of a new building is a complex process which requires a team of professionals working together to produce the desired results. We can call this team of professionals the building team, and their interrelationship can be expressed by Fig. 2.1.

The client is the person or group of people with the actual need for the building, such as a new house, office or factory. The client is responsible for financing all the work and, therefore, in effect, employs the entire building team.

The architect is the client's agent and is considered to be the leader of the building team. The architect must interpret the client's requirements and produce working drawings. During the building process the architect will supervise all aspects of the work until the building is handed over to the client.

The quantity surveyor measures the quantities of labour and material necessary to complete the building work from drawings supplied by the architect.

Specialist engineers advise the architect during the design stage. They will prepare drawings and calculations on specialist areas of work.

The clerk of works is the architect's 'on-site' representative. He or she will make sure that the contractors carry out the work in accordance with the drawings and other contract documents. They can also agree general matters directly with the building contractor as the architect's representative.

The local authority will ensure that the proposed building conforms to the relevant planning and building legislation.

The health and safety inspectors will ensure that the government's legislation concerning health and safety is fully implemented by the building contractor.

The building contractor will enter into a contract with the client to carry out the construction work in accordance with contract documents. The building

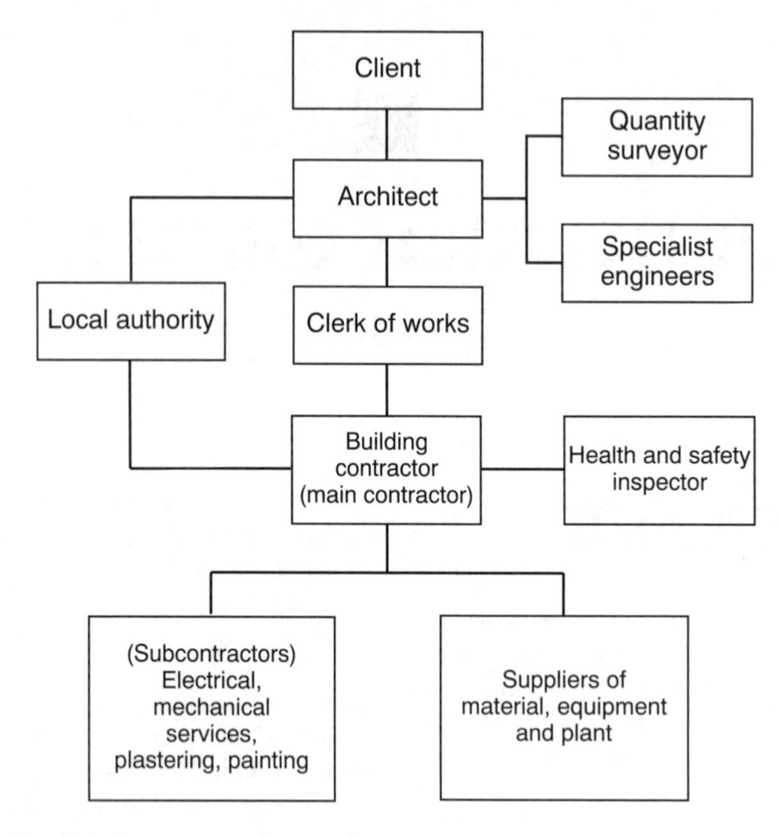

Fig. 2.1 The building team.

contractor is the main contractor and he or she, in turn, may engage subcontractors to carry out specialist services such as electrical installation, mechanical services, plastering and painting.

The electrical team

The electrical contractor is the subcontractor responsible for the installation of electrical equipment within the building. An electrical contracting firm is made up of a group of individuals with varying duties and responsibilities; (see Fig. 2.2). There is often no clear distinction between the duties of the individuals, and the responsibilities carried by an employee will vary from one employer to another. If the firm is to be successful, the individuals must work together to meet the requirements of their customers. Good customer relationships are important for the success of the firm and the continuing employment of the employee.

The customer or his representatives will probably see more of the electrician and the electrical trainee than the managing director of the firm and, therefore, the image presented by them is very important. They should always be polite and be seen to be capable and in command of the situation. This gives a customer confidence in the firm's ability to meet his or her needs. The electrician and his trainee should be appropriately dressed for the job in hand, which probably means an overall of some kind. Footwear is also important, but sometimes a difficult consideration for a journeyman electrician. For example, if working in a factory, the safety regulations insist that protective footwear be worn, but rubber boots may be most appropriate for a building site. However, neither of these would be the most suitable footwear for an electrician fixing a new light fitting in the home of the managing director!

The electrical installation in a building is often carried out alongside other trades. It makes sound sense to help other trades where possible and to develop good working relationships with other employees.

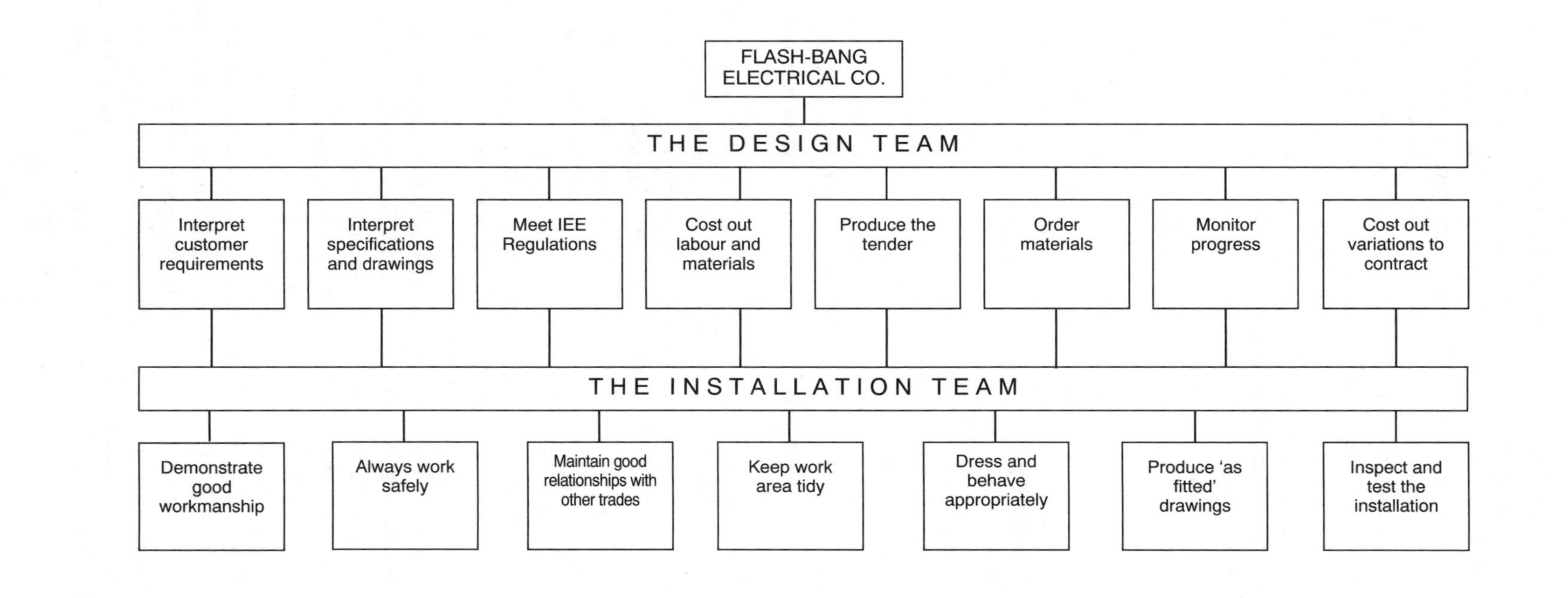

Fig. 2.2 The electrical team.

The employer has the responsibility of finding sufficient work for his employees, paying government taxes and meeting the requirements of the Health and Safety at Work Act described in Chapter 3. The rates of pay and conditions for electricians and trainees are determined by negotiation between the Joint Industry Board and the Amalgamated Engineering and Electrical Trades Union, which will also represent their members in any disputes. Electricians are usually paid at a rate agreed for their grade as an electrician, approved electrician or technician electrician; movements through the grades are determined by a combination of academic achievement and practical experience.

Designing an electrical installation

The designer of an electrical installation must ensure that the design meets the requirements of the IEE Wiring Regulations for electrical installations and any other regulations which may be relevant to a particular installation. The designer may be a professional technician or engineer whose job it is to design electrical installations for a large contracting firm. In a smaller firm, the designer may also be the electrician who will carry out the installation to the customer's requirements. The designer of any electrical installation is the person who interprets the electrical requirements of the customer within the regulations, identifies the appropriate types of installation, the most suitable methods of protection and control and the size of cables to be used.

A large electrical installation may require many meetings with the customer and his professional representatives in order to identify a specification of what is required. The designer can then identify the general characteristics of the electrical installation and its compatibility with other services and equipment, as indicated in Part 3 of the Regulations. The protection and safety of the installation, and of those who will use it, must be considered, with due regard to Part 4 of the Regulations. An assessment of the frequency and quality of the maintenance to be expected (Regulation 341–01–01) will give an indication of the type of installation which is most appropriate.

The size and quantity of all the materials, cables, control equipment and accessories can then be determined. This is called a 'bill of quantities'.

It is common practice to ask a number of electrical contractors to tender or submit a price for work specified by the bill of quantities. The contractor must cost all the materials, assess the labour cost required to install the materials and add on profit and overhead costs in order to arrive at a final estimate for the work. The contractor tendering the lowest cost is usually, but not always, awarded the contract.

To complete the contract in the specified time the electrical contractor must use the management skills required by any business to ensure that men and materials are on site as and when they are required. If alterations or modifications are made to the electrical installation as the work proceeds which are outside the original specification, then a variation order must be issued so that the electrical contractor can be paid for the additional work.

The specification for the chosen wiring system will be largely determined by the building construction and the activities to be carried out in the completed building.

An industrial building, for example, will require an electrical installation which incorporates flexibility and mechanical protection. This can be achieved by a conduit, tray or trunking installation.

In a block of purpose-built flats, all the electrical connections must be accessible from one flat without intruding upon the surrounding flats. A loop-in conduit system, in which the only connections are at the light switch and outlet positions, would meet this requirement.

For a domestic electrical installation an appropriate lighting scheme and multiple socket outlets for the connection of domestic appliances, all at a reasonable cost, are important factors which can usually be met by a PVC insulated and sheathed wiring system.

The final choice of a wiring system must rest with those designing the installation and those ordering the work, but whatever system is employed, good workmanship is essential for compliance with the regulations. The necessary skills can be acquired by an electrical trainee who has the correct attitude and dedication to his craft.

Legal contracts

Before work commences, some form of legal contract should be agreed between the two parties, that is, those providing the work (e.g. the subcontracting electrical company) and those asking for the work to be carried out (e.g. the main building company).

A contract is a formal document which sets out the terms of agreement between the two parties. A standard form of building contract typically contains four sections:

1 The articles of agreement – this names the parties, the proposed building and the date of the contract period.
2 The contractual conditions – this states the rights and obligations of the parties concerned, e.g. whether there will be interim payments for work or a penalty if work is not completed on time.
3 The appendix – this contains details of costings, e.g. the rate to be paid for extras as daywork, who will be responsible for defects, how much of the contract tender will be retained upon completion and for how long.
4 The supplementary agreement – this allows the electrical contractor to recoup any value-added tax paid on materials at interim periods.

In signing the contract, the electrical contractor has agreed to carry out the work to the appropriate standards in the time stated and for the agreed cost. The other party, say the main building contractor, is agreeing to pay the price stated for that work upon completion of the installation.

If a dispute arises the contract provides written evidence of what was agreed and will form the basis for a solution.

For smaller electrical jobs, a verbal contract may be agreed, but if a dispute arises there is no written evidence of what was agreed and it then becomes a matter of one person's word against another's.

Exercises

1 The person responsible for financing the building team is the:
   (a) main contractor
   (b) subcontractor
   (c) client
   (d) architect.
2 The person responsible for interpreting the client's requirements to the building team is the:
   (a) main contractor
   (b) subcontractor
   (c) client
   (d) architect.
3 The building contractor is also called the:
   (a) main contractor
   (b) subcontractor
   (c) client
   (d) architect.
4 The electrical contractor is also called the:
   (a) main contractor
   (b) subcontractor
   (c) client
   (d) architect.
5 The people responsible for interpreting the architect's electrical specifications and drawings are the:
   (a) building team
   (b) electrical design team
   (c) electrical installation team
   (d) construction industry.
6 The people responsible for demonstrating good workmanship and maintaining good relationships with other trades are the:
   (a) building team
   (b) electrical design team
   (c) electrical installation team
   (d) construction industry.
7 Briefly describe the duties of each of the following people:
   (a) the clerk of works
   (b) the health and safety inspector
   (c) the electrician
   (d) the foreman electrician.
8 Describe the importance of a correct attitude towards the customer by an apprentice electrician and other members of the installation team.
9 State eight separate tasks carried out by the electrical design team.
10 State seven separate tasks carried out by the electrical installation team.
11 State the purpose of a 'variation' order.
12 State the advantages of a written legal contract as compared to a verbal contract.

3

SAFETY AT WORK

Introduction

Every day in the UK many hundreds of workers become victims of industrial accidents which require professional treatment by a doctor or nurse. Some accidents are inconvenient as well as painful, such as stepping on a nail which penetrates the foot, requiring treatment in a hospital casualty department. Many other accidents lead to a long-term stay in hospital, and about 100 people die each year as a result of an industrial accident. Most accidents could have been avoided and are the result of workers' ignorance, neglect, forgetfulness or recklessness. The construction industry has one of the worst accident records. An electrician is part of the construction industry and is therefore more likely to have an accident than a worker in any other industry.

The Health and Safety at Work Act 1974

Many governments have passed laws aimed at improving safety at work but the most important recent legislation has been the Health and Safety at Work Act 1974. The purpose of the Act is to provide the legal framework for stimulating and encouraging high standards of health and safety at work; the Act puts the responsibility for safety at work on both workers and managers.

The employer has a duty to care for the health and safety of employees (Section 2 of the Act). To do this he must ensure that

- the working conditions and standard of hygiene are appropriate;
- the plant, tools and equipment are properly maintained;
- the necessary safety equipment – such as personal protective equipment, dust and fume extractors and machine guards – is available and properly used;
- the workers are trained to use equipment and plant safely.

Employees have a duty to care for their own health and safety and that of others who may be affected by their actions (Section 7 of the Act). To do this they must

- take reasonable care to avoid injury to themselves or others as a result of their work activity;
- co-operate with their employer, helping him or her to comply with the requirements of the Act;
- not interfere with or misuse anything provided to protect their health and safety.

Failure to comply with the Health and Safety at Work Act is a criminal offence and any infringement of the law can result in heavy fines, a prison sentence or both.

ENFORCEMENT

Laws and rules must be enforced if they are to be effective. The system of control under the Health and Safety at Work Act comes from the Health and Safety Executive (HSE) which is charged with enforcing the law. The HSE is divided into a number of specialist inspectorates or sections which operate from local offices throughout the UK. From the local offices the inspectors visit individual places of work.

The HSE inspectors have been given wide-ranging powers to assist them in the enforcement of the law. They can:

1 enter premises unannounced and carry out investigations, take measurements or photographs;
2 take statements from individuals;
3 check the records and documents required by legislation;
4 give information and advice to an employee or employer about safety in the workplace;
5 demand the dismantling or destruction of any equipment, material or substance likely to cause immediate serious injury;
6 issue an improvement notice which will require an employer to put right, within a specified period of time, a minor infringement of the legislation;
7 issue a prohibition notice which will require an employer to stop immediately any activity likely to result in serious injury, and which will be enforced until the situation is corrected;
8 prosecute all persons who fail to comply with their safety duties, including employers, employees, designers, manufacturers, suppliers and the self-employed.

SAFETY DOCUMENTATION

Under the Health and Safety at Work Act, the employer is responsible for ensuring that adequate instruction and information is given to employees to make them safety-conscious. Part 1, section 3 of the Act instructs all employers to prepare a written health and safety policy statement and to bring this to the notice of all employees.

To promote adequate health and safety measures the employer must consult with the employees' safety representatives. In companies which employ more than 20 people this is normally undertaken by forming a safety committee which is made up of a safety officer and employee representatives, usually nominated by a trade union. The safety officer is usually employed full-time in that role. Small companies might employ a safety supervisor, who will have other duties within the company, or alternatively they could join a 'safety group'. The safety group then shares the cost of employing a safety adviser or safety officer, who visits each company in rotation. An employee who identifies a dangerous situation should initially report to his site safety representative. The safety representative should then bring the dangerous situation to the notice of the safety committee for action which will remove the danger. This may mean changing company policy or procedures or making modifications to equipment. All actions of the safety committee should be documented and recorded as evidence that the company takes seriously its health and safety policy.

Under the general protective umbrella of the Health and Safety at Work Act, other pieces of legislation also affect those working in the electrical contracting industry.

The Control of Substances Hazardous to Health Regulations 1988 (COSHH)

These Regulations control people's exposure to hazardous substances in the workplace. Regulation 6 requires employers to assess the risks to health from working with hazardous substances, to train employees in techniques which will reduce the risk or provide personal protective equipment so that employees will not endanger themselves or others through exposure to hazardous substances.

Employees should also know what cleaning, storage and disposal procedures are required and what emergency procedures to take.

All this information must be available to anyone using hazardous substances and the documentation made available to a visiting HSE inspector.

Hazardous substances include:

1 any substance which gives off fumes causing headaches or respiratory irritation;
2 man-made fibres which might cause skin or eye irritation (e.g. loft insulation);
3 acids causing skin burns and breathing irritation (e.g. car batteries, which contain dilute sulphuric acid);
4 solvents causing skin and respiratory irritation

(strong solvents are used to cement together PVC conduit fittings and tube);
5 fumes and gases causing asphyxiation (burning PVC gives off toxic fumes);
6 cement and wood dust causing breathing problems and eye irritation.

Where personal protective equipment is provided by an employer, employees have a duty to use it to safeguard themselves.

The Electricity at Work Regulations 1989 (EWR)

This legislation came into force in 1990 and replaced earlier regulations such as the Electricity (Factories Act) Special Regulations 1944. The purpose of the Regulations is to 'require precautions to be taken against the risk of death or personal injury from electricity in work activities'.

Section 4 of the EWR tells us that 'all systems must be constructed so as to prevent danger . . ., and be properly maintained. . . . Every work activity shall be carried out in a manner which does not give rise to danger. . . . In the case of work of an electrical nature, it is preferable that the conductors be made dead before work commences'.

The EWR do not tell us specifically how to carry out our work activities and ensure compliance but if proceedings were brought against an individual for breaking the EWR, the only acceptable defence would be 'to prove that all reasonable steps were taken and all diligence exercised to avoid the offence' (Regulation 29). An electrical contractor could reasonably be expected to have 'exercised all diligence' if the installation was wired according to the IEE Wiring Regulations (see below).

The Electricity Supply Regulations 1988

These Regulations impose requirements upon the regional electricity companies regarding the installation and use of electric lines and equipment. The Regulations are administered by the Engineering Inspectorate of the Electricity Division of the Department of Energy and will not normally concern the electrical contractor except that it is these Regulations which lay down the earthing requirement of the electrical supply at the meter position.

The IEE Wiring Regulations

The Institution of Electrical Engineers Requirements for Electrical Installations (the IEE Regulations) are non-statutory regulations. They relate principally to the design, selection, erection, inspection and testing of electrical installations, whether permanent or temporary, in and about buildings generally and to agricultural and horticultural premises, construction sites and caravans and their sites. Paragraph 7 of the introduction to the EWR says: 'the IEE Wiring Regulations is a code of practice which is widely recognised and accepted in the United Kingdom and compliance with them is likely to achieve compliance with all relevant aspects of the Electricity at Work Regulations'. The IEE Wiring Regulations only apply to installations operating at a voltage up to 1000 V a.c. They do not apply to electrical installations in mines and quarries, where special regulations apply because of the adverse conditions experienced there.

The current edition of the IEE Wiring Regulations, is the 16th edition incorporating amendment numbers 1 (1994) and 2 (1997). The main reason for incorporating the IEE Wiring Regulations into British Standard BS 7671 was to create harmonization with European standards.

To assist electricians in their understanding of the Regulations a number of guidance notes have been published. The guidance notes which I will frequently make reference to in this book are those contained in the *On Site Guide*. Six other guidance notes booklets are also currently available. These are:

- *Selection and Erection*;
- *Isolation and Switching*;
- *Inspection and Testing*;
- *Protection against Fire*;
- *Protection against Electric Shock*;
- *Protection against Overcurrent*.

These guidance notes are intended to be read in conjunction with the Regulations.

Safety signs

The rules and regulations of the working environment are communicated to employees by written instructions, signs, symbols and codes and by other employees as they go about their work.

All signs in the working environment are intended to inform. They give warning of possible dangers, and should be obeyed. At first there were many different safety signs, but British Standard BS 5378 Part 1 (1980) introduced a standard system which gives health and safety information with a minimum use of words. Its purpose is to establish an internationally understood system of safety signs and safety colours which draws attention to objects and situations that do, or could, affect health and safety. Signs fall into four categories: prohibited activities; warnings; mandatory instructions; and safe conditions.

PROHIBITION SIGNS

These are circular white signs with a red border and red cross bar, and are given in Fig. 3.1. They indicate an activity which *must not* be done.

WARNING SIGNS

These are triangular yellow signs with a black border and symbol, and are given in Fig. 3.2. They *give warning* of a hazard or danger.

MANDATORY SIGNS

These are circular blue signs with a white symbol, and are given in Fig. 3.3. They *give instructions* which must be obeyed.

SAFE CONDITION SIGNS

These are square or rectangular green signs with a white symbol, and are given in Fig. 3.4. They *give information* about safety provision.

Accidents at work

Despite new legislation, improved information, education and training, accidents at work do still happen. An accident may be defined as an uncontrolled event causing injury or damage to an individual or property. An accident can nearly always be avoided if correct procedures and methods of working are followed. Any accident which results in an absence from work for more than 3 days, causes a major injury or death, is notifiable to the HSE. There are more than 40 000 accidents reported to the HSE each year which occur as a result of some building-related activity. To avoid having an accident you should:

1 follow all safety procedures (e.g. fit safety signs when isolating supplies and screen off work areas from the general public);

No entry

No smoking

Fig. 3.1 Prohibition signs.

Fig. 3.2 Warning signs.

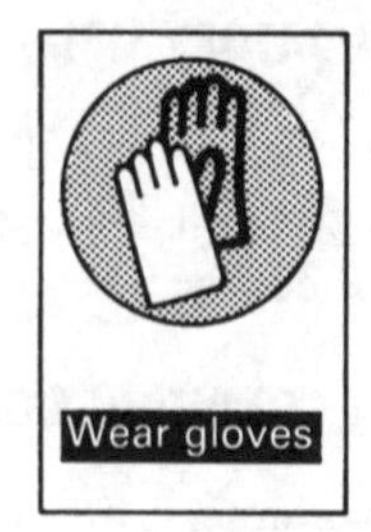

Fig. 3.3 Mandatory signs.

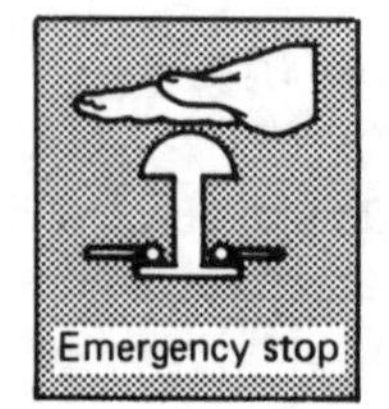

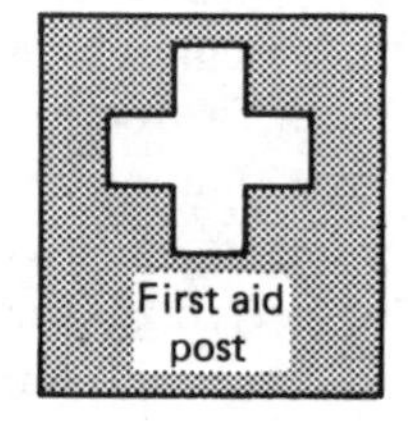

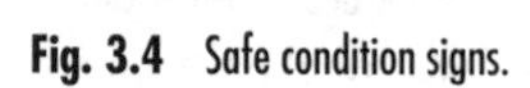

Fig. 3.4 Safe condition signs.

2 not misuse or interfere with equipment provided for health and safety;
3 dress appropriately and use personal protective equipment (PPE) when appropriate;
4 behave appropriately and with care;
5 avoid over-enthusiasm and foolishness;
6 stay alert and avoid fatigue;
7 not use alcohol or drugs at work;
8 work within your level of competence;
9 attend safety courses and read safety literature;
10 take a positive decision to act and work safely.

If you observe a hazardous situation at work, first make the hazard safe, using an appropriate method, or screen it off, but only if you can do so without putting yourself or others at risk, then report the situation to your safety representative or supervisor.

Fire control

A fire is a chemical reaction which will continue if fuel, oxygen and heat are present. To eliminate a fire *one* of these components must be removed. This is often expressed by means of the fire triangle shown in Fig. 3.5; all three corners of the triangle must be present for a fire to burn.

FUEL

Fuel is found in the construction industry in many forms: petrol and paraffin for portable generators and heaters; bottled gas for heating and soldering. Most solvents are flammable. Rubbish also represents a source of fuel: off-cuts of wood, roofing felt, rags, empty solvent cans and discarded packaging will all provide fuel for a fire.

To eliminate fuel as a source of fire, all flammable liquids and gases should be stored correctly, usually in an outside locked store. The working environment should be kept clean by placing rags in a metal bin with a lid. Combustible waste material should be removed from the work site or burned outside under controlled conditions by a competent person.

OXYGEN

Oxygen is all around us in the air we breathe, but can be eliminated from a small fire by smothering with a fire blanket, sand or foam. Closing doors and windows but not locking them will limit the amount of oxygen available to a fire in a building and help to prevent it spreading.

Most substances will burn if they are at a high enough temperature and have a supply of oxygen. The minimum temperature at which a substance will

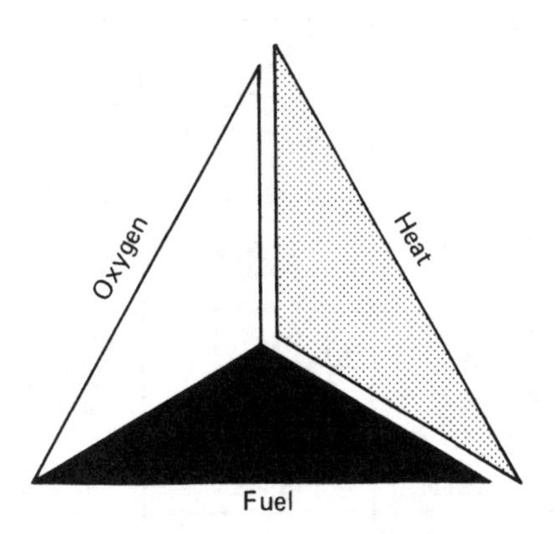

Fig. 3.5 The fire triangle.

burn is called the 'minimum ignition temperature' and for most materials this is considerably higher than the surrounding temperature. However, a danger does exist from portable heaters, blow torches and hot air guns which provide heat and can cause a fire by raising the temperature of materials placed in their path above the minimum ignition temperature. A safe distance must be maintained between heat sources and all flammable materials.

HEAT

Heat can be removed from a fire by dousing with water, but water must not be used on burning liquids since the water will spread the liquid and the fire. Some fire extinguishers have a cooling action which removes heat from the fire.

Fires in industry damage property and materials, injure people and sometimes cause loss of life. Everyone should make an effort to prevent fires, but those which do break out should be extinguished as quickly as possible.

In the event of fire you should

- raise the alarm;
- turn off machinery, gas and electricity supplies in the area of the fire;
- close doors and windows but without locking or bolting them;
- remove combustible materials and fuels away from the path of the fire, if the fire is small, and if this can be done safely;
- attack small fires with the correct extinguisher.

Only attack the fire if you can do so without endangering your own safety in any way. Those not involved in fighting the fire should walk to a safe area or assembly point.

Fires are divided into four classes or categories:

- Class A are wood, paper and textile fires.
- Class B are liquid fires such as paint, petrol and oil.
- Class C are fires involving gas or spilled liquefied gas.
- Class D are very special types of fire involving burning metal.

Electrical fires do not have a special category because, once started, they can be identified as one of the four above types.

Fire extinguishers are for dealing with small fires, and different types of fire must be attacked with a different type of extinguisher. Using the wrong type of extinguisher could make matters worse. For example, water must not be used on a liquid or electrical fire. The normal procedure when dealing with electrical fires is to cut off the electrical supply and use an extinguisher which is appropriate to whatever is burning. Figure 3.6 shows the correct type of extinguisher to be used on three of the categories of fire.

Electrical safety

Electrical supplies at voltages above extra low voltages (ELV) – that is, above 50 V a.c. – can kill human beings and livestock and should therefore be treated with the greatest respect. As an electrician working on electrical installations and equipment, you should always make sure that the supply is first switched off. Every circuit must be provided with a means of isolation (Regulation 130–06–01) and you should isolate and lock off before work begins. In order to deter anyone from reconnecting the supply, a 'Danger Electrician at Work' sign should be displayed on the isolation switch. Where a test instrument or voltage indicator such as that shown in Fig. 3.7 is used to prove conductors dead, Regulation 4(3) of the Electricity at Work Regulations 1989 recommends that the following procedure be adopted so that the device itself is 'proved':

1. Connect the test device to the supply which is to be isolated; this should indicate mains voltage.
2. Isolate the supply and observe that the test device now reads zero volts.

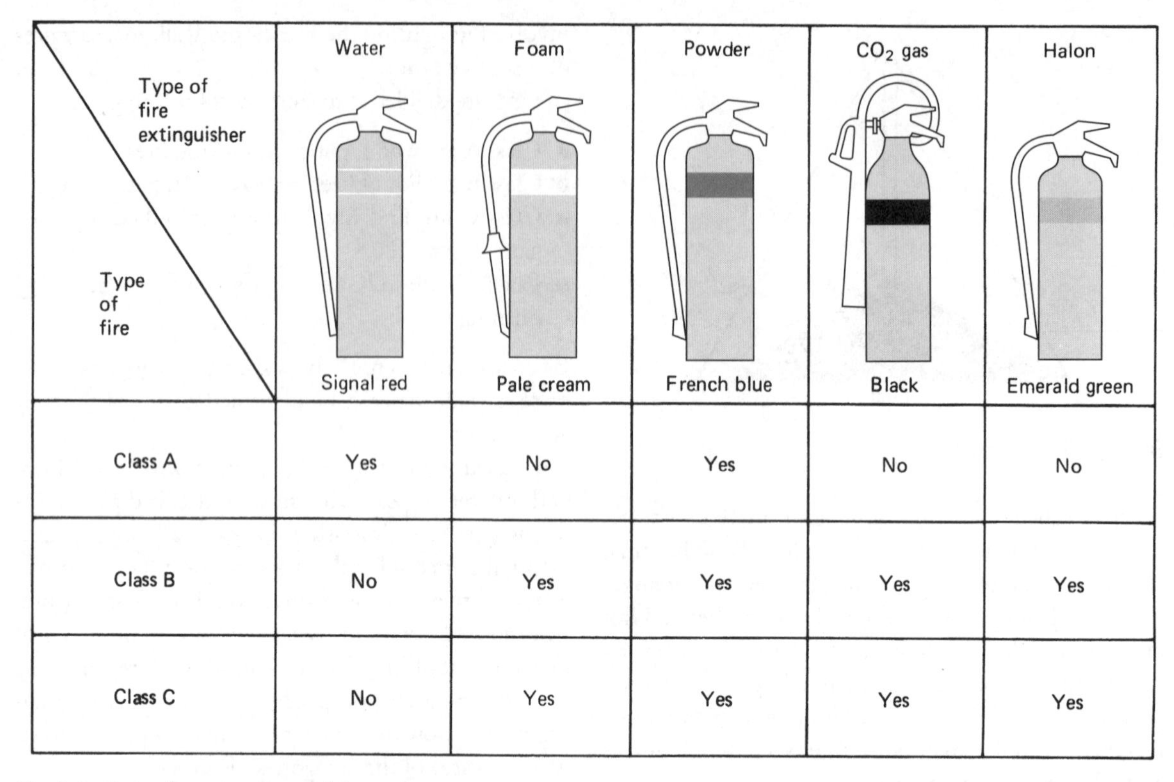

Type of fire \ Type of fire extinguisher	Water	Foam	Powder	CO_2 gas	Halon
	Signal red	Pale cream	French blue	Black	Emerald green
Class A	Yes	No	Yes	No	No
Class B	No	Yes	Yes	Yes	Yes
Class C	No	Yes	Yes	Yes	Yes

Fig. 3.6 Various fire extinguishers and their applications.

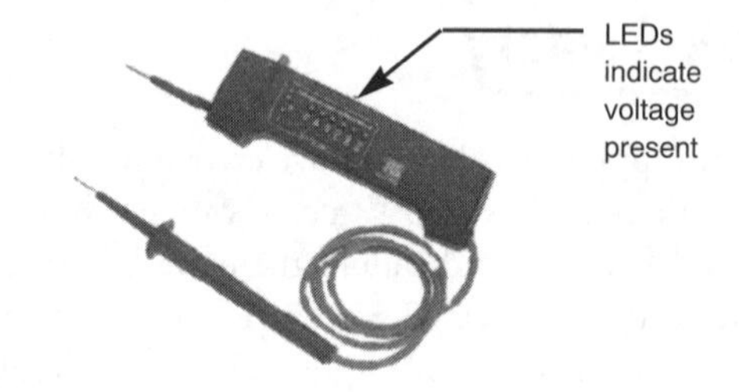

Fig. 3.7 A voltage indicator.

3 Connect the test device to another source of supply to 'prove' that the device is still working correctly.
4 Lock off the supply and place warning notices. Only then should work commence on the 'dead' supply.

The test device must incorporate fused test leads to comply with HSE Guidance Note GS 38, *Electrical Test Equipment Used by Electricians.*

Temporary electrical supplies on construction sites can save many person-hours of labour by providing energy for fixed and portable tools and lighting. However, as stated previously in this chapter, construction sites are dangerous places and the temporary electrical supplies must be safe. IEE Regulation 110–01 tells us that the regulations apply to temporary electrical installations such as construction sites. The frequency of inspection of construction sites is increased to every 3 months because of the inherent dangers. See also Chapter 4 of *Advanced Electrical Installation Work.* Regulation 604–02–02 recommends the following voltages for distributing to plant and equipment on construction sites:

400 V – fixed plant such as cranes
230 V – site offices and fixed floodlighting robustly installed
110 V – portable tools and hand lamps
50 V or 25 V – potable lamps used in damp or confined places.

Portable tools must be fed from a 110 V socket outlet unit (see Fig. 3.8(a)) incorporating splash-proof sockets and plugs with a keyway which prevents a tool from one voltage being connected to the socket outlet of a different voltage.

Socket outlet and plugs are also colour-coded for voltage identification: 25 V violet, 50 V white, 110 V

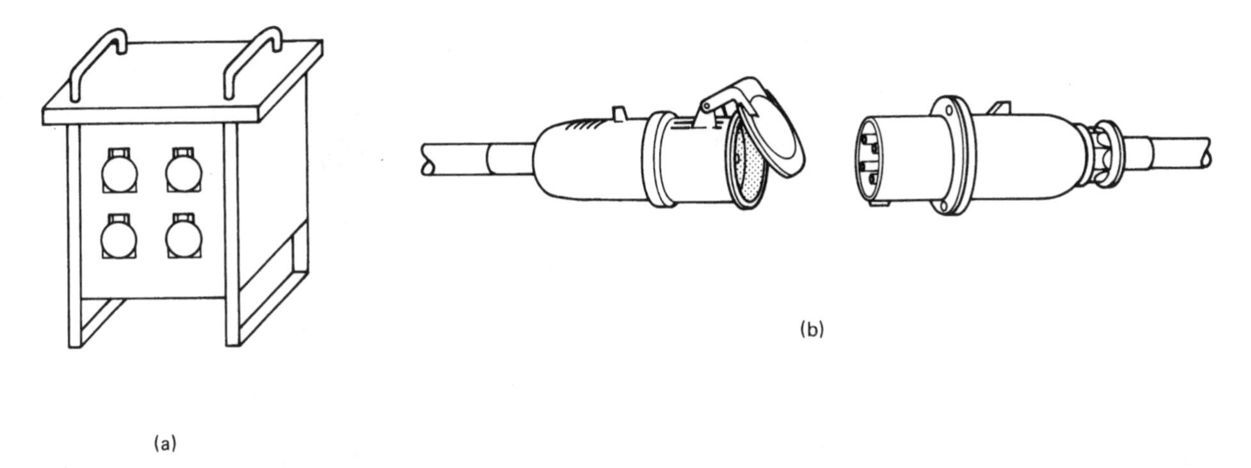

Fig. 3.8 110 V distribution unit and cable connector, suitable for construction site electrical supplies: (a) reduced-voltage distribution unit incorporating industrial sockets to BS 4343; (b) industrial plug and connector.

yellow, 230 V blue and 400 V red, as shown in Fig. 3.8(b).

ELECTRIC SHOCK

Electric shock occurs when a person becomes part of the electrical circuit, as shown in Fig. 3.9. The level or intensity of the shock will depend upon many factors, such as age, fitness and the circumstances in which the shock is received. The lethal level is approximately 50 mA, above which muscles contract, the heart flutters and breathing stops. A shock above the 50 mA level is therefore fatal unless the person is quickly separated from the supply. Below 50 mA only an unpleasant tingling sensation may be experienced or you may be thrown across a room, roof or ladder, but the resulting fall may lead to serious injury.

To prevent people receiving an electric shock accidentally, all circuits contain protective devices. All exposed metal is earthed, fuses and miniature circuit breakers (MCBs) are designed to trip under fault conditions and residual current devices (RCDs) are designed to trip below the fatal level as described in Chapter 6.

Construction workers and particularly electricians do receive electric shocks, usually as a result of carelessness or unforeseen circumstances. When this happens it is necessary to act quickly to prevent the electric shock becoming fatal. Actions to be taken upon finding a workmate receiving an electric shock are as follows:

- Switch off the supply if possible.
- Alternatively, remove the person from the supply *without touching him*, e.g. push him off with a piece of wood, pull him off with a scarf, dry towel or coat.
- If breathing or heart has stopped, immediately call professional help by dialling 999 or 112 and asking for the ambulance service. Give precise directions to the scene of the accident. The casualty stands the best chance of survival if the emergency services can get a rapid-response paramedic team quickly to the scene. They have extensive training and will have specialist equipment with them.
- Only then should you apply resuscitation or cardiac massage until the patient recovers, or help arrives.
- Treat for shock.

First aid

Despite all the safety precautions taken on construction sites to prevent injury to the workforce, accidents do happen and *you* may be the only other person able to take action to assist a workmate. If you are not a qualified first aider limit your help to obvious common-sense assistance and call for help *but* do remember that if a workmate's heart or breathing has stopped as a result of an accident he has only minutes to live unless you act quickly.

There now follows a description of some first aid procedures which should be practised under expert guidance before they are required in an emergency.

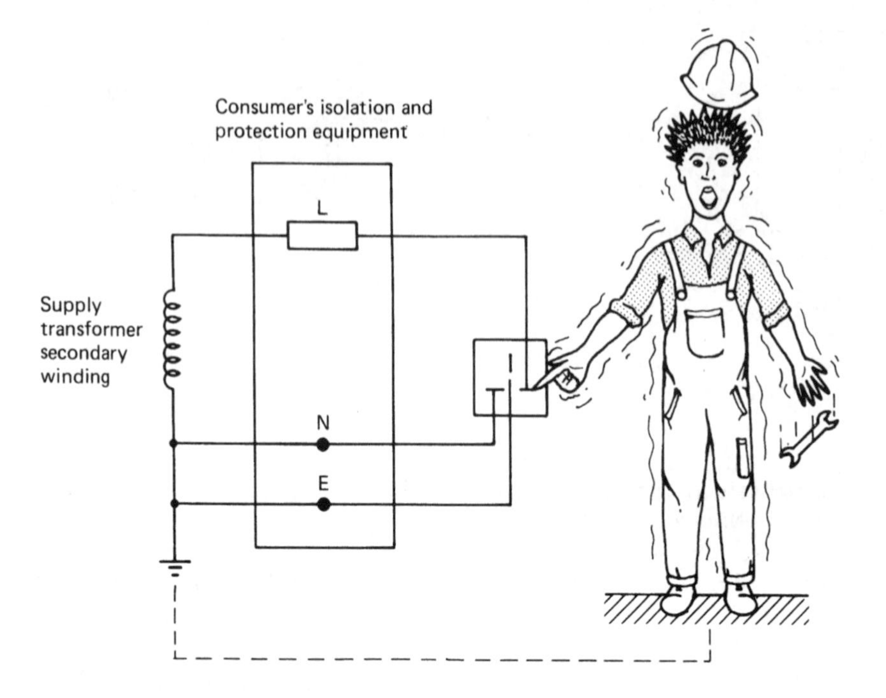

Fig. 3.9 Touching live and earth or live and neutral makes a person part of the electrical circuit and can lead to an electric shock.

Bleeding

If the wound is dirty, rinse it under clean running water. Clean the skin around the wound and apply a plaster, pulling the skin together.

If the bleeding is severe apply direct pressure to reduce the bleeding and raise the limb if possible. Apply a sterile dressing or pad and bandage firmly before obtaining professional advice.

To avoid possible contact with hepatitis or the AIDS virus, when dealing with open wounds, first aiders should avoid contact with fresh blood by wearing plastic or rubber protective gloves, or by allowing the casualty to apply pressure to the bleeding wound.

Burns

Remove heat from the burn to relieve the pain by placing the injured part under clean cold water. Do not remove burnt clothing sticking to the skin. Do not apply lotions or ointments. Do not break blisters or attempt to remove loose skin. Cover the injured area with a clean dry dressing.

Broken bones

Make the casualty as comfortable as possible by supporting the broken limb either by hand or with padding. Do not move the casualty unless by remaining in that position he is likely to suffer further injury. Obtain professional help as soon as possible.

Contact with chemicals

Wash the affected area very thoroughly with clean cold water. Remove any contaminated clothing. Cover the affected area with a clean sterile dressing and seek expert advice. It is a wise precaution to treat all chemical substances as possibly harmful; even commonly used substances can be dangerous if contamination is from concentrated solutions. When handling dangerous substances it is also good practice to have a neutralizing agent to hand.

Disposal of dangerous substances must not be into the main drains since this can give rise to an environmental hazard, but should be undertaken in accordance with local authority regulations.

Exposure to toxic fumes

Get the casualty into fresh air quickly and encourage deep breathing if conscious. Resuscitate if breathing has stopped. Obtain expert medical advice as fumes may cause irritation of the lungs.

Sprains and bruising

A cold compress can help to relieve swelling and pain. Soak a towel or cloth in cold water, squeeze it out and place it on the injured part. Renew the compress every few minutes.

Breathing stopped

Remove any restrictions from the face and any vomit, loose or false teeth from the mouth. Loosen tight clothing around the neck, chest and waist. To ensure a good airway, lay the casualty on his back and support the shoulders on some padding. Tilt the head backwards and open the mouth. If the casualty is faintly breathing, lifting the tongue clear of the airway may be all that is necessary to restore normal breathing. However, if the casualty does not begin to breathe, open your mouth wide and take a deep breath, close the casualty's nose by pinching with your fingers, and, sealing your lips around his mouth, blow into his lungs until the chest rises. Remove your mouth and watch the casualty's chest fall. Continue this procedure at your natural breathing rate. If the mouth is damaged or you have difficulty making a seal around the casualty's mouth, close his mouth and inflate the lungs through his nostrils. Give artificial respiration until natural breathing is restored or until professional help arrives.

Heart stopped beating

This sometimes happens following a severe electric shock. If the casualty's lips are blue, the pupils of his eyes widely dilated and the pulse in his neck cannot be felt, then he may have gone into cardiac arrest. Act quickly and lay the casualty on his back. Kneel down beside him and place the heel of one hand in the centre of his chest. Cover this hand with your other hand and interlace the fingers. Straighten your arms and press down on his chest sharply with the heel of your hands and then release the pressure. Continue to do this 15 times at the rate of one push per second. Check the casualty's pulse. If none is felt, give two breaths of artificial respiration and then a further 15 chest compressions. Continue this procedure until the heartbeat is restored and the artificial respiration until normal breathing returns. Pay close attention to the condition of the casualty while giving heart massage. When a pulse is restored the blueness around the mouth will quickly go away and you should stop the heart massage. Look carefully at the rate of breathing. When this is also normal, stop giving artificial respiration. Treat the casualty for shock, place him in the recovery position and obtain professional help.

Shock

Everyone suffers from shock following an accident. The severity of the shock depends upon the nature and extent of the injury. In cases of severe shock the casualty will become pale and his skin become clammy from sweating. He may feel faint, have blurred vision, feel sick and complain of thirst. Reassure the casualty that everything that needs to be done is being done. Loosen tight clothing and keep him warm and dry until help arrives. *Do not* move him unnecessarily or give him anything to drink.

EMPLOYER'S RESPONSIBILITIES

An employer must provide adequate first aid equipment and, on large sites, appoint a qualified first aider. Where only a few people are employed someone should be nominated to take charge in the event of an accident occurring.

ACCIDENT REPORTS

Every accident should be reported to the employer and the details of the accident and treatment given entered in an 'accident book'. Failure to do so may influence the payment of compensation later.

Exercises

1 For any fire to continue to burn three components must be present. These are:
   (a) fuel, wood, cardboard
   (b) petrol, oxygen, bottled gas
   (c) flames, fuel, heat
   (d) fuel, oxygen, heat.

2 A CO_2 gas fire extinguisher, colour-coded black, is suitable on:
(a) class A fires only
(b) class A and B fires only
(c) class B and C fires only
(d) class A, B and C fires.

3 The recommended voltage for portable hand tools on construction sites is:
(a) 50 V
(b) 110 V
(c) 230 V
(d) 400 V.

4 A small blow-torch burn to the arm of a workmate should be treated by:
(a) immersing in cold water before applying a clean dry dressing
(b) pricking blisters before applying a clean dry dressing
(c) covering burned skin with cream or petroleum jelly to exclude the air before applying a clean dry dressing
(d) applying direct pressure to the burned skin to remove the heat from the burn and relieve the pain.

5 State the responsibilities under the Health and Safety at Work Act of:
(a) an employer to his employees
(b) an employee to his employer and fellow workers.

6 Safety signs are used in the working environment to give information and warnings. Describe the purpose of the four categories of signs and state their colour code and shape. You may use sketches to illustrate your answer.

7 A trainee electrician discovers a large well-established fire in a store-room of the building in which he is working. The building is an office block which is under construction but almost complete. There are six offices on each of six floors and the store-room and fire are in an office on the fourth floor of the building. The trainee knows that there are between 10 and 20 other construction workers somewhere in the building and that the fire alarm system is not connected. Describe the actions which the trainee should take to prevent this emergency becoming a disaster.

8 Describe a suitable electrical distribution system for a construction site comprising
(a) heavy current using fixed machines
(b) site cabins
(c) robustly installed site lighting
(d) portable tools.
Identify suitable voltages and how the various voltages may be obtained from the mains input position which is at 400 V.

9 Describe the action to be taken upon finding a workmate apparently dead on the floor and connected to a live electrical supply.

10 Briefly identify the main difference between the Electricity at Work Regulations 1989 and the IEE Regulations, 16th edition (BS 7671).

4

MOVING LOADS – GAINING ACCESS

Any heavy object which needs to be moved can be classified as a 'load'. The load might be a 110 V transformer, a box of tools, a pipe vice or boxes of accessories. Whatever the heavy object is, it must be moved carefully and thoughtfully using appropriate lifting techniques, if personal pain and injury are to be avoided. In the construction industry many hundreds of man-hours are lost as a result of workers injuring themselves by using incorrect lifting procedures.

Manual lifting

When lifting heavy loads correct lifting procedures *must* be adopted to avoid back injuries. Figure 4.1 demonstrates the technique.

Fig. 4.1 Correct manual lifting and carrying procedures.

Do not lift objects from the floor with the back bent and the legs straight as this causes excessive stress on the spine. *Always* lift with the *back straight* and the *legs bent* so that the powerful leg muscles do the lifting work. Bend at the hips and knees to get down to the level of the object being lifted, positioning the body as close to the object as possible. Grasp the object firmly and, keeping the back straight and head erect, use the leg muscles to raise in a smooth movement. Carry the load close to the body. When putting the object down keep the back straight and bend at the hips and knees, reversing the lifting procedure.

There have been too many injuries over the years resulting from bad manual handling techniques. The problem has become so serious that the Health and Safety Executive has introduced new legislation under the Health and Safety at Work Act 1974, the Manual Handling Operations Regulations 1992.

Where a job involves considerable manual handling, employers must now train employees in the correct lifting procedures and provide the appropriate equipment necessary to promote the safe manual handling of loads.

Consider some 'good practice' when lifting loads:

1 Do not lift the load manually if it is more appropriate to use a mechanical aid.
2 Plan ahead to avoid unnecessary or repeated movement of loads.
3 Take account of the centre of gravity of the load when lifting – the weight acts through the centre of gravity (see Chapter 8).
4 Never leave a suspended load unsupervised.
5 Always lift and lower loads gently.
6 Clear obstacles out of the lifting area.

7 Use the manual lifting techniques described above and avoid sudden or jerky movements.
8 Use gloves when manual handling to avoid injury from rough or sharp edges.
9 Take special care when moving loads wrapped in grease or bubble-wrap.
10 Never move a load over other people or walk under suspended loads.

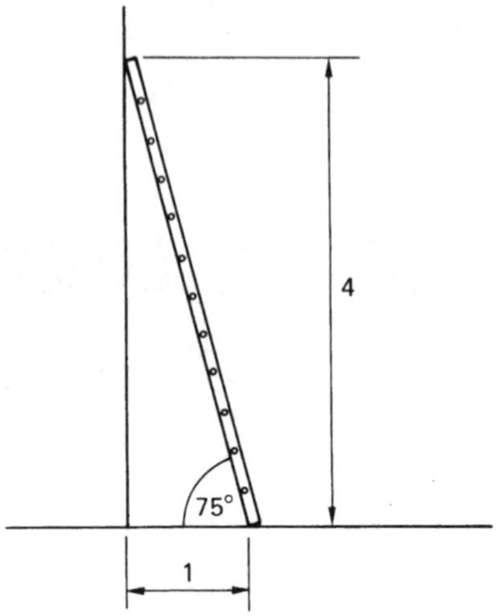

Working above ground level

Working above ground level creates added dangers and slows down the work rate of the electrician. Every precaution should be taken to ensure that the working platform is appropriate for the purpose and in good condition.

Fig. 4.2 A correctly erected ladder.

LADDERS

It is advisable to inspect the ladder before climbing it. It should be straight and firm. All rungs and tie rods should be in position and there should be no cracks in the stiles. The ladder should not be painted since the paint may be hiding defects.

Extension ladders should be erected in the closed position and extended one section at a time. Each section should overlap by at least the number of rungs indicated below:

- Ladder up to 4.8 m length – 2 rungs overlap.
- Ladder up to 6.0 m length –3 rungs overlap.
- Ladder over 6.0 m length – 4 rungs overlap.

The angle of the ladder to the building should be in the proportion 4 up to 1 out or 75° as shown in Fig. 4.2. The ladder should be lashed at the top and bottom to prevent unwanted movement and placed on firm and level ground. When ladders provide access to a roof or working platform the ladder must extend at least 1 m or 5 rungs above the landing place.

Short ladders may be carried by one person resting the ladder on the shoulder, but longer ladders should be carried by two people, one at each end, to avoid accidents when turning corners.

Long ladders or extension ladders should be erected by two people as shown in Fig. 4.3. One person stands on or 'foots' the ladder, while the other person lifts and walks under the ladder towards the walls. When the ladder is upright it can be positioned in the correct place, at the correct angle and secured before being climbed.

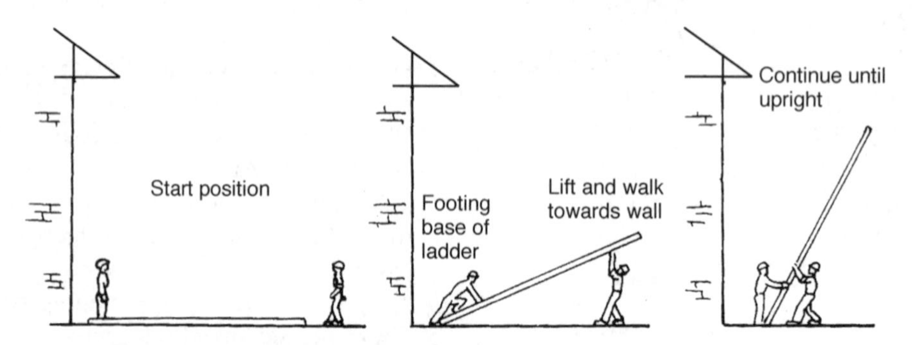

Fig. 4.3 Correct procedure for erecting long or extension ladder.

TRESTLE SCAFFOLD

Figure 4.4 shows a trestle scaffold. Two pairs of trestles spanned by scaffolding boards provide a simple working platform. The platform must be at least two boards or 450 mm wide. At least one-third of the trestle must be above the working platform. If the platform is more than 2 m above the ground, toe-boards and guardrails must be fitted, and a separate ladder provided for access. The boards which form the working platform should be of equal length and not overhang the trestles by more than four times their own thickness. The maximum span of boards between trestles is:

- 1.3 m for boards 40 mm thick
- 2.5 m for boards 50 mm thick.

Trestles which are higher than 3.6 m must be tied to the building to give them stability. Where anyone can fall more than 4.5 m from the working platform, trestles may not be used.

MOBILE SCAFFOLD TOWERS

Mobile scaffold towers may be constructed of basic scaffold components or made from light alloy tube. The tower is built up by slotting the sections together until the required height is reached. A scaffold tower is shown in Fig. 4.5.

If the working platform is above 2 m from the ground it must be closed-boarded and fitted with guardrails and toeboards. When the platform is being used, all four wheels must be locked. The platform must not be moved unless it is clear of tools, equipment and workers and should be pushed at the base of the tower and not at the top.

Fig. 4.4 A trestle scaffold.

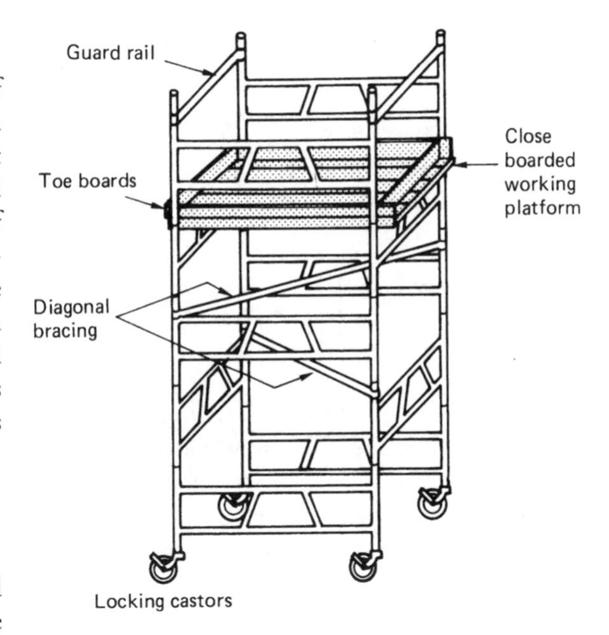

Fig. 4.5 A mobile scaffold tower.

The stability of the tower depends upon the ratio of the base width to tower height. A ratio of base to height of 1 : 3 gives good stability. Outriggers can be used to increase stability by effectively increasing the base width. If outriggers are used then they must be fitted diagonally across all four corners of the tower and not on one side only. The tower must not be built more than 12 m high unless it has been specially designed for that purpose. Any tower higher than 9 m should be secured to the structure of the building to increase stability.

Access to the working platform of a scaffold tower should be by a ladder securely fastened vertically to the tower. Ladders must never be leaned against a tower since this might push the tower over.

Simple machines

Our physical abilities in the field of lifting and moving heavy objects are limited. However, over the centuries we have used our superior intelligence to design tools, mechanisms and machines which have overcome this physical inadequacy. This concept is shown in Fig. 4.6.

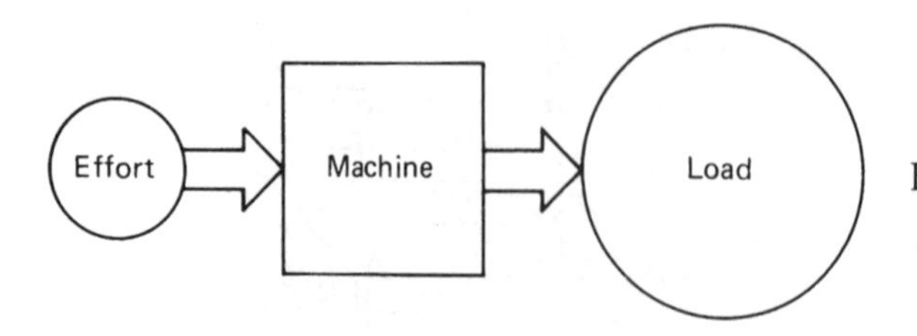

Fig. 4.6 Simple machine concept.

By definition, a machine is an assembly of parts, some fixed, others movable, by which motion and force are transmitted. With the aid of a machine we are able to magnify the effort exerted at the input and lift or move large loads at the output.

MECHANICAL ADVANTAGE (*MA*)

This is the advantage given by the machine and is defined as the ratio of the load to the effort.

$$MA = \frac{\text{Load}}{\text{Effort}} \text{ (no units)}$$

VELOCITY RATIO (*VR*)

This is the ratio of the distance moved by the effort to the distance moved by the load. Because the distance moved by the load and effort are dependent upon the construction of a particular machine, the velocity ratio is usually a constant for that machine.

$$VR = \frac{\text{Distance moved by effort}}{\text{Distance moved by load}} \text{ (no units)}$$

EFFICIENCY

In all machines the power available at the output is less than that which is put in because losses occur in the machine. These losses may result from friction in the bearings, wind resistance to moving parts, heat, noise or vibrations.

The ratio of the output power to the input power is known as the *efficiency* of the machine. The symbol for efficiency is the Greek letter 'eta' (η). In general,

$$\text{Efficiency} = \frac{\text{Work output}}{\text{Work input}}$$

But

$$\text{Work Output} = \text{Load} \times \text{Distance moved by the load}$$

and

$$\text{Work Input} = \text{Effort} \times \text{Distance moved by the effort}$$

If we divide these two equations we see that

$$\frac{\text{Load}}{\text{Effort}} = MA$$

$$\frac{\text{Distance moved by load}}{\text{Distance moved by effort}} = \frac{1}{VR}$$

and therefore

$$\text{Efficiency} = MA \times \frac{1}{VR}$$

or

$$\eta = \frac{MA}{VR}$$

FORCE

The presence of a force can only be detected by its effect on a body. A force may cause a stationary body to move or bring a moving body to rest.

Gravitational force causes objects to fall towards the earth. A spring balance extends when a mass is attached because the force of gravity acts upon the mass and extends the spring. Because the force of gravity acts upon all masses on earth, these masses tend to accelerate and exert a force which is dependent upon the mass of the body and the acceleration due to gravity. The standard acceleration due to gravity is internationally accepted as 9.81 m/s^2. Thus a mass of one kilogram will exert a force of 9.81 N.

The SI unit of force is called the newton (symbol N) to commemorate the great English scientist Sir Isaac Newton (1642–1727).

$$\text{Force} = \text{Mass} \times \text{Acceleration (N)}$$

$$\text{Acceleration due to gravity } g = 9.81 \text{ m/s}^2$$

PULLEYS

A pulley is simply a wheel with a grooved rim around which a rope is passed. A load is attached to one end of the rope and an effort applied to the other end.

A single pulley offers no mechanical advantage because the effort applied must equal the load, but it does have uses in raising small loads. They are often used on construction sites, for example, to lift buckets of mortar up to the top of a building scaffold.

Pulley blocks which contain more than one pulley do offer a mechanical advantage since the total load is shared equally by each vertical rope; see Fig. 4.7.

If the load is raised by 1 cm, each length of rope in the system will shorten by 1 cm. Therefore, the effort applied at

- system A will raise the load by 1 cm
- system B will raise the load by 2 cm
- system C will raise the load by 4 cm.

Now

$$VR = \frac{\text{Distance moved by effort}}{\text{Distance moved by load}}$$

so

$$\text{for system A} \quad VR = \frac{1}{1} = 1$$

$$\text{for system B} \quad VR = \frac{2}{1} = 2$$

$$\text{for system C} \quad VR = \frac{4}{1} = 4$$

The velocity ratio of any pulley system, such as a block and tackle, can be determined by counting the number of pulley wheels.

Pulley blocks are used for lifting heavy loads such as car engines from the body of the vehicle, and passenger lifts and cranes operate using this principle. The limitations of this machine are that the upper pulley must be secured at a higher level than the height through which the load will travel.

The following 'good practice' should be borne in mind when using any pulley or block and tackle system:

1. Ensure that the lifting equipment is suspended from a 'secure' point.
2. Do not exceed the maximum safe working load (SWL) indicated on the lifting equipment.
3. Use the lifting 'eye' fitted to the load. Heavy loads often have a lifting point or points cast into the outer casing.
4. If the load does not have a lifting eye, place slings under the load and adjust the lifting point to above the centre of gravity. Remember, the total weight acts through the centre of gravity.
5. Avoid 'shock loading' the lifting equipment.
6. Prevent the load from swinging and twisting while being lifted.
7. Finally, always look at the lifting problems first before taking any action. Ask yourself these questions:
   (a) Do I have the appropriate equipment for the job?
   (b) Do I have the necessary skill and experience for moving the load?

If you do not feel confident to tackle the job, seek help and guidance from your supervisor.

EXAMPLE

A transformer having a mass of 100 kg is lifted on to a lorry by means of a pulley block containing three pulleys in the upper block and three in the lower block. If the effort required is 200 N, calculate the efficiency of the system. (Take g as 10 m/s^2.)

$$\text{Force exerted by transfomer} = \text{Mass} \times \text{Acceleration (N)}$$

$$\text{Force} = 100 \text{ kg} \times 10 \text{ m/s}^2 = 1000 \text{ N}$$

$$MA = \frac{\text{Load}}{\text{Effort}} = \frac{1000 \text{ N}}{200 \text{ N}} = 6$$

$$VR = \text{Total number of pulley wheels} = 5$$

$$\eta = \frac{MA}{VR} = \frac{5}{6} = 0.833$$

$$\text{or Percentage efficiency} = 0.833 \times 100 = 83.3\%$$

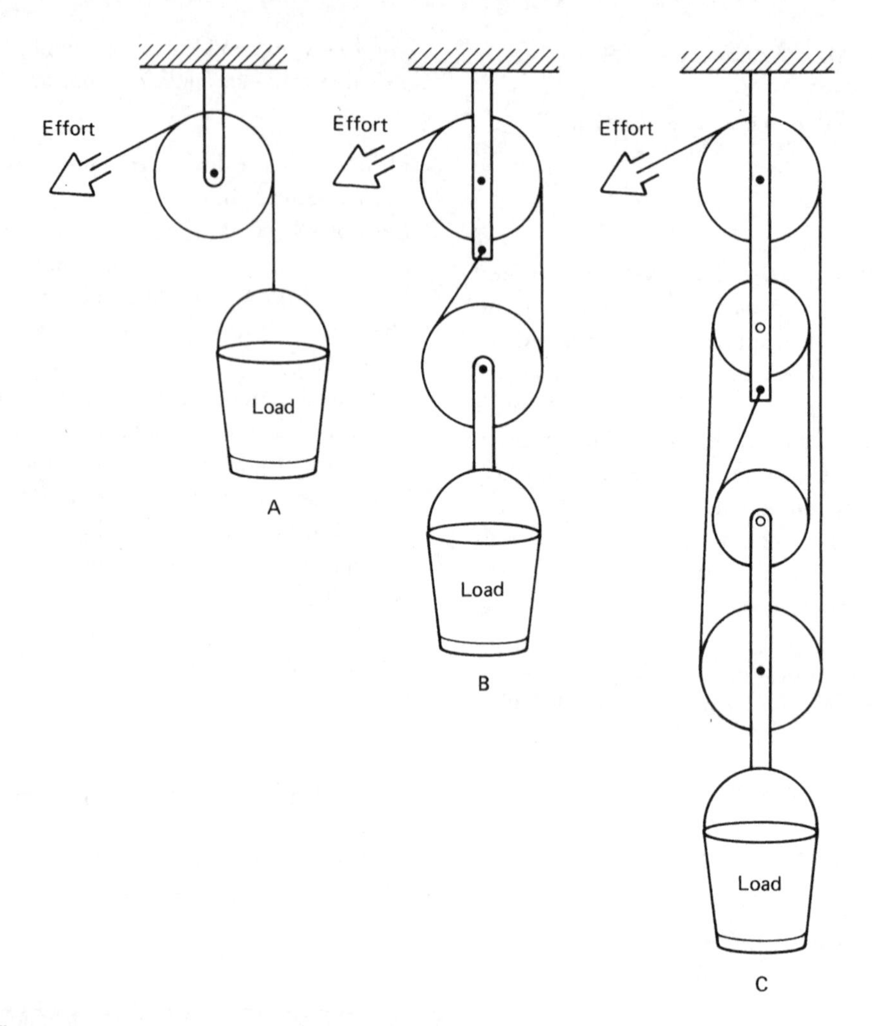

Fig. 4.7 Pulley systems.

THE WEDGE

The wedge provides a means of converting motion in one direction to motion in another at right angles. Driving the wedge under the load causes the load to move up an inclined plane, thus changing the direction of the force by 90°. This is shown in Fig. 4.8.

The wedge is used to prevent horizontal motion, for example when placing a wedge under a door. The early Egyptians used the inclined plane of the wedge to raise the huge blocks of stone with which they built the pyramids.

An electrician might use the mechanical advantage of an inclined plane by pulling a heavy object up the sloping surface formed by a plank into, say, the loading bay of his or her van.

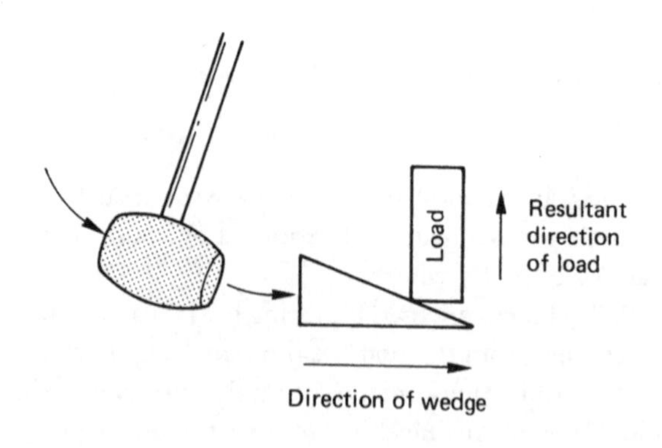

Fig. 4.8 A wedge.

This uses the lever principle discussed in Chapter 8. A small effort exerted over a greater distance (the length of the plank) can overcome a heavy load travelling a shorter distance. (The load only travels through a vertical distance equal to the height of the van loading bay.)

THE SCREW JACK

The screw jack is a simple machine which makes use of a screw thread to lift a large load with a small effort. The principle is similar to that of a wedge in that the load moves up an inclined plane, in this case, the screw thread. Figure 4.9 shows the screw jack.

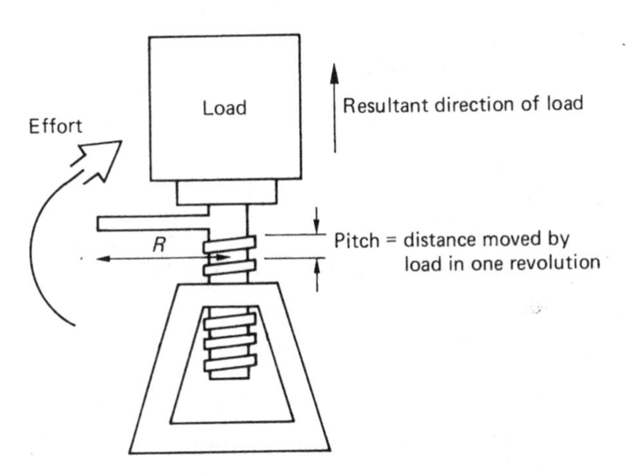

Fig. 4.9 A screw jack.

For this machine consider *one* revolution of the jack arm:

$$VR = \frac{\text{Distance moved by effort}}{\text{Distance moved by load}}$$

$$= \frac{\text{Circumference of jack arm}}{\text{Pitch}}$$

$$VR = \frac{2\pi R}{\text{Pitch}}$$

The screw jack is used extensively as a motor car jack, but the principle is also used for roof supports, an engineer's vice and the tipping mechanism of trucks.

EXAMPLE

A screw jack is used to raise a motor car of mass 1 tonne (1000 kg). The length of the jack arm is 250 mm and the pitch of the thread 5 mm. If an effort of 250 N is required to raise the arm, calculate the efficiency of this machine. (Take g as 10 m/s^2.) We must first change the mass into a force:

$$\text{Force} = \text{Mass} \times \text{Acceleration (N)}$$

$$\therefore \text{Force} = 1000\text{ kg} \times 10\text{ m/s}^2 = 10\,000\text{ N}$$

$$MA = \frac{\text{Load}}{\text{Effort}} = \frac{10\,000\text{ N}}{250\text{ N}} = 40$$

$$VR = \frac{2\pi R}{\text{Pitch}} = \frac{2 \times 3.142 \times 0.25\text{ m}}{0.005\text{ m}} = 314.2$$

$$\eta = \frac{MA}{VR} = \frac{40}{314.2} = 0.127$$

$$\text{or Percentage efficiency} = 0.127 \times 100 = 12.7\%$$

We usually expect a machine to have a high efficiency (better than about 80%), but the efficiency of the screw jack is very poor, as you can see from the example. This is because friction on the moving surfaces, the inclined plane of the screw thread, is very high. A mechanically more efficient car jack would use the liquid properties of hydraulic fluid in a hydraulic jack.

Transporting loads

If the load is to be moved after being lifted, first make sure it is secure.

Some quite simple machines can ease the burden of moving very heavy loads.

A sack truck is an effective means for one person to move very heavy loads. The mechanical advantage of the sack truck, which allows a little human effort to overcome a large load, is based on the principle of levers and is discussed in Chapter 8.

A flat bed truck has a sturdy horizontal base of approximately 1.5 m × 0.75 m supported in an angle iron frame with four heavy castors on each corner. This is ideal for moving heavy loads on hard smooth surfaces. You can often see them being used in supermarkets to move boxes of canned products and in DIY stores for moving building materials.

A fork-lift truck can be powered by batteries or bottled gas for inside use or by combustion engine for outside use. The operator will need training and must hold a certificate of competence. If the machine is to be used on the public highway the operator will also need a driving licence.

Exercises

1 To avoid back injuries when manually lifting heavy weights from ground level a worker should:
(a) bend both legs and back
(b) bend legs but keep back straight
(c) keep legs straight but bend back
(d) keep both legs and back straight.

2 The angle of a ladder to the building upon which it is resting should be in the proportions of:
(a) 1 up to 4 out
(b) 4 up to 75 out
(c) 4 up to 1 out
(d) 75 up to 4 out.

3 The angle which a correctly erected ladder should make with level ground is:
(a) 41°
(b) 45°
(c) 57°
(d) 75°.

4 The working platform of a trestle scaffold which is 1.5 m above the ground should be:
(a) at least two boards wide
(b) fitted with toeboards
(c) fitted with toeboards and guardrails
(d) secured to the building to give stability.

5 For good stability mobile towers must have a base width to tower height ratio of:
(a) 1 : 2
(b) 1 : 3
(c) 1 : 4
(d) 1 : 5.

6 A block and tackle pulley system contains three pulleys in the upper sheath and two in the lower sheath to which the load is attached. The velocity ratio of the system is:
(a) 2
(b) 3
(c) 5
(d) 6

7 The *MA* and *VR* of a particular machine were found to be 3 and 4, respectively. The efficiency of the machine is:
(a) 12%
(b) 13.3%
(c) 70%
(d) 75%.

8 A machine of 60% efficiency has an *MA* of 3 and therefore the *VR* is:
(a) 5
(b) 6
(c) 7
(d) 8.

9 The efficiency of an 8 pulley wheel system is found to be 75% and therefore the *MA* is:
(a) 3
(b) 4
(c) 5
(d) 6.

10 Which particular machine would be most suitable for raising a 200 kg lift motor 9 m up a 12 m lift shaft:
(a) a screw jack
(b) a lever
(c) a pulley system
(d) a wedge.

11 Which, if any, of the following machines use the principle of an inclined plane in their operation:
(a) the lever
(b) the screw jack
(c) the pulley system
(d) none of the above.

5

ELECTRICAL INSTALLATION – WORK PROCEDURES

Tools and equipment

A craftsman earns his living by hiring out his skills or selling products made using his skills and expertise. He shapes his environment, mostly for the better, improving the living standards of himself and others.

Tools extend the limited physical responses of the human body and therefore good-quality, sharp tools are important to a craftsman. An electrician is no less a craftsman than a wood carver. Both must work with a high degree of skill and expertise and both must have sympathy and respect for the materials which they use. Modern electrical installations using new materials are lasting longer than 50 years. Therefore they must be properly installed. Good design, good workmanship and the use of proper materials are essential if the installation is to comply with the relevant regulations, and reliably and safely meet the requirements of the customer for over half a century.

An electrician must develop a number of basic craft skills particular to his own trade, but he also requires some of the skills used in many other trades. An electrician's tool-kit will reflect both the specific and general nature of the work.

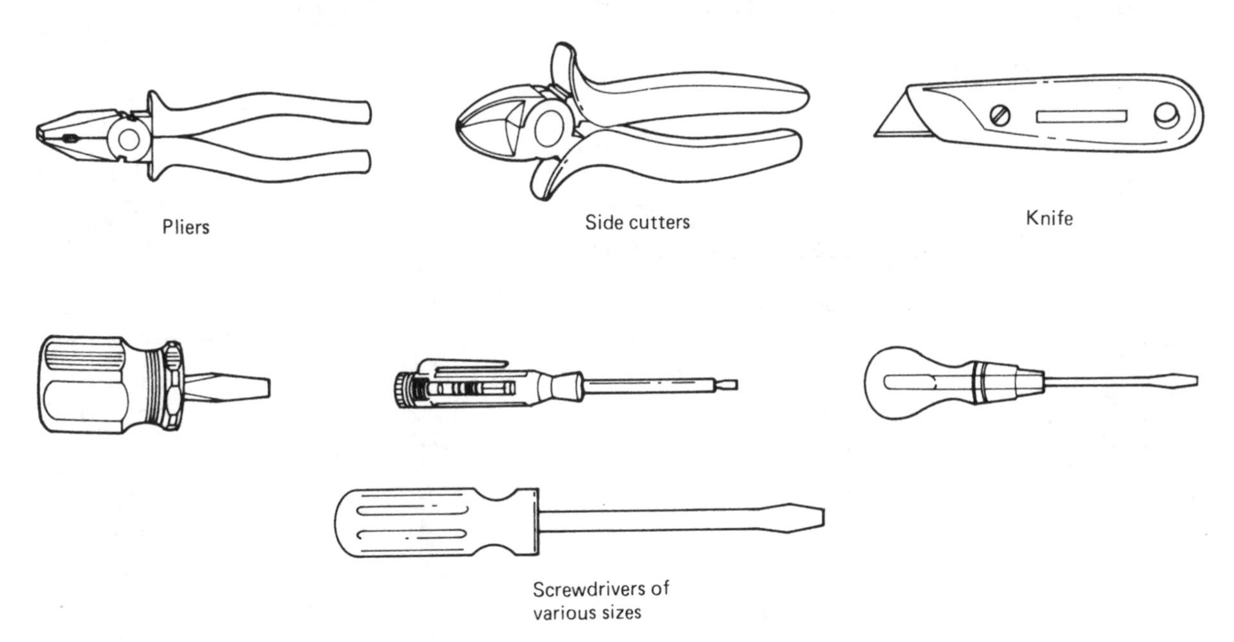

Fig. 5.1 The tools used for making electrical connections.

The basic tools required by an electrician are those used in the stripping and connecting of conductors. These are pliers, side cutters, knife and an assortment of screwdrivers, as shown in Fig. 5.1.

The tools required in addition to these basic implements will depend upon the type of installation work being undertaken. When wiring new houses or rewiring old ones, the additional tools required are those usually associated with a bricklayer and joiner. Examples are shown in Fig. 5.2.

When working on industrial installations, installing conduit and trunking, the additional tools required by an electrician would more normally be those associated with a fitter or sheet-metal fabricator, and examples are shown in Fig. 5.3.

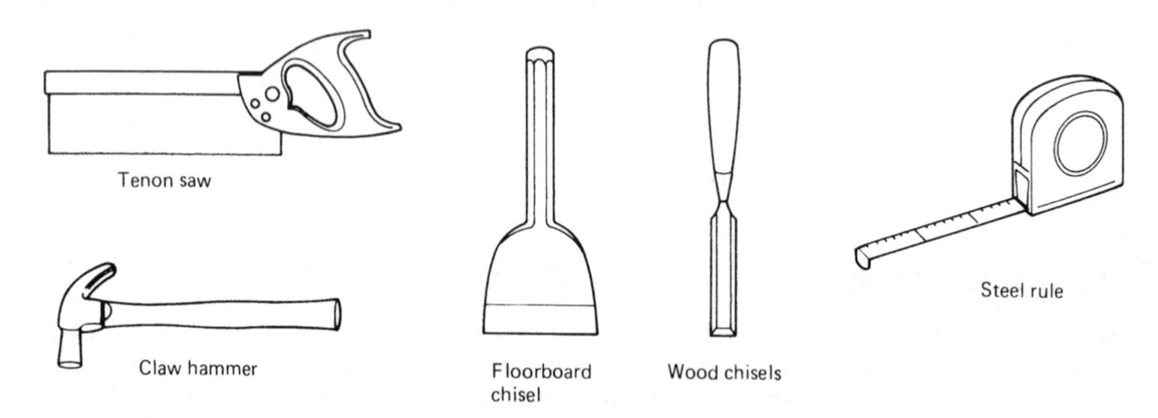

Fig. 5.2 Some additional tools required by an electrician engaged in house wiring.

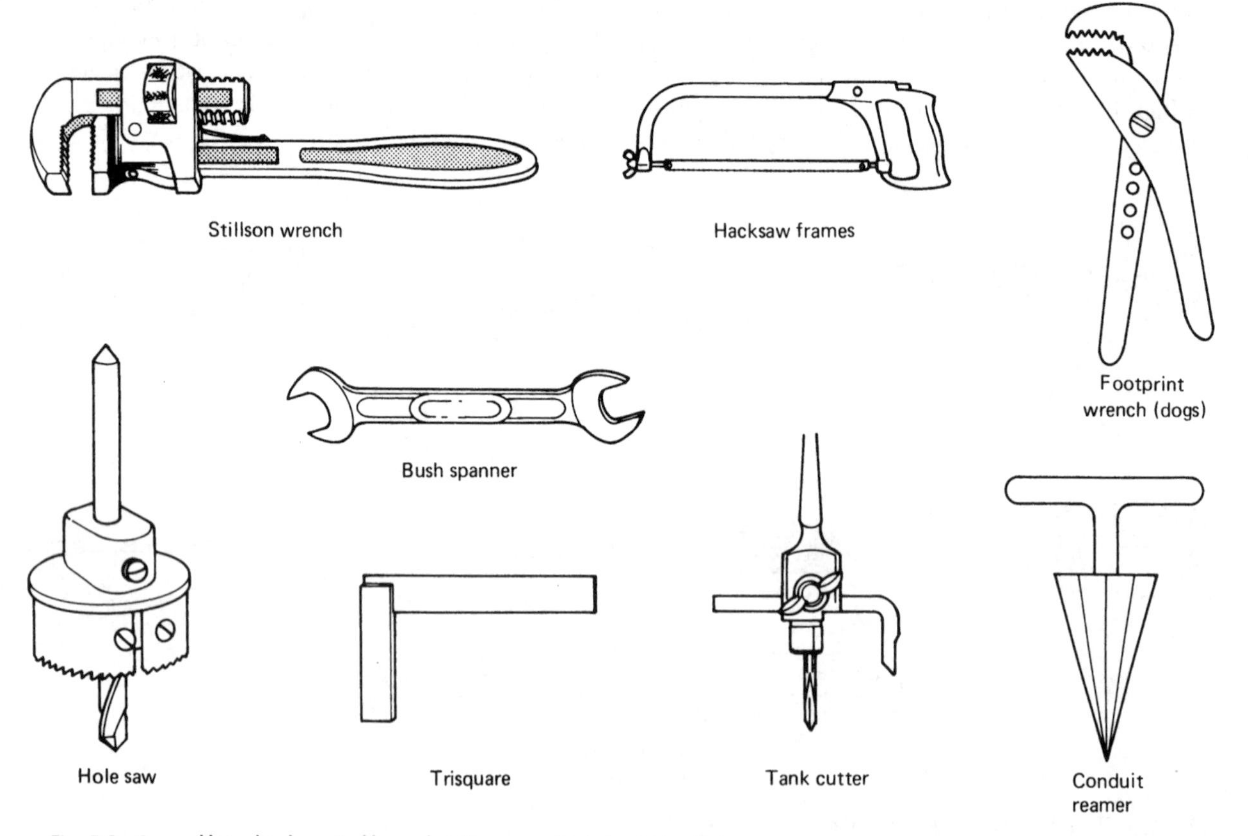

Fig. 5.3 Some additional tools required by an electrician engaged in industrial installations.

Where special tools are required, for example those required to terminate mineral insulated (MI) cables or the bending and cutting tools for conduit and cable trays as shown in Fig. 5.4, they will often be provided by an employer but most hand-tools are provided by the electrician himself.

In general, good-quality tools last longer and stay sharper than those of inferior quality, but tools are very expensive to buy. A good set of tools can be assembled over the training period if the basic tools are bought first and the extended tool-kit acquired one tool at a time.

Another name for an installation electrician is a 'journeyman' electrician and, as the name implies, an electrician must be mobile and prepared to carry his tools from one job to another. Therefore, a good tool-box is an essential early investment, so that the right tools for the job can be easily transported.

Tools should be cared for and maintained in good condition if they are to be used efficiently and remain serviceable. Screwdrivers should have a flat squared off end and wood chisels should be very sharp. Access to a grindstone will help an electrician to maintain his tools in first-class condition. Additionally, wood chisels will require sharpening on an oilstone to give them a very sharp edge.

ELECTRICAL TOOLS

Portable electrical tools can reduce much of the hard work for any tradesman and increase his productivity. Electrical tools should be maintained in a good condition and be appropriate for the purpose for which they are used. The use of reduced voltage double insulation or an RCD can further increase safety without any loss of productivity. Some useful electrical tools are shown in Fig. 5.5.

Electric drills are probably used most frequently of all electrical tools. They may be used to drill metal or wood. Wire brushes are made which fit into the drill chuck for cleaning the metal. Variable-speed electric drills, which incorporate a vibrator, will also drill brick and concrete as easily as wood when fitted with a masonry drill bit.

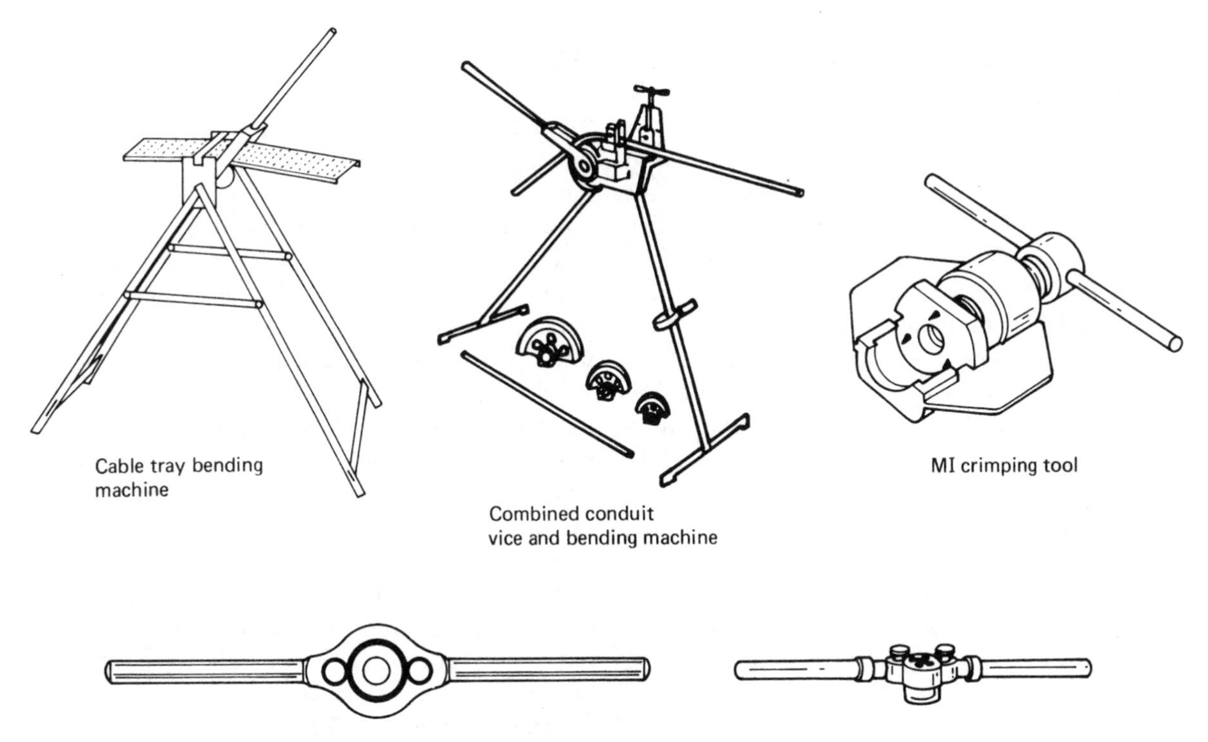

Fig. 5.4 Some special tools required by an electrician engaged in industrial installations.

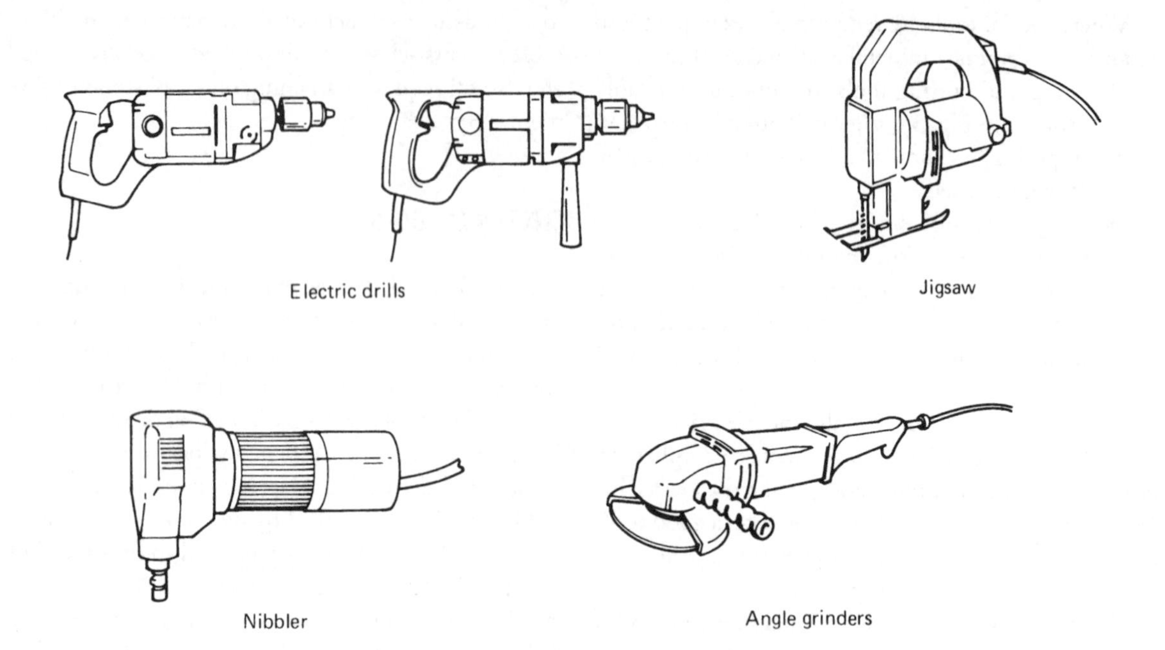

Fig. 5.5 Electrical hand-tools.

Hammer drills give between two and three thousand impacts per minute and are used for drilling concrete walls and floors.

Cordless electric drills are also available which incorporate a rechargeable battery, usually in the handle. They offer the convenience of electric drilling when an electrical supply is not available or if an extension cable is impractical.

Angle grinders are useful for cutting chases in brick or concrete. The discs are interchangeable. Silicon carbide discs are suitable for cutting slate, marble, tiles, brick and concrete, and aluminium oxide discs for cutting iron and steel such as conduit and trucking.

Jigsaws can be fitted with wood or metal cutting blades. With a wood cutting blade fitted they are useful for cutting across floorboards and skirting boards or any other application where a pad saw would be used. With a metal cutting blade fitted they may be used to cut trunking.

When a lot of trunking work is to be undertaken, an electric nibbler is a worthwhile investment. This nibbles out the sheet metal, is easily controllable and is one alternative to the jigsaw.

STEEL TOOLS

To understand why tools need to undergo different processes to make them suitable for a variety of purposes, it is necessary to look at the processes and the properties that they impart to the tools.

The hand-tools used by a craftsman are often made from steel. To cut through wood, brick or cable conductors, the cutting edge must be sharp, hard and tough, but the head of, say, a brick chisel, which is struck by a hammer, must be soft and tough in order to prevent cracking and splintering. Consequently, the tools we use require different properties of softness and hardness depending upon the tool and its application.

Steel tools are alloys of iron and carbon. They contain between 0.7% and 1.7% of carbon and, to a large extent, the proportion of carbon contained determines the properties of the steel; the higher the carbon content, the stronger and harder the steel.

Steels undergo chemical and structural changes when heated. If a piece of carbon steel is heated steadily its temperature will rise at a uniform rate until it reaches approximately 700°C. At this point, even though the heating is continued, the temperature of the steel remains constant for a short period and then continues to rise at a slower rate until it reaches about 780°C. This pause in the temperature rise and the slowing down of the rate of temperature increase indicates that energy is being absorbed by the steel, bringing about chemical and structural changes. If heating is continued beyond 780°C the temperature will rise at the same rate as when first heated, indicating that the chemical and structural changes are completed.

If the same piece of steel is allowed to cool naturally after heating, the action is reversed and the metal returns to its normal composition.

Table 5.1 Colour indication of temperature

Colour	Approximate temperature °C	Type of article to be tempered to this temperature
Pale straw	220–230	Metal turning tools
Dark straw	240–245	Taps, dies, drills
Yellow brown	250–255	Wood turning tools
Brown, just turning purple	260–265	Chisels and axes
Purple	270–280	Cold chisels, punches and knives
Blue	290–300	Screwdrivers

Hardening

When the temperature of a steel tool is raised chemical and structural changes take place within the steel as described above. If the temperature of the heated steel is lowered quickly, by quenching in clean water or oil, the chemical and structural changes cannot return to normal but are 'frozen' or locked in their new configuration. The steel then has the properties of being hard and brittle.

Annealing

The purpose of annealing is to soften steel or relieve internal stresses and strains set up by previous working or use, making the finished tool more malleable. Annealing is achieved by raising the temperature of the steel and then allowing it to cool very slowly, probably in the hearth of the forge used to heat the metal.

Tempering

Hardened steel is too brittle for most hand-tools and is tempered to give the steel back some of its normal toughness and ductility. To temper small articles, one surface of the hardened steel is polished and then heated slowly. The polished surface will change colour as heat is absorbed because thin films of oxide form at different temperatures. Table 5.1 shows the connection between temperature and colour as the steel is quenched from its tempering temperature.

Some tools, cold chisels and punches for example, require a fairly soft tough head but a hardened and tempered cutting edge or point. In these cases the hardening and tempering can be carried out at the same time.

The working or cutting end and half the tool length are heated to the hardening temperature, which is indicated when the metal glows cherry red. The tool is then removed from the heat and the working end quenched up to half the distance heated. After quenching, the end of the tool is quickly polished and the tempering colours observed as the heat from the unquenched portion travels through the metal by conduction. When the required colour reaches the end, the whole tool is quenched.

Whether oil or water is used for quenching depends on the type of tool being tempered. Water quenching produces a very hard steel but it is liable to cause cracks and distortion. Oil quenching is less liable to cause these defects but produces a slightly softer steel. A faster and more even rate of cooling can be obtained if the steel is moved about in the cooling liquid. If only part of the steel is to be hardened, that part should be moved up and down in the liquid to avoid a sharp boundary between the soft and hard portions.

Joining materials

Plastic can be joined with an appropriate solvent. Metals may be welded, brazed or soldered, but the most popular method of on-site joining of metals on electrical installations is by nuts and bolts or rivets.

A nut and bolt joint may be considered a temporary fastening since the parts can easily be separated if required by unscrewing the nut and removing the bolt. A rivet is a permanent fastening since the parts riveted together cannot be easily separated.

Two pieces of metal joined by a bolt and nut and by a machine screw and nut are shown in Fig. 5.6. The nut is tightened to secure the joint. When joining trunking or cable trays, a round head machine screw should be used with the head inside to reduce the risk of damage to cables being drawn into the trunking or tray.

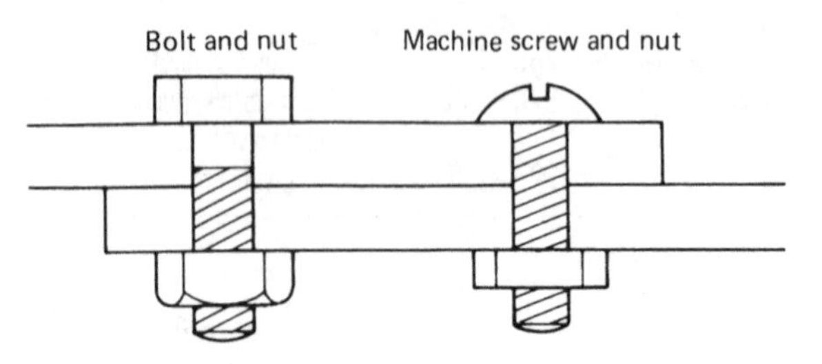

Fig. 5.6 Joining of metals.

Thin sheet material such as trunking is often joined using a pop riveter. Special rivets are used with a hand-tool, as shown in Fig. 5.7. Where possible, the parts to be riveted should be clamped and drilled together with a clearance hole for the rivet. The stem of the rivet is pushed into the nose bush of the riveter until the alloy sleeve of the rivet is flush with the nose bush (a). The rivet is then placed in the hole and the handles squeezed together (b). The alloy sleeve is compressed and the rivet stem will break off when the rivet is set and the joint complete (c). To release the broken-off stem piece, the nose bush is turned upwards and the handles opened sharply. The stem will fall out and is discarded (d).

BRACKET SUPPORTS

Conduit and trunking may be fixed directly to a surface such as a brick wall or concrete ceiling, but where cable runs are across girders or other steel framework, spring steel clips may be used but support brackets or clips often require manufacturing.

The brackets are usually made from flat iron, which is painted after manufacturing to prevent corrosion. They may be made on-site by the electrician or, if many brackets are required, the electrical contractor may make a working sketch with dimensions and have the items manufactured by a blacksmith or metal fabricator.

The type of bracket required will be determined by the installation, but Fig. 5.8 gives some examples of brackets which may be modified to suit particular circumstances.

Fixing methods

PVC insulated and sheathed wiring systems are usually fixed with PVC clips in order to comply with IEE Regulation 522–08 and Table 4A of the *On Site Guide*. The clips are supplied in various sizes to hold the cable firmly, and the fixing nail is a hardened masonry nail. Figure 5.9 shows a cable clip of this type. The use of a masonry nail means that fixings to wood, plaster, brick or stone can be made with equal ease.

When heavier cables, trunking, conduit or luminaires have to be fixed a screw fixing is often needed. Wood screws may be screwed directly into wood but when fixing to brick, stone, plaster or concrete it is necessary to drill a hole in the masonry material,

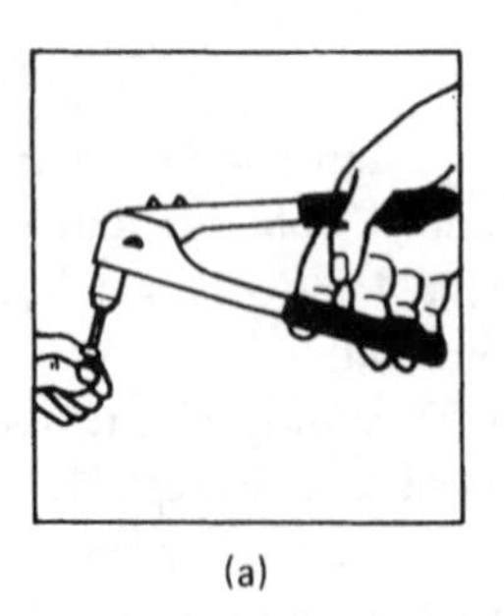

(a)

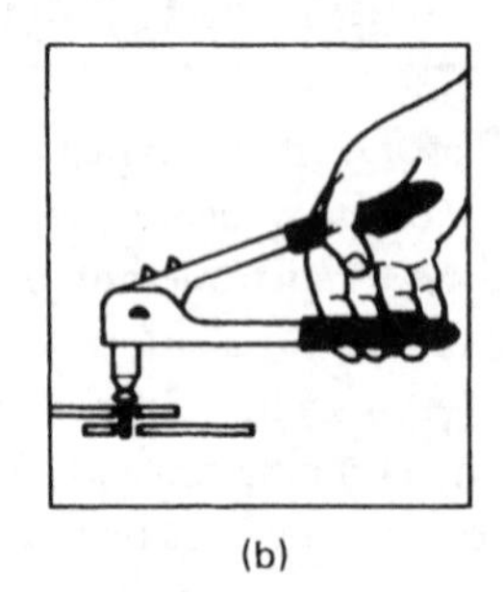

(b)

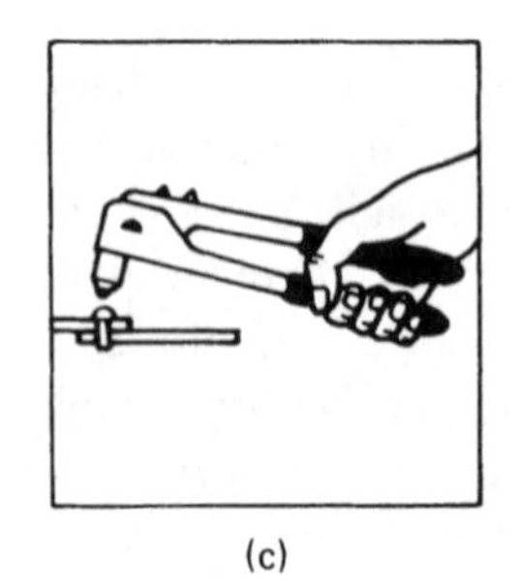

(c)

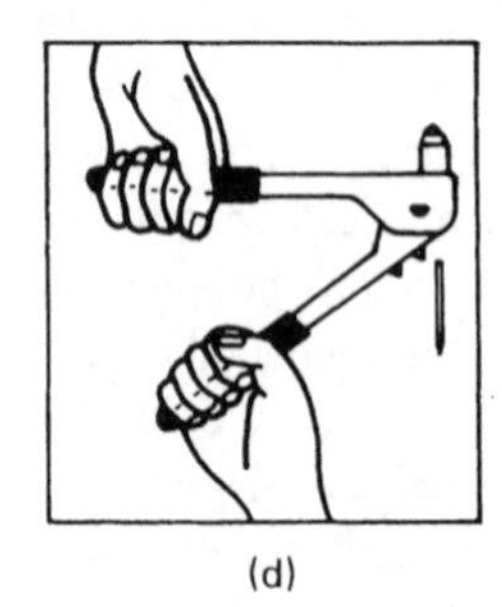

(d)

Fig. 5.7 Metal joining with pop rivets.

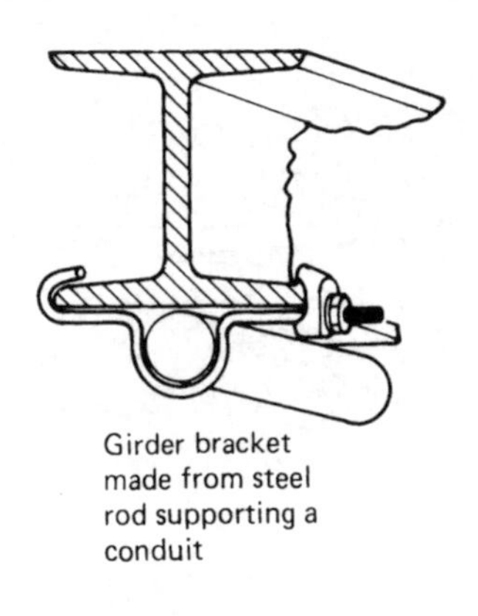

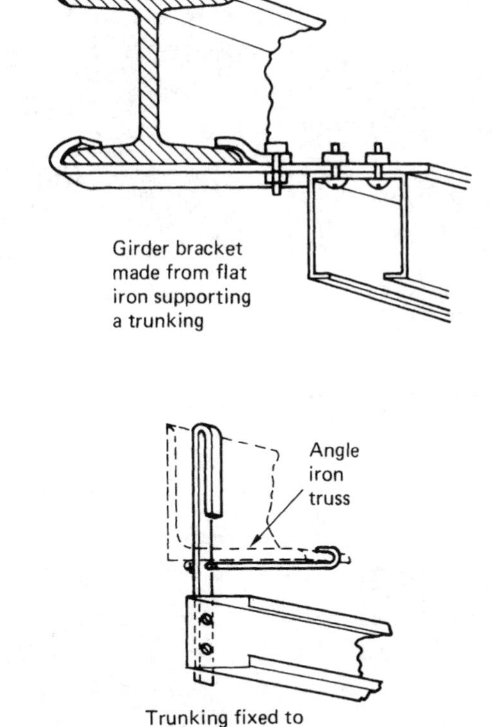

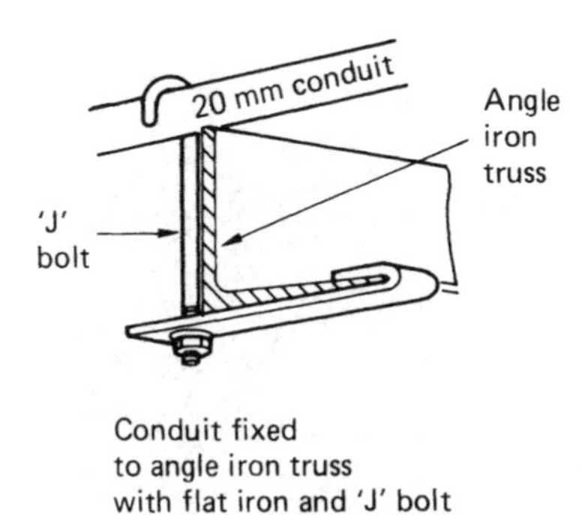

Fig. 5.8 Bracket supports for conduits and trunking.

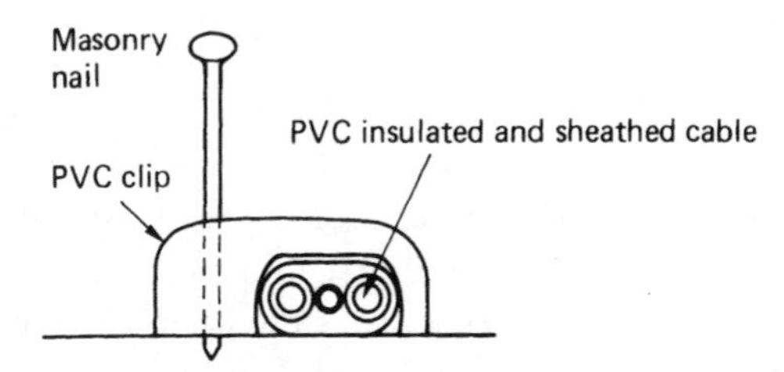

Fig. 5.9 PVC insulated and sheathed cable clip.

which is then plugged with a material to which the screw can be secured.

Plastic plugs

A plastic plug is made of a hollow plastic tube split up to half its length to allow for expansion. Each size of plastic plug is colour-coded to match a wood screw size.

A hole is drilled into the masonry, using a masonry drill of the same diameter, to the length of the plastic plug (see Fig. 5.10). The plastic plug is inserted into the hole and tapped home until it is level with the surface of the masonry. Finally, the fixing screw is driven into the plastic plug until it becomes tight and the fixture is secure.

Expansion bolts

The most common expansion bolt is made by Rawlbolt and consists of a split iron shell held together at one end by a steel ferrule and a spring wire clip at the other end. Tightening the bolt draws up an expanding bolt inside the split iron shell, forcing the iron to expand and grip the masonry. Rawlbolts are for heavy-duty masonry fixings (see Fig. 5.11).

A hole is drilled in the masonry to take the iron shell and ferrule. The iron shell is inserted with the spring wire clip end first so that the ferrule is at the

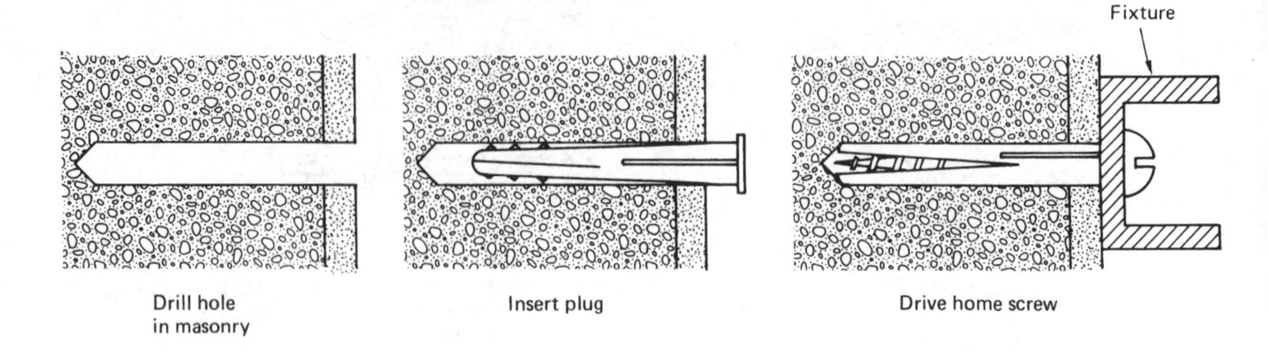

Fig. 5.10 Screw fixing to plastic plug.

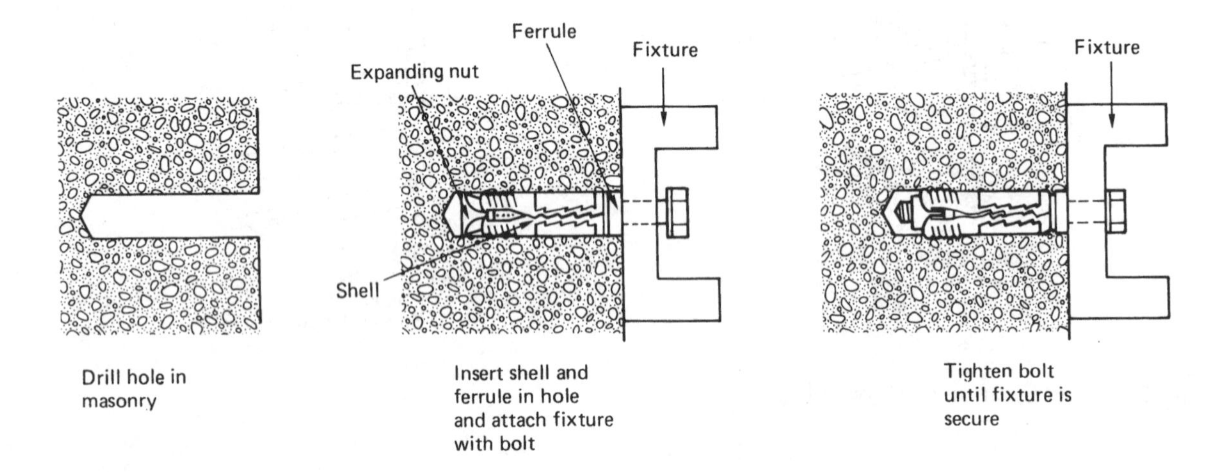

Fig. 5.11 Expansion bolt fixing.

outer surface. The bolt is passed through the fixture, located in the expanding nut and tightened until the fixing becomes secure.

Spring toggle bolts

A spring toggle bolt provides one method of fixing to hollow partition walls which are usually faced with plasterboard and a plaster skimming. Plasterboard and plaster wall or ceiling surfaces are not strong enough to support a load fixed directly into the plasterboard, but the spring toggle spreads the load over a larger area, making the fixing suitable for light loads (see Fig. 5.12).

A hole is drilled through the plasterboard and into the cavity. The toggle bolt is passed through the fixture and the toggle wings screwed into the bolt. The toggle wings are compressed and passed through the hole in the plasterboard and into the cavity where they spring apart and rest on the cavity side of the plasterboard. The bolt is tightened until the fixing becomes firm. The bolt of the spring toggle cannot be removed after fixing without loosening the toggle wings. If it becomes necessary to remove and refix the fixture a new toggle bolt will have to be used.

Making good

An electrician spends most of his working life in other people's premises, which might be part of a construc-

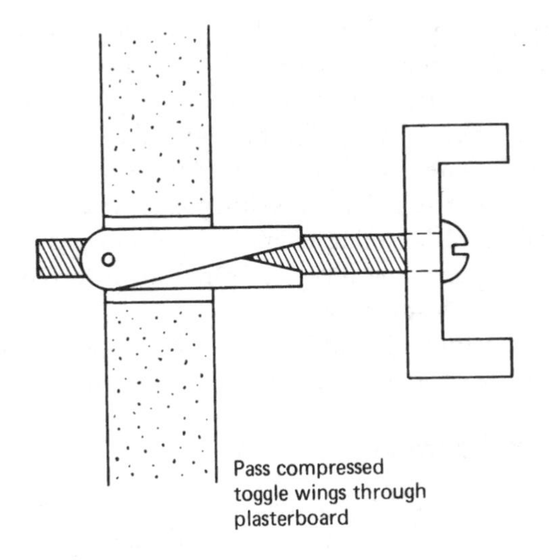

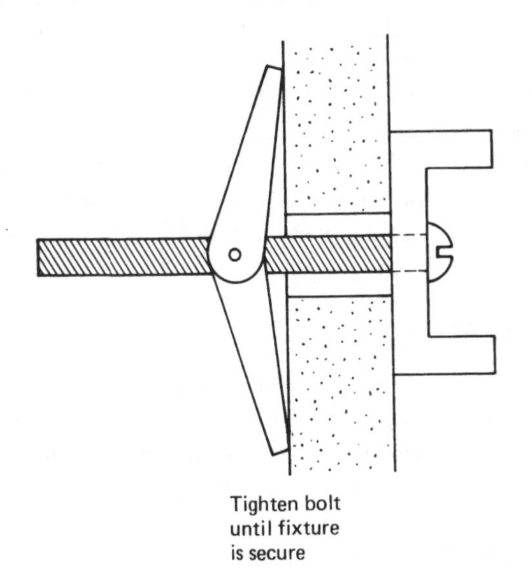

Fig. 5.12 Spring toggle bolt fixing.

tion site or a building where the customer's normal business is being carried out. If at all possible, the intrusion of an electrician to carry out installation work should not interfere with other workers or the general occupants of the building. Every effort should be made not to damage the customer's property, but where this cannot be avoided the damage should be made good (Regulation 527–02). This might involve plastering up a chase in plaster to a switch drop or making good brickwork where trunking passes through a wall or floor. If damage to the fabric of the building is anticipated as a result of the installation of electrical equipment it is always better to discuss the extent of the damage with the customer before carrying out the work. An electrician should work neatly and tidily, always cleaning up when the work is completed, or at the end of the day, and at all times give consideration to others in an effort to maintain the best relationship with everyone concerned.

Protection of materials

All the materials used in an electrical installation must be in perfect condition when installed if they are to give good service during the lifetime of the installation. It is the electrical contractor's responsibility to ensure that the materials which he installs meet the design specifications and are in perfect condition.

Sufficient materials must be stored on-site to ensure continuity of the installation work, thereby avoiding time loss due to waiting for material deliveries. Materials which are transported to the site should be checked upon arrival to ensure that they are exactly the materials ordered, and that they are in good condition. The materials should be stored as close as possible to the point of installation, so that the electrician's time is spent installing the materials and not transporting them.

The materials store must be appropriate for the type of material being stored. In general, the store should be dry in order to prevent corrosion of equipment, tools and materials, and maintained at a temperature well above 0°C, because PVC cables become brittle at low temperatures. Fragile materials such as switches, sockets, lampholders, thermostats, consumer units and luminaires must be handled and stored in a way which will prevent damage occurring.

Electrical materials are expensive and attractive to thieves, and pilfering by other site workers can occur. These losses reduce profit margins and causes inconvenience when further materials have to be reordered. There might also be long delivery times on such materials. To ensure the security of site materials, the store must be locked with a good-quality padlock and the key held by a responsible person. All materials, tools and equipment which are not in use *must* be locked away, otherwise they may be stolen.

Removal of materials

Upon completion of the job, the site should be thoroughly cleaned up and all unused plant and materials returned to head office. Waste materials must be disposed of appropriately and responsibly either by placing them in the main contractor's skip or by transporting them to the local authority's refuse tip.

On-site communications

Good communication is about transferring information from one person to another. Electricians and other professionals in the construction trades communicate with each other and the general public by means of drawings, sketches and symbols, in addition to what we say and do.

DRAWINGS AND DIAGRAMS

Many different types of electrical drawing and diagram can be identified: layout, schematic, block, wiring and circuit diagrams. The type of diagram to be used in any particular application is the one which most clearly communicates the desired information.

Layout drawings

These are scale drawings based upon the architect's site plan of the building and show the positions of the electrical equipment which is to be installed. The electrical equipment is identified by a graphical symbol.

The standard symbols used by the electrical contracting industry are those recommended by the British Standard EN 60617, *Graphical Symbols for Electrical Power, Telecommunications and Electronic Diagrams*. Some of the more common electrical installation symbols are given in Fig. 5.13.

A layout drawing of a small domestic extension is shown in Fig. 5.14. It can be seen that the mains intake position, probably a consumer's unit, is situated in the store-room which also contains one light controlled by a switch at the door. The bathroom contains one lighting point controlled by a one-way switch at the door. The kitchen has two doors and a switch is installed at each door to control the fluorescent luminaire. There are also three double sockets situated around the kitchen. The sitting room has a two-way switch at each door controlling the centre lighting point. Two wall lights with built-in switches are to be wired, one at each side of the window. Two double sockets and one switched socket are also to be installed in the sitting room. The bedroom has two lighting points controlled independently by two one-way switches at the door.

The wiring diagrams and installation procedures for all these circuits can be found in the next chapter.

As-fitted drawings

When the installation is completed a set of drawings should be produced which indicate the final positions of all the electrical equipment. As the building and electrical installation progresses, it is sometimes necessary to modify the positions of equipment indicated on the layout drawing because, for example, the position of a doorway has been changed. The layout drawings indicate the original intentions for the positions of equipment, while the 'as-fitted' drawing indicates the actual positions of equipment upon completion of the job.

Detail drawings

These are additional drawings produced by the architect to clarify some point of detail. For example, a drawing might be produced to give a fuller description of the suspended ceiling arrangements.

Schematic diagrams

A schematic diagram is a diagram in outline of, for example, a motor starter circuit. It uses graphical symbols to indicate the interrelationship of the electrical elements in a circuit. These help us to understand the working operation of the circuit.

An electrical schematic diagram looks very like a circuit diagram. A mechanical schematic diagram gives a more complex description of the individual elements in the system, indicating, for example, acceleration, velocity, position, force sensing and viscous damping.

Block diagrams

A block diagram is a very simple diagram in which the various items or pieces of equipment are represented

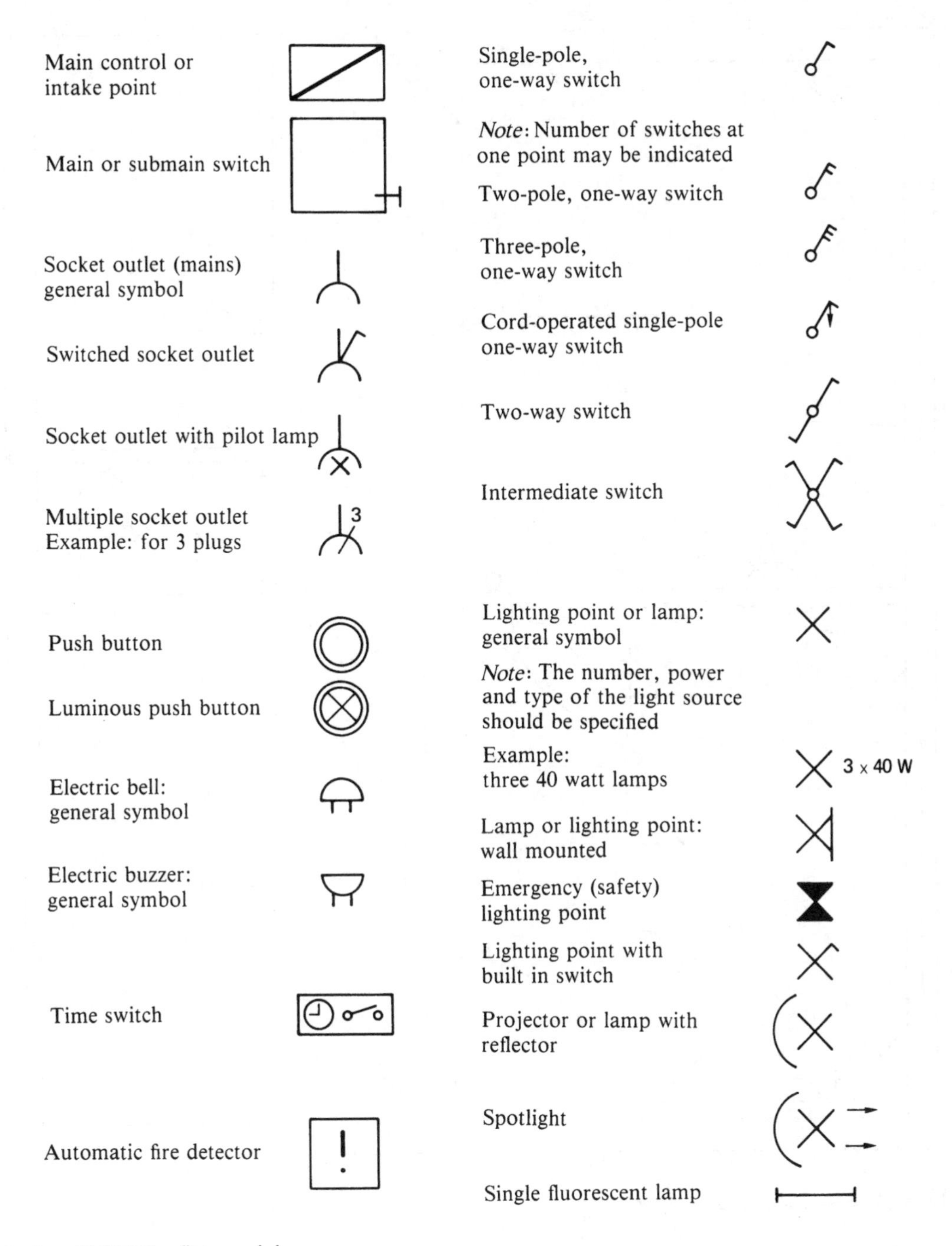

Fig. 5.13 Some EN 60617 installation symbols.

by a square or rectangular box. The purpose of the block diagram is to show how the components of the circuit relate to each other, and therefore the individual circuit connections are not shown. Figure 5.15 shows the block diagram of a space heating control system.

Wiring diagrams

A wiring diagram or connection diagram shows the detailed connections between components or items of equipment. They do not indicate how a piece of equipment or circuit works. The purpose of a wiring

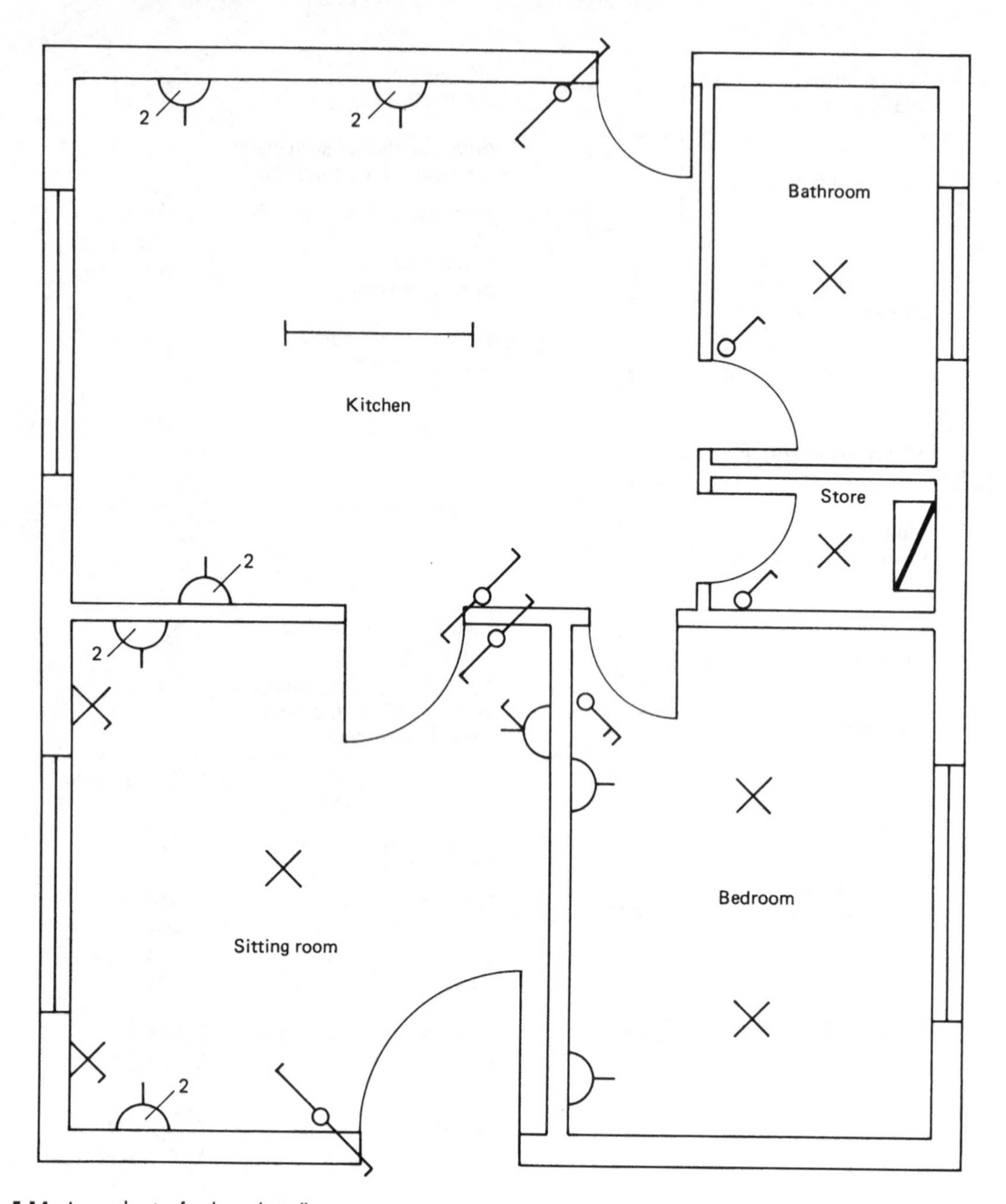

Fig. 5.14 Layout drawing for electrical installation.

diagram is to help someone with the actual wiring of the circuit. Figure 5.16 shows the wiring diagram for a space heating control system. Other wiring diagrams can be seen in Figs 6.23–6.26 in the next chapter.

Circuit diagrams

A circuit diagram shows most clearly how a circuit works. All the essential parts and connections are represented by their graphical symbols. The purpose of a circuit diagram is to help our understanding of the circuit. It will be laid out as clearly as possible, without regard to the physical layout of the actual components, and therefore it may not indicate the most convenient way to wire the circuit. Figure 5.17 shows the circuit diagram of our same space heating control system. Most of the diagrams in Chapters 8 and 10 are circuit diagrams.

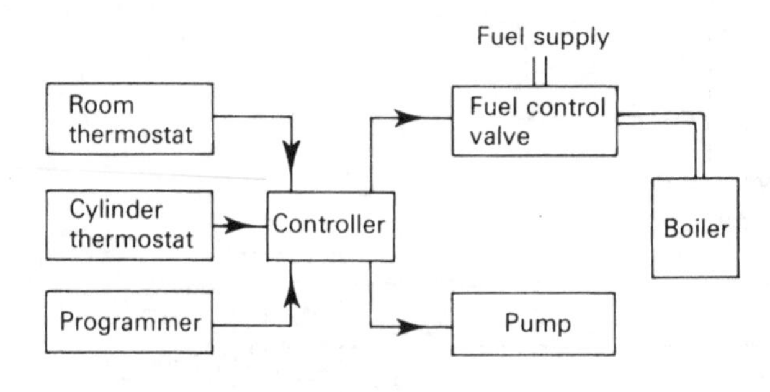

Fig. 5.15 Space heating control system.

Supplementary diagrams

A supplementary diagram conveys additional information in a way which is usually a mixture of the other categories of drawings. Figure 5.18 shows the supplementary diagram for our space heating control system and is probably the most useful diagram for initially setting out the wiring for the heating system.

Freehand working diagrams

Freehand working drawings or sketches are another important way in which we communicate our ideas. The drawings of the brackets earlier in this chapter in

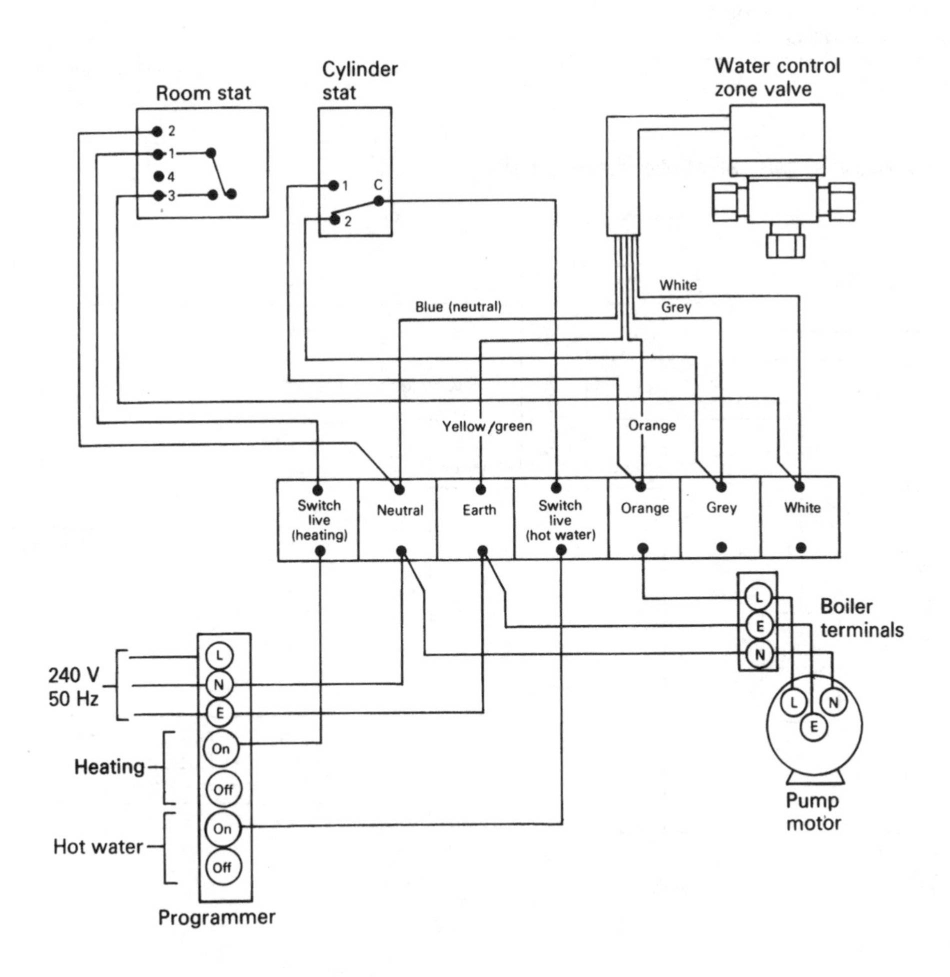

Fig. 5.16 Wiring diagram for space heating control (Honeywell 'Y' plan).

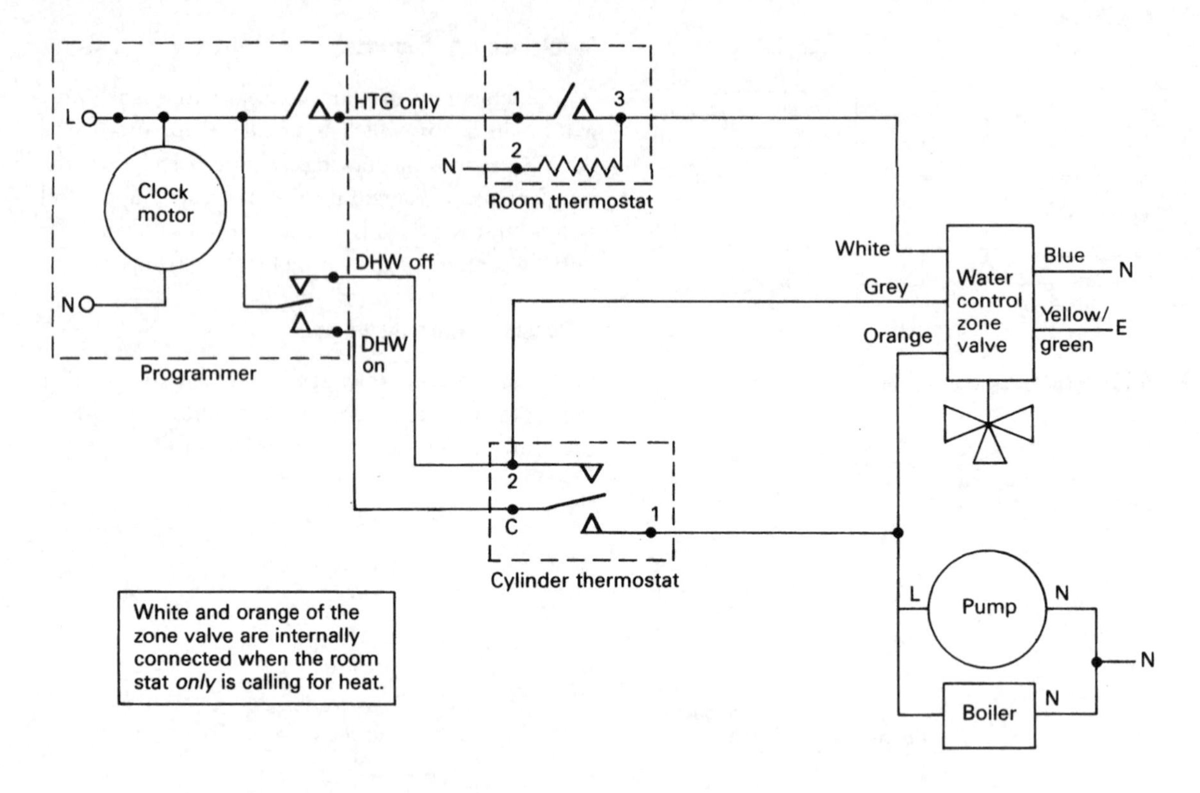

Fig. 5.17 Circuit diagram for space heating control (Honeywell 'Y' plan).

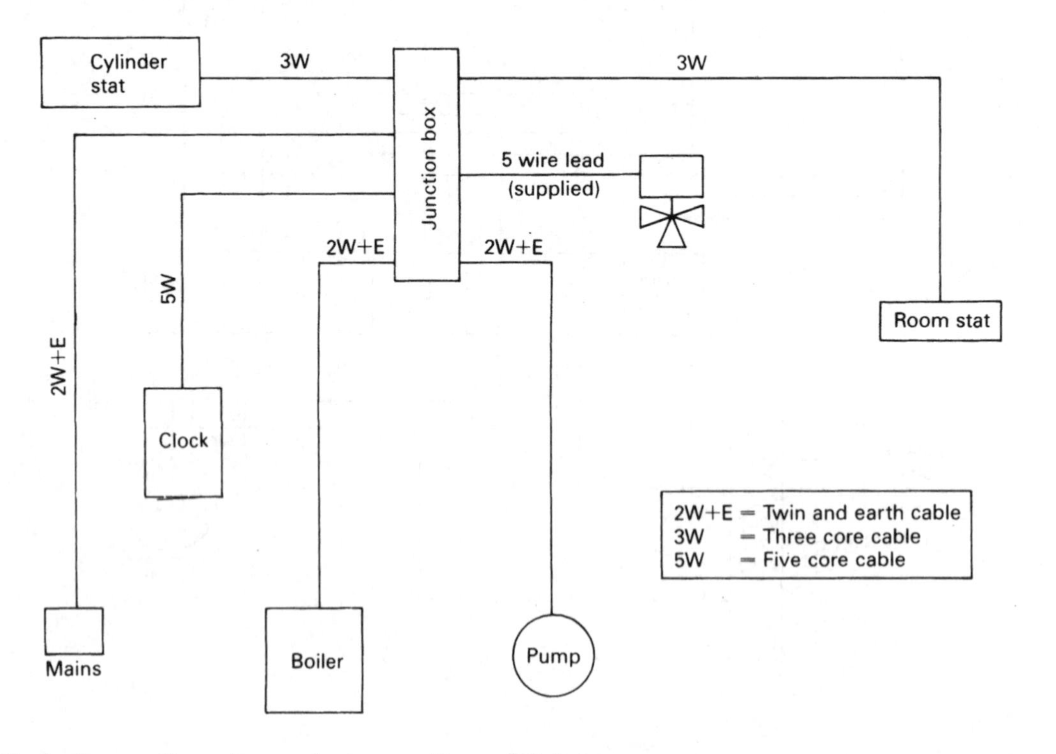

Fig. 5.18 Supplementary diagram for a space heating system (Honeywell 'Y' plan).

Fig. 5.8 were done from freehand sketches. A freehand sketch may be done as an initial draft of an idea before a full working drawing is made or, in the case of Fig. 5.8, it may be used to enable someone actually to make the bracket. It is often much easier to produce a sketch of your ideas or intentions than to describe them or produce a list of instructions.

To convey the message or information clearly it is better to make your sketch large rather than too small. It should also contain all the dimensions necessary to indicate clearly the size of the finished object depicted by the sketch.

You could practise freehand sketching by drawing some of the tools and equipment used in our trade and shown in Figs 5.1 and 5.4.

TELEPHONE MESSAGES

Telephones today play one of the most important roles in enabling people to communicate with each other. The advantage of a telephone message over a written message is its speed; the disadvantage is that no record is kept of an agreement made over the telephone. Therefore, business agreements made on the telephone are often followed up by written confirmation.

When *taking* a telephone call, remember that you cannot be seen and, therefore, gestures and facial expressions will not help to make you understood. Always be polite and helpful when answering your company's telephone – you are your company's most important representative at that moment. Speak clearly and loud enough to be heard without shouting, sound cheerful and write down messages if asked. Always read back what you have written down to make sure that you are passing on what the caller intended.

Many companies now use standard telephone message pads such as that shown in Fig. 5.19 because they prompt people to collect all the relevant information. In this case John Gall wants Dave Twem to pick up

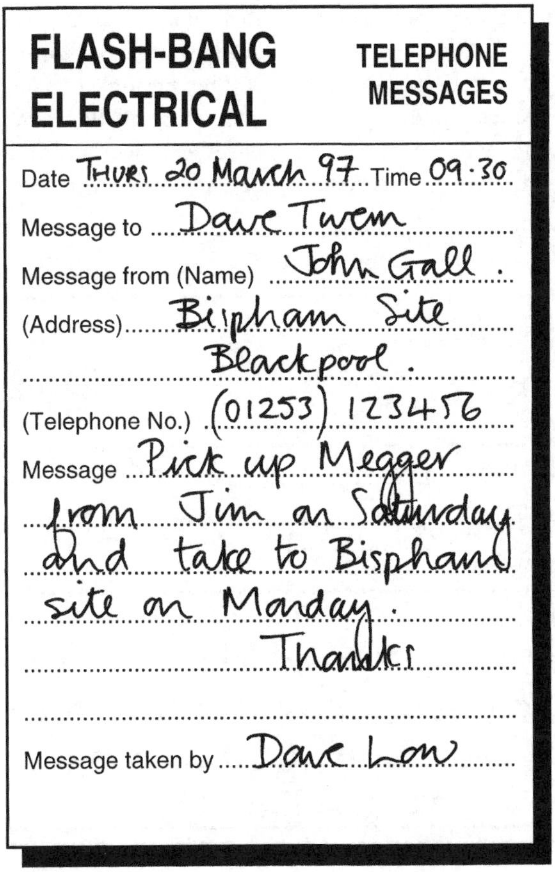

FLASH-BANG ELECTRICAL — TELEPHONE MESSAGES

Date Thurs 20 March 97 Time 09:30

Message to Dave Twem

Message from (Name) John Gall.

(Address) Bispham Site
Blackpool.

(Telephone No.) (01253) 123456

Message Pick up Megger from Jim on Saturday and take to Bispham site on Monday.
Thanks

Message taken by Dave Low

Fig. 5.19 Typical standard telephone message pad.

the Megger from Jim on Saturday and take it to the Bispham site on Monday. The person taking the call and relaying the message is Dave Low.

When *making* a telephone call, make sure you know what you want to say or ask. Make notes so that you have times, dates and any other relevant information ready before you make the call.

WRITTEN MESSAGES

A lot of communications between and within larger organizations take place by completing standard forms or sending internal memos. Written messages have the advantage of being 'auditable'. An auditor can follow the paperwork trail to see, for example, who was responsible for ordering certain materials.

When completing standard forms, follow the instructions given and ensure that your writing is legible. Do not leave blank spaces on the form, always specifying 'not applicable' or 'N/A' whenever necessary. Sign or give your name and the date as asked for on the form. Finally, read through the form again to make sure you have answered all the relevant sections correctly.

Internal memos are forms of written communication used within an organization; they are not normally used for communicating with customers or suppliers. Figure 5.20 shows the layout of a typical standard memo form used by Dave Twem to notify John Gall that he has ordered the hammer drill.

Letters provide a permanent record of communications between organizations and individuals. They may be handwritten, but formal business letters give a better impression of the organization if they are typewritten. A letter should be written using simple concise language, and the tone of the letter should always be polite even if it is one of complaint. Always include the date of the correspondence. The greeting on a formal letter should be 'Dear Sir/Madam' and concluded with 'Yours faithfully'. A less formal greeting would be 'Dear Mr Smith' and concluded 'Yours sincerely'. Your name and status should be typed below your signature.

DELIVERY NOTES

When materials are delivered to site, the person receiving the goods is required to sign the driver's 'delivery note'. This record is used to confirm that goods have been delivered by the supplier, who will then send out an invoice requesting payment, usually at the end of the month.

The person receiving the goods must carefully check that all the items stated on the delivery note have been delivered in good condition. Any missing or damaged items must be clearly indicated on the

FLASH-BANG ELECTRICAL — internal **MEMO**

From Dave Twem To John Gall

Subject Power Tools Date FRI 21 March 97

Message

Have today ordered Hammer Drill from P.S. Electrical – should be with you end of next week – Hope this is OK. Dave.

Fig. 5.20 Typical standard memo form.

delivery note before signing, because, by signing the delivery note the person is saying 'yes, these items were delivered to me as my company's representative on that date and in good condition and I am now responsible for these goods.' Copies of delivery notes are sent to head office so that payment can be made for the goods received.

TIME SHEETS

A time sheet is a standard form completed by each employee to inform the employer of the actual time spent working on a particular contract or site. This helps the employer to bill the hours of work to an individual job. It is usually a weekly document and includes the number of hours worked, the name of the job and any travelling expenses claimed.

JOB SHEETS

A job sheet or job card carries information about a job which needs to be done, usually a small job. It gives the name and address of the customer, contact telephone numbers, often a job reference number and a brief description of the work to be carried out. A typical job sheet work description might be:

- Job 1 Upstairs lights not working
- Job 2 Funny fishy smell from kettle socket in kitchen

An electrician might typically have a 'jobbing day' where he picks up a number of job sheets from the office and carries out the work specified.

Job 1, for example, might be the result of a blown fuse which is easily rectified, but the electrician must search a little further for the fault which caused the fuse to blow in the first place. The actual fault might, for example, be a decayed flex on a pendant drop which has become shorted out, blowing the fuse. The pendant drop would be re-flexed or replaced, along with any others in poor condition. The installation would then be tested for correct operation and the customer given an account of what has been done to correct the fault. General information and assurances about the condition of the installation as a whole might be requested and given before setting off to job 2.

The kettle socket outlet at job 2 is probably getting warm and, therefore, giving off that 'fishy' bakelite smell because loose connections are causing the bakelite socket to burn locally. A visual inspection would confirm the diagnosis. A typical solution would be to replace the socket and repair any damage to the conductors inside the socket box. Check the kettle plug top for damage and loose connections. Make sure all connections are tight before reassuring the customer that all is well; then, off to the next job or back to the office.

The time spent on each job and the materials used are sometimes recorded on the job sheet, but alternatively a daywork sheet can be used. This will depend upon what is normal practice for the particular electrical company. This information can then be used to 'bill' the customer for work carried out.

DAYWORK SHEETS

Daywork is one way of recording variations to a contract, that is, work done which is outside the scope of the original contract. If daywork is to be carried out, the site supervisor must first obtain a signature from the client's representative, for example, the architect, to authorize the extra work. A careful record must then be kept on the daywork sheets of all extra time and materials used so that the client can be billed for the extra work.

REPORTS

On large jobs, the foreman or supervisor is often required to keep a report of the relevant events which happen on the site – for example, how many people from your company are working on site each day, what goods were delivered, whether there were any breakages or accidents, and records of site meetings attended. Some firms have two separate documents, a site diary to record daily events and a weekly report which is a summary of the week's events extracted from the site diary. The site diary remains on-site and the weekly report is sent to head office to keep managers informed of the work's progress.

PERSONAL COMMUNICATIONS

Remember that it is the customers who actually pay the wages of everyone employed in your company. You should always be polite and listen carefully to their wishes. They may be elderly or of a different religion or cultural background than you. In a domestic situation, the playing of loud music on a radio may not be approved of. Treat the property in which you are working with the utmost care. When working in houses, shops and offices use dust sheets to protect floor coverings and furnishings. Clean up periodically and made a special effort when the job is completed.

Dress appropriately: an unkempt or untidy appearance will encourage the customer to think that your work will be of poor quality.

The electrical installation in a building is often carried out alongside other trades. It makes good sense to help other trades where possible and to develop good working relationships with other employees. The customer will be most happy if the workers give an impression of working together as a team for the successful completion of the project.

Finally, remember that the customer will probably see more of the electrician and the electrical trainee than the managing director of your firm and, therefore, the image presented by you will be assumed to reflect the policy of the company. You are, therefore, your company's most important representative. Always give the impression of being capable and in command of the situation, because this gives customers confidence in the company's ability to meet their needs. However, if a problem does occur which is outside your previous experience and you do not feel confident to solve it successfully, then contact your supervisor for professional help and guidance. It is not unreasonable for a young member of the company's team to seek help and guidance from those employees with more experience. This approach would be preferred by most companies rather than having to meet the cost of an expensive blunder.

Exercises

1 'A scale drawing showing the position of equipment by graphical symbols' is a description of a:
(a) block diagram
(b) layout diagram
(c) wiring diagram
(d) circuit diagram.

2 'A diagram which shows the detailed connections between individual items of equipment' is a description of a:
(a) block diagram
(b) layout diagram
(c) wiring diagram
(d) circuit diagram.

3 'A diagram which shows most clearly how a circuit works, with all items represented by graphical symbols' is a description of a:
(a) block diagram
(b) layout diagram
(c) wiring diagram
(d) circuit diagram.

4 A common method of joining together lengths of PVC conduit is by:
(a) pop rivets
(b) nuts and bolts
(c) solvent adhesives
(d) soldering.

5 A common method of joining together lengths of cable tray is by:
(a) pop rivets
(b) nuts and bolts
(c) solvent adhesives
(d) soldering.

6 A common method of permanently joining together thin sheet metal is by:
(a) pop rivets
(b) nuts and bolts
(c) solvent adhesives
(d) machine screws.

7 The best method of fixing a luminaire to a brick wall would be by:
(a) expansion bolts
(b) woodscrews into a plastic plug
(c) spring toggle bolts
(d) masonry nails.

8 The best method of fixing a luminaire to a plasterboard wall would be by:
(a) expansion bolts
(b) woodscrews into a plastic plug
(c) spring toggle bolts
(d) masonry nails.

9 The best method of fixing a large electric motor to a concrete floor would be by:
(a) expansion bolts
(b) woodscrews into a plastic plug
(c) spring toggle bolts
(d) masonry nails.

10 A record of work done which is outside the scope of the original contract would be kept on a:
(a) memo
(b) daywork sheet

(c) time sheet
(d) delivery note.

11 A record of goods delivered to site is recorded on a:
(a) memo
(b) daywork sheet
(c) time sheet
(d) delivery note.

12 Describe how you would arrange for electrical accessories, switchgear and cable to be stored on a large construction site.

13 Describe, with the aid of a sketch, a flat iron bracket suitable for fixing three 20 mm conduits to the bottom of an 'I' section rolled steel joist.

6

ELECTRICAL INSTALLATION – THEORY

Properties of materials

Let us first of all define some technical terms and discuss the properties of materials used in electrical installation work.

Conductor A material (usually a metal) which allows heat and electricity to pass easily through it.
Insulator A material (usually a non-metal) which will *not* allow heat and electricity to pass easily through it.
Ferrous A word used to describe all metals in which the main constituent is iron. The word 'ferrous' comes from the Latin word *ferrum* meaning iron. Ferrous metals have magnetic properties. Cast iron, wrought iron and steel are all ferrous metals.
Non-ferrous Metals which *do not* contain iron are called non-ferrous. They are non-magnetic and resist rusting. Copper, aluminium, tin, lead, zinc and brass are examples of non-ferrous metals.
Alloy An alloy is a mixture of two or more metals. Brass is an alloy of copper and zinc, usually in the ratio 70% to 30% or 60% to 40%.
Corrosion The destruction of a metal by chemical action. Most corrosion takes place when a metal is in contact with moisture (see also mild steel and zinc).
Thermoplastic polymers These may be repeatedly warmed and cooled without appreciable changes occurring in the properties of the material. They are good insulators, but give off toxic fumes when burned. They have a flexible quality when operated up to a maximum temperature of 70°C but should not be flexed when the air temperature is near 0°C, otherwise they may crack. Polyvinylchloride (PVC) used for cable insulation is a thermoplastic polymer.

Thermosetting polymers Once heated and formed, products made from thermosetting polymers are fixed rigidly. Plug tops, socket outlets and switch plates are made from this material.
Rubber is a tough elastic substance made from the sap of tropical plants. It is a good insulator, but degrades and becomes brittle when exposed to sunlight.
Synthetic rubber is manufactured, as opposed to being produced naturally. Synthetic or artificial rubber is carefully manufactured to have all the good qualities of natural rubber – flexibility, good insulation and suitability for use over a wide range of temperatures.
Silicon rubber Introducing organic compounds into synthetic rubber produces a good insulating material which is flexible over a wide range of temperatures and which retains its insulating properties even when burned. These properties make it ideal for cables used in fire alarm installations such as FP200 cables.
Magnesium oxide The conductors of mineral insulated metal sheathed (MICC) cables are insulated with compressed magnesium oxide, a white chalk-like substance which is heat-resistant and a good insulator and lasts for many years. The magnesium oxide insulation, copper conductors and sheath, often additionally manufactured with various external sheaths to provide further protection from corrosion and weather, produce a cable designed for long-life and high-temperature installations. However, the magnesium oxide is very hygroscopic, which means that it attracts moisture and, therefore, the cable must be terminated with a special moisture-excluding seal, as shown in Fig. 6.3.

COPPER

Copper is extracted from an ore which is mined in South Africa, North America, Australia and Chile. For electrical purposes it is refined to about 98.8% pure copper, the impurities being extracted from the ore by smelting and electrolysis. It is a very good conductor, is non-magnetic and offers considerable resistance to atmospheric corrosion. Copper toughens with work, but may be annealed, or softened, by heating to dull red before quenching.

Copper forms the largest portion of the alloy brass, and is used in the manufacture of electrical cables, domestic heating systems, refrigerator tubes and vehicle radiators. An attractive soft reddish brown metal, copper is easily worked and is also used to manufacture decorative articles and jewellery.

ALUMINIUM

Aluminium is a grey-white metal obtained from the mineral bauxite which is found in the USA, Germany and the Russian Federation. It is a very good conductor, is non-magnetic, offers very good resistance to atmospheric corrosion and is notable for its extreme softness and lightness. It is used in the manufacture of power cables. The overhead cables of the National Grid are made of an aluminium conductor reinforced by a core of steel. Copper conductors would be too heavy to support themselves between the pylons. Lightness and resistance to corrosion make aluminium an ideal metal for the manufacture of cooking pots and food containers.

Aluminium alloys retain the corrosion resistance properties of pure aluminium with an increase in strength. The alloys are cast into cylinder heads and gearboxes for motorcars, and switch-boxes and luminaires for electrical installations. Special processes and fluxes have now been developed which allow aluminium to be welded and soldered.

BRASS

Brass is a non-ferrous alloy of copper and zinc which is easily cast. Because it is harder than copper or aluminium it is easily machined. It is a good conductor and is highly resistant to corrosion. For these reasons it is often used in the electrical and plumbing trades. Taps, valves, pipes, electrical terminals, plug top pins and terminal glands for steel wire armour (SWA) and MI cables are some of the many applications.

Brass is an attractive yellow metal which is also used for decorative household articles and jewellery. The combined properties of being an attractive metal which is highly resistant to corrosion make it a popular metal for ships' furnishings.

CAST STEEL

Cast steel is also called tool steel or high carbon steel. It is an alloy of iron and carbon which is melted in airtight crucibles and then poured into moulds to form ingots. These ingots are then rolled or pressed into various shapes from which the finished products are made. Cast steel can be hardened and tempered and is therefore ideal for manufacturing tools (see also Chapter 5). Hammer heads, pliers, wire cutters, chisels, files and many machine parts are also made from cast steel.

MILD STEEL

Mild steel is also an alloy of iron and carbon but contains much less carbon than cast steel. It can be filed, drilled or sawn quite easily and may be bent when hot or cold, but repeated cold bending may cause it to fracture. In moist conditions corrosion takes place rapidly unless the metal is protected. Mild steel is the most widely used metal in the world, having considerable strength and rigidity without being brittle. Ships, bridges, girders, motorcar bodies, bicycles, nails, screws, conduit, trunking, tray and SWA are all made of mild steel.

ZINC

Zinc is a non-ferrous metal which is used mainly to protect steel against corrosion and in making the alloy brass. Mild steel coated with zinc is sometimes called *galvanized steel*, and this coating considerably improves steel's resistance to corrosion. Conduit, trunking, tray, steel wire armour, outside luminaires and electricity pylons are made of galvanized steel.

Construction of cables

Most cables can be considered to be constructed in three parts: the *conductor* which must be of a suitable cross-section to carry the load current; the *insulation*, which has a colour or number code for identification; and the *outer sheath* which may contain some means of providing protection from mechanical damage.

The conductors of a cable are made of either copper or aluminium and may be stranded or solid. Solid conductors are only used in fixed wiring installations and may be shaped in larger cables. Stranded conductors are more flexible and conductor sizes from 4.0 mm^2 to 25 mm^2 contain seven strands. A 10 mm^2 conductor, for example, has seven 1.35 mm diameter strands which collectively make up the 10 mm^2 cross-sectional area of the cable. Conductors above 25 mm^2 have more than seven strands, depending upon the size of the cable. Flexible cords have multiple strands of very fine wire, as fine as one strand of human hair. This gives the cable its very flexible quality.

PVC INSULATED AND SHEATHED CABLES

Domestic and commercial installations use this cable, which may be clipped direct to a surface, sunk in plaster or installed in conduit or trunking. It is the simplest and least expensive cable. Figure 6.1 shows a sketch of a twin and earth cable.

The conductors are covered with a colour-coded PVC insulation and then contained singly or with others in a PVC outer sheath.

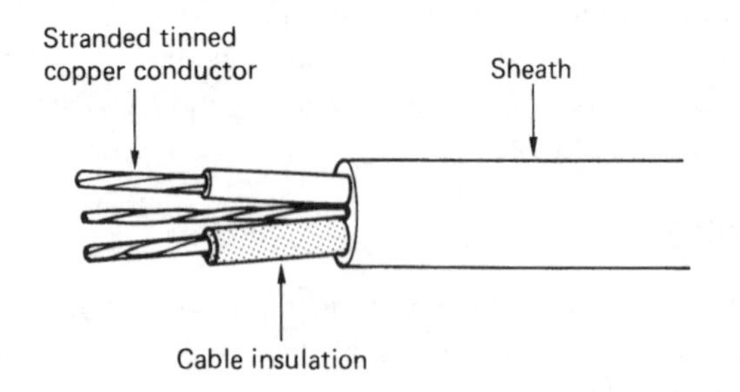

Fig. 6.1 A twin and earth PVC insulated and sheathed cable.

PVC/SWA cable

PVC insulated steel wire armour cables are used for wiring underground between buildings, for main supplies to dwellings, rising submains and industrial installations. They are used where some mechanical protection of the cable conductors is required.

The conductors are covered with colour-coded PVC insulation and then contained either singly or with others in a PVC sheath (see Fig. 6.2). Around this sheath is placed an armour protection of steel wires twisted along the length of the cable, and a final PVC sheath covering the steel wires protects them from corrosion. The armour sheath also provides the circuit protective conductor (CPC) and the cable is simply terminated using a compression gland.

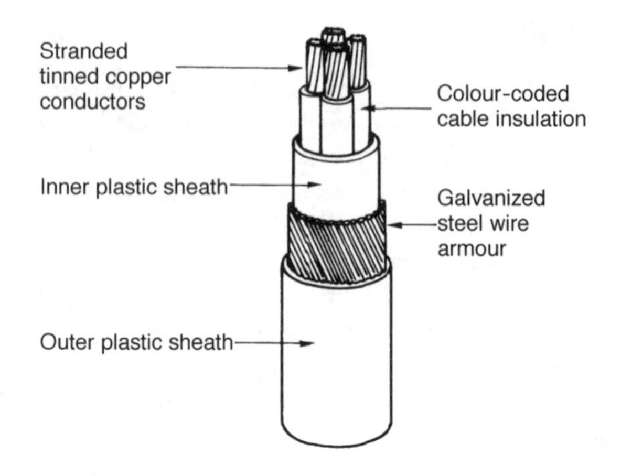

Fig. 6.2 A four-core PVC/SWA cable.

MI CABLE

A mineral insulated (MI) cable has a seamless copper sheath which makes it waterproof and fire- and corrosion-resistant. These characteristics often make it the only cable choice for hazardous or high-temperature installations such as oil refineries and chemical works, boiler-houses and furnaces, petrol pump and fire alarm installations.

The cable has a small overall diameter when compared to alternative cables and may be supplied as bare copper or with a PVC oversheath. It is colour-coded orange for general electrical wiring, white for emergency lighting or red for fire alarm wiring. The copper outer sheath provides the CPC, and the cable is terminated with a pot and sealed with compound and a compression gland (see Fig. 6.3).

The copper conductors are embedded in a white powder, magnesium oxide, which is non-ageing and non-combustible, but which is hygroscopic, which means that it readily absorbs moisture from the surrounding air, unless adequately terminated. The termination of an MI cable is a complicated process requiring the electrician to demonstrate a high level of practical skill and expertise for the termination to be successful.

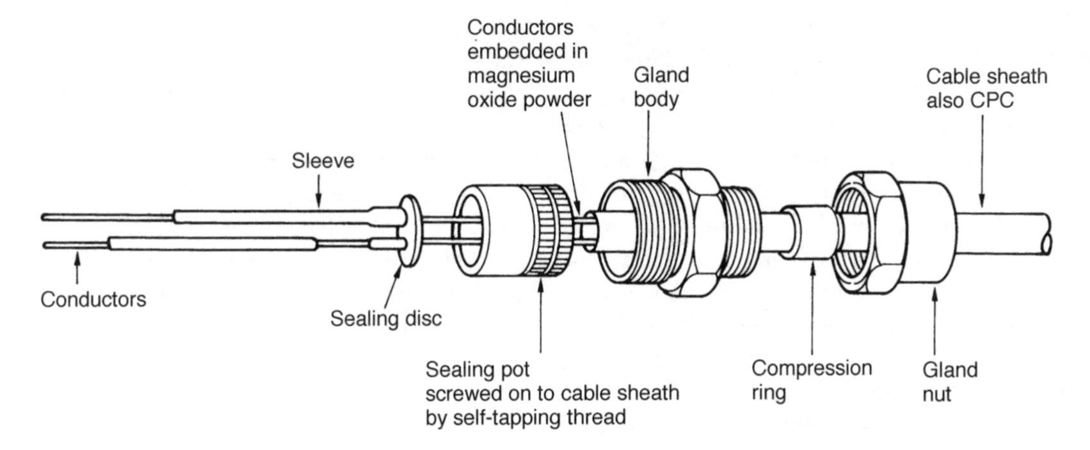

Fig. 6.3 MI cable with terminating seal and gland.

HIGH-VOLTAGE POWER CABLES

The cables used for high-voltage power distribution require termination and installation expertise beyond the normal experience of a contracting electrician. The regulations covering high-voltage distribution are beyond the scope of the IEE Regulations for electrical installations. Operating at voltages in excess of 33 kV and delivering thousands of kilowatts, these cables are either suspended out of reach on pylons or buried in the ground in carefully constructed trenches.

HIGH-VOLTAGE OVERHEAD CABLES

Suspended from cable towers or pylons, overhead cables must be light, flexible and strong.

The cable is constructed of stranded aluminium conductors formed around a core of steel stranded conductors (see Fig. 6.4). The aluminium conductors carry the current and the steel core provides the tensile strength required to suspend the cable between pylons. The cable is not insulated since it is placed out of reach and insulation would only add to the weight of the cable.

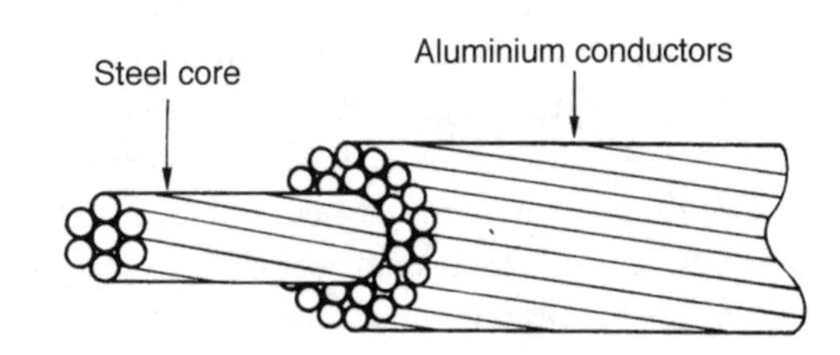

Fig. 6.4 132 kV overhead cable construction.

HIGH-VOLTAGE UNDERGROUND CABLES

High-voltage cables are only buried underground in special circumstances when overhead cables would be unsuitable, for example, because they might spoil a view of natural beauty (see Chapter 1). Underground cables are very expensive because they are much more complicated to manufacture than overhead cables. In transporting vast quantities of power, heat is generated within the cable. This heat is removed by passing oil through the cable to expansion points, where the

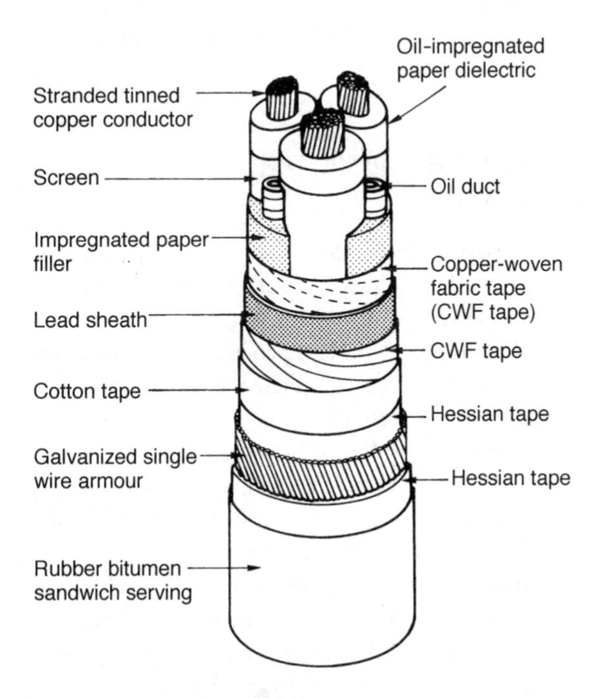

Fig. 6.5 132 kV underground cable construction.

oil is cooled. The system is similar to the water cooling of an internal combustion engine. Figure 6.5 shows a typical high voltage cable construction.

The conductors may be aluminium or copper, solid or stranded. They are insulated with oil-impregnated brown paper wrapped in layers around the conductors. The oil ducts allow the oil to flow through the cable, removing excess heat. The whole cable within the lead sheath is saturated with oil, which is a good insulator. The lead sheath keeps the oil in and moisture out of the cable, and this is supported by the copper-woven fabric tape. The cable is protected by steel wire armouring, which has bitumen or PVC serving over it to protect the armour sheath from corrosion. The termination and installation of these cables is a very specialized job, undertaken by the supply authorities only.

Installing wiring systems

The final choice of a wiring system must rest with those designing the installation and those ordering the work, but whatever system is employed, good workmanship is essential for compliance with the Regulations. The necessary skills can be acquired by an electrical trainee who has the correct attitude and dedication to his craft.

PVC INSULATED AND SHEATHED CABLE INSTALLATIONS

PVC insulated and sheathed wiring systems are used extensively for lighting and socket installations in domestic dwellings. Mechanical damage to the cable caused by impact, abrasion, penetration, compression or tension must be minimized during installation (Regulation 522–06–01). The cables are generally fixed using plastic clips incorporating a masonry nail, which means the cables can be fixed to wood, plaster or brick with almost equal ease. Cables should be run horizontally or vertically, not diagonally, down a wall. All kinks should be removed so that the cable is run straight and neatly between clips fixed at equal distances providing adequate support for the cable so that it does not become damaged by its own weight (Regulation 522–08–04 and Table 4A of the *On Site Guide*). Where cables are bent, the radius of the bend should not cause the conductors to be damaged (Regulation 522–08–03 and Table 4E of the *On Site Guide*).

Terminations or joints in the cable may be made in ceiling roses, junction boxes, or behind sockets or switches, provided that they are enclosed in a non-ignitable material, are properly insulated and are mechanically and electrically secure (IEE Regulation 526). All joints must be accessible for inspection testing and maintenance when the installation is completed.

Where PVC insulated and sheathed cables are concealed in walls, floors or partitions, they must be provided with a box incorporating an earth terminal at each outlet position. PVC cables do not react chemically with plaster, as do some cables, and consequently PVC cables may be buried under plaster. Further protection by channel or conduit is only necessary if mechanical protection from nails or screws is required. However, Regulation 522–06–06 now tells us that where PVC cables are to be embedded in wet plaster, they should be covered in capping to protect them from the plasterer's trowel. Figure 6.6 shows a typical PVC installation. To identify the most probable cable routes, Regulation 522–06–06 tells us that outside a zone formed by a 150 mm border all around a wall edge, cables can only be run horizontally or vertically to a point or accessory unless they are contained in a substantial earthed enclosure, such as a conduit, which can withstand nail penetration, as shown in Fig. 6.7.

Where cables pass through walls, floors and ceilings the hole should be made good with incombustible material such as mortar or plaster to prevent the spread of fire (Regulation 527–02–01). Cables passing through metal boxes should be bushed with a rubber grommet to prevent abrasion of the cable. Holes drilled in floor joists through which cables are run should be 50 mm below the top or 50 mm above the bottom of the joist to prevent damage to the cable by nail penetration (Regulation 522–06–05), as shown in Fig. 6.8. PVC cables should not be installed when the surrounding temperature is below 0°C or when the cable temperature has been below 0°C for the previous 24 hours because the insulation becomes brittle at low temperatures and may be damaged during installation.

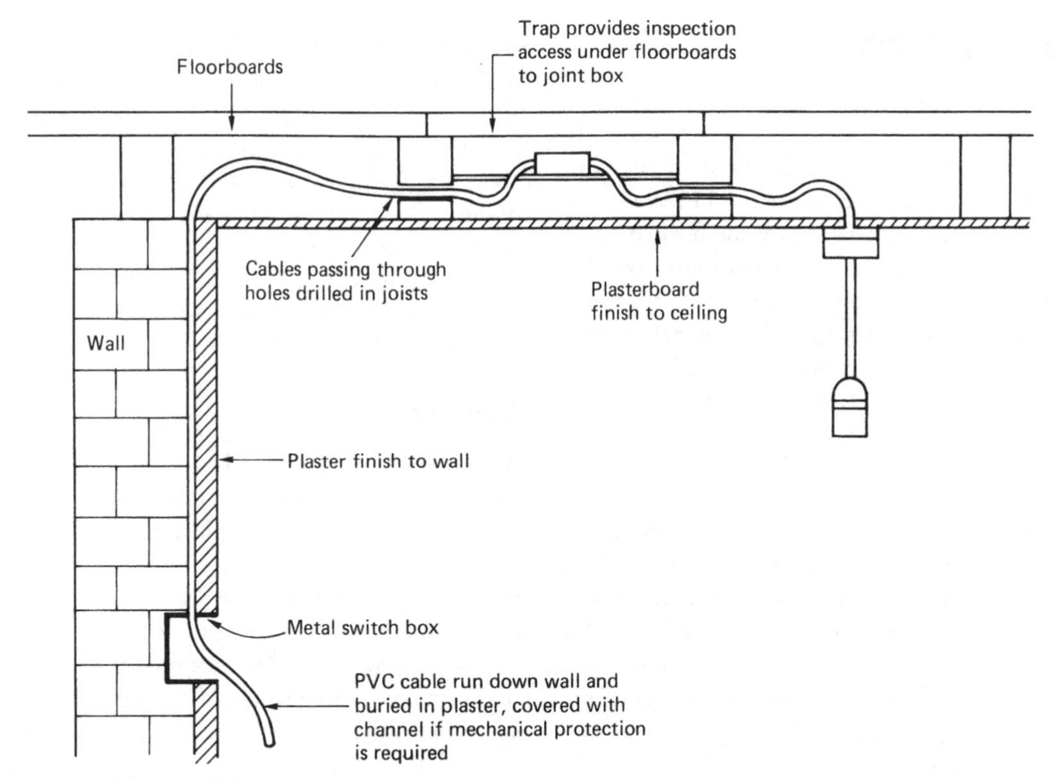

Fig. 6.6 A concealed PVC sheathed wiring system.

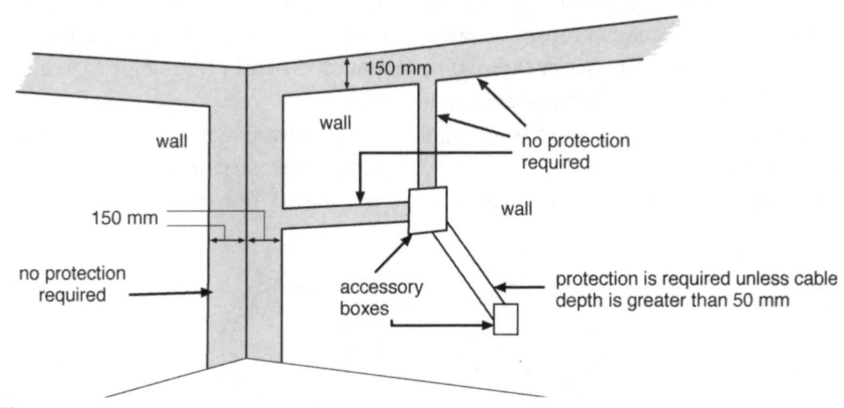

Fig. 6.7 Permitted cable routes.

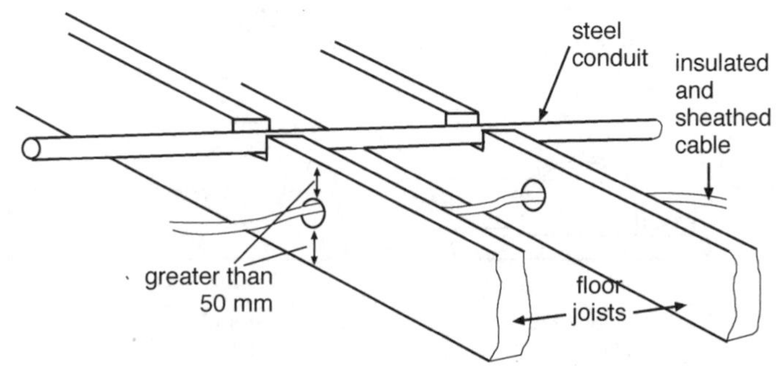

Fig. 6.8 Correct installation of conductors in floor joists.

CONDUIT INSTALLATIONS

A conduit is a tube, channel or pipe in which insulated conductors are contained. The conduit, in effect, replaces the PVC outer sheath of a cable, providing mechanical protection for the insulated conductors. A conduit installation can be rewired easily or altered at any time, and this flexibility, coupled with mechanical protection, makes conduit installations popular for commercial and industrial applications. There are three types of conduit used in electrical installation work: steel, PVC and flexible.

Steel conduit

Steel conduits are made to a specification defined by BS 4568 and are either heavy gauge welded or solid drawn. Heavy gauge is made from a sheet of steel welded along the seam to form a tube and is used for most installation work. Solid drawn conduit is a seamless tube which is much more expensive and only used for special gas-tight, explosion-proof or flame-proof installations.

Conduit is supplied in 3.75 m lengths and typical sizes are 16, 20, 25 and 32 mm. Conduit tubing and fittings are supplied in a black enamel finish for internal use or hot galvanized finish for use on external or damp installations. A wide range of fittings are available and the conduit is fixed using saddles or pipe hooks, as shown in Fig. 6.9.

Metal conduits are threaded with stocks and dies and bent using special bending machines. The metal conduit is also utilized as the circuit protective conductor and, therefore, all connections must be screwed up tightly and all burrs removed so that cables will not be damaged as they are drawn into the conduit. Metal conduits containing a.c. circuits must contain phase and neutral conductors in the same conduit to prevent eddy currents flowing, which would result in the metal conduit becoming hot (Regulation 521–02–01).

PVC conduit

PVC conduit used on typical electrical installations is heavy gauge standard impact tube manufactured to BS 4607. The conduit size and range of fittings are the same as those available for metal conduit. PVC conduit is most often joined by placing the end of the conduit into the appropriate fitting and fixing with a PVC solvent adhesive. PVC conduit can be bent by hand using a bending spring of the same diameter as the inside of the conduit. The spring is pushed into the conduit to the point of the intended bend and the conduit then bent over the knee. The spring ensures that the conduit keeps its circular shape. In cold weather, a little warmth applied to the point of the intended bend often helps to achieve a more successful bend.

The advantages of a PVC conduit system are that it may be installed much more quickly than steel conduit and is non-corrosive, but it does not have the

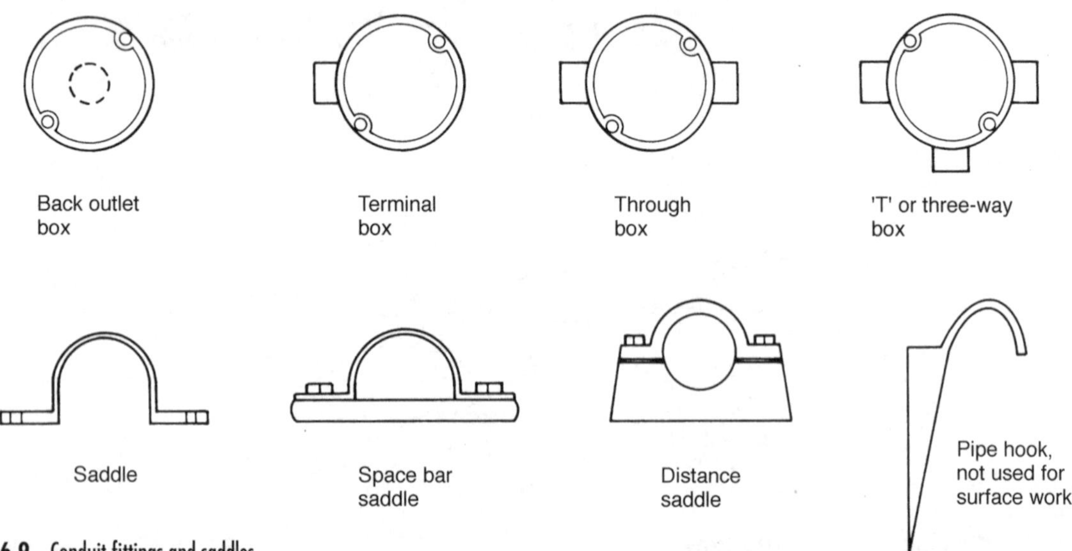

Fig. 6.9 Conduit fittings and saddles.

mechanical strength of steel conduit. Since PVC conduit is an insulator it cannot be used as the CPC and a separate earth conductor must be run to every outlet. It is not suitable for installations subjected to temperatures below –5°C or above 60°C. Where luminaires are suspended from PVC conduit boxes, precautions must be taken to ensure that the lamp does not raise the box temperature or that the mass of the luminaire supported by each box does not exceed the maximum recommended by the manufacturer (IEE Regulation 522–01). PVC conduit also expands much more than metal conduit and so long runs require an expansion coupling to allow for conduit movement and help to prevent distortion during temperature changes.

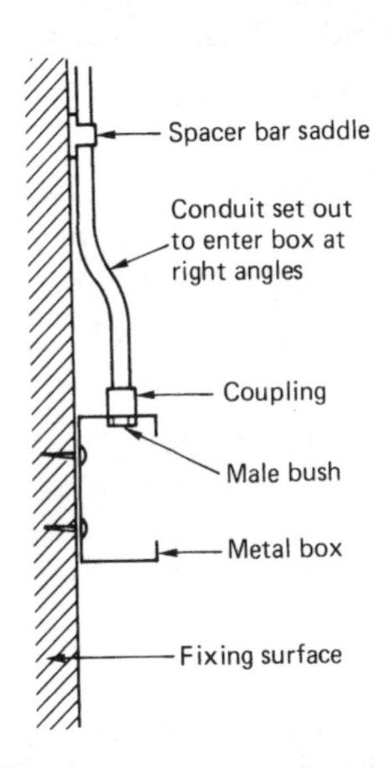

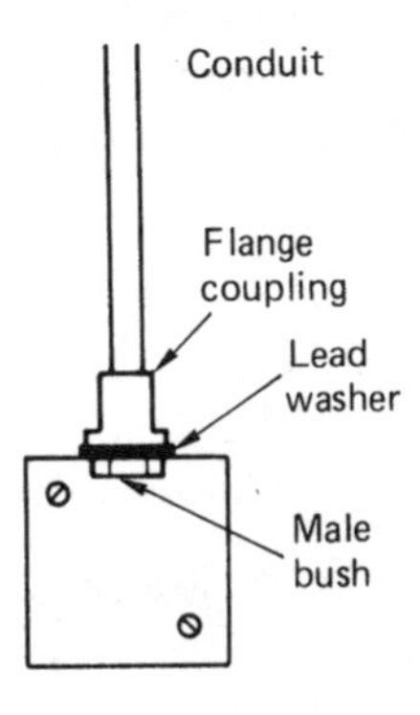

Fig. 6.10 Terminating conduits.

All conduit installations must be erected first before any wiring is installed (IEE Regulation 522–08–02). The radius of all bends in conduit must not cause the cables to suffer damage, and therefore the minimum radius of bends given in Table 4E of the *On Site Guide* applies (IEE Regulation 522–08–03). All conduits should terminate in a box or fitting and meet the boxes or fittings at right angles, as shown in Fig. 6.10. Any unused conduit box entries should be blanked off and all boxes covered with a box lid, fitting or accessory to provide complete enclosure of the conduit system. Conduit runs should be separate from other services, unless intentionally bonded, to prevent arcing occurring from a faulty circuit within the conduit, which might cause the pipe of another service to become punctured.

When drawing cables into conduit they must first be *run off* the cable drum. That is, the drum must be rotated as shown in Fig. 6.11 and not allowed to *spiral off*, which will cause the cable to twist.

Cables should be fed into the conduit in a manner which prevents any cable crossing over and becoming twisted inside the conduit. The cable insulation must not be damaged on the metal edges of the draw-in box. Cables can be pulled in on a draw wire if the run is a long one. The draw wire itself may be drawn in on a fish tape, which is a thin spring steel or plastic tape.

A limit must be placed on the number of bends between boxes in a conduit run and the number of cables which may be drawn into a conduit to prevent the cables being strained during wiring. Appendix 5 of the *On Site Guide* gives a guide to the cable capacities of conduits and trunking.

Flexible conduit

Flexible conduit is made of interlinked metal spirals often covered with a PVC sleeving. The tubing must not be relied upon to provide a continuous earth path and, consequently, a separate CPC must be run either inside or outside the flexible tube (Regulation 543–02–01).

Flexible conduit is used for the final connection to motors so that the vibrations of the motor are not transmitted throughout the electrical installation and to allow for modifications to be made to the final motor position and drive belt adjustments.

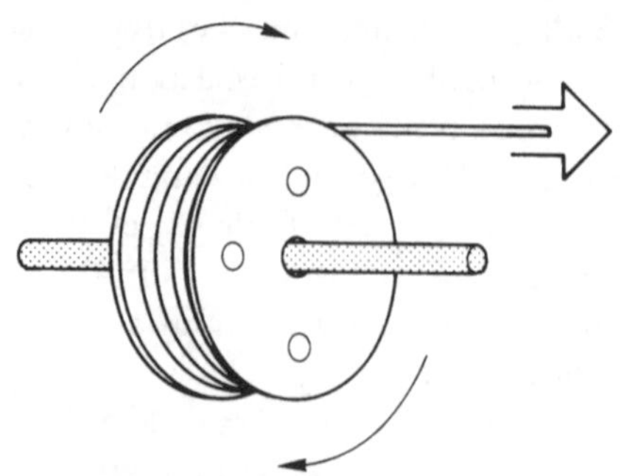

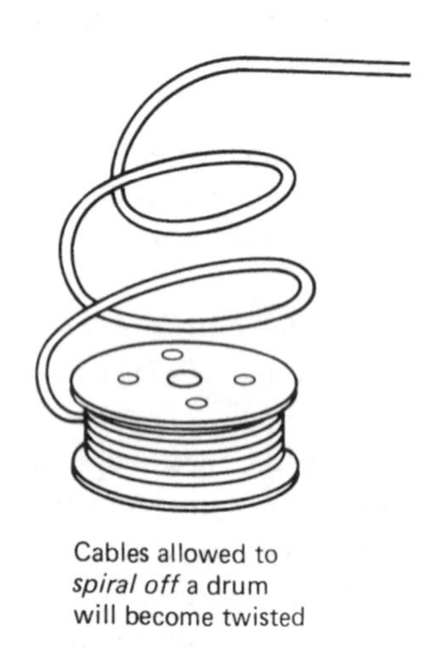

Fig. 6.11 Running off cable from a drum.

Conduit capacities

Single PVC insulated conductors are usually drawn into the installed conduit to complete the installation. Having decided upon the type, size and number of cables required for a final circuit, it is then necessary to select the appropriate size of conduit to accommodate those cables.

The tables in Appendix 5 of the *On Site Guide* describe a 'factor system' for determining the size of conduit required to enclose a number of conductors. The method is as follows:

- Identify the cable factor for the particular size of conductor. (This is given in Table 5A for straight conduit runs and Table 5C for cables run in conduits which incorporate bends.)
- Multiply the cable factor by the number of conductors, to give the sum of the cable factors.
- Identify the appropriate part of the conduit factor table given by the length of run and number of bends. (For straight runs of conduit less than 3 m in length, the conduit factors are given in Table 5B. For conduit runs in excess of 3 m or incorporating bends, the conduit factors are given in Table 5D.)
- The correct size of conduit to accommodate the cables is that conduit which has a factor equal to or greater than the sum of the cable factors.

EXAMPLE 1

Six 2.5 mm² PVC insulated cables are to be run in a conduit containing two bends between boxes 10 m apart. Determine the minimum size of conduit to contain these cables.

From Table 5C,
The factor for one 2.5 mm² cable $= 30$
The sum of the cable factors $= 6 \times 30$
$= 180$

From Table 5D, a 25 mm conduit, 10 m long and containing two bends, has a factor of 260. A 20 mm conduit containing two bends only has a factor of 141 which is less than 180, the sum of the cable factors and, therefore, 25 mm conduit is the minimum size to contain these cables.

EXAMPLE 2

Ten 1.0 mm² PVC insulated cables are to be drawn into a plastic conduit which is 6 m long between boxes and contains one bend. A 4.0 mm PVC insulated CPC is also included. Determine the minimum size of conduit to contain these conductors.

From Table 5C,
the factor for one 1.0 mm cable $= 16$
the factor for one 4.0 mm cable $= 43$.
The sum of the cable factors $= (10 \times 16) + (1 \times 43) = 203$.

From Table 5D, a 20 mm conduit, 6 m long and containing one bend, has a factor of 233. A 16 mm conduit containing one bend only has a factor of 143 which is less than 203, the sum of the cable factors and, therefore, 20 mm conduit is the minimum size to contain these cables.

TRUNKING INSTALLATIONS

A trunking is an enclosure provided for the protection of cables which is normally square or rectangular in cross-section, having one removable side. Trunking may be thought of as a more accessible conduit system and for industrial and commercial installations it is replacing the larger conduit sizes. A trunking system can have great flexibility when used in conjunction with conduit; the trunking forms the background or framework for the installation, with conduits running from the trunking to the point controlling the current using apparatus. When an alteration or extension is required it is easy to drill a hole in the side of the trunking and run a conduit to the new point. The new wiring can then be drawn through the new conduit and the existing trunking to the supply point.

Trunking is supplied in 3 m lengths and various cross-sections measured in millimetres from 50 × 50 up to 300 × 150. Most trunking is available in either steel or plastic.

Metallic trunking

Metallic trunking is formed from mild steel sheet, coated with grey or silver enamel paint for internal use or a hot-dipped galvanized coating where damp conditions might be encountered. A wide range of accessories are available, such as 45° bends, 90° bends, tee and four-way junctions, for speedy on-site assembly. Alternatively, bends may be fabricated in lengths of trunking, as shown in Fig. 6.12. This may be necessary or more convenient if a bend or set is non-standard, but it does take more time to fabricate bends than merely to bolt on standard accessories.

When fabricating bends the trunking should be supported with wooden blocks for sawing and filing, in order to prevent the sheet steel vibrating or becoming deformed. Fish plates must be made and riveted or bolted to the trunking to form a solid and secure bend. When manufactured bends are used, the continuity of the earth path must be ensured across the joint by making all fixing screw connections very tight, or fitting a separate copper strap between the

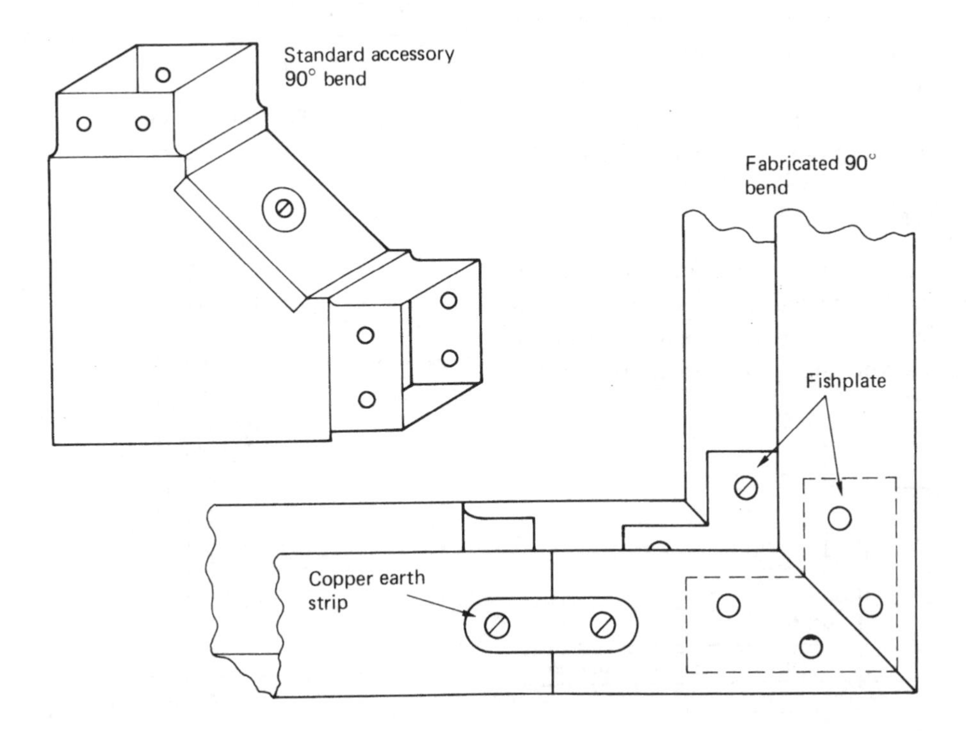

Fig. 6.12 Alternative trunking bends.

trunking and the standard bend. If an earth continuity test on the trunking is found to be unsatisfactory, an insulated CPC must be installed inside the trunking. The size of the protective conductor will be determined by the largest cable contained in the trunking, as described by Table 54G of the IEE Regulations.

Non-metallic trunking

Trunking and trunking accessories are also available in high-impact PVC. The accessories are usually secured to the lengths of trunking with a PVC solvent adhesive. PVC trunking, like PVC conduit, is easy to install and is non-corrosive. A separate CPC will need to be installed and non-metallic trunking may require more frequent fixings because it is less rigid than metallic trunking. All trunking fixings should use round-headed screws to prevent damage to cables since the thin sheet construction makes it impossible to countersink screw heads.

Mini-trunking

Mini-trunking is very small PVC trunking, ideal for surface wiring in domestic and commercial installations such as offices. The trunking has a cross-section of 16 mm × 16 mm, 25 mm × 16 mm, 38 mm × 16 mm or 38 mm × 25 mm and is ideal for switch drops or for housing auxiliary circuits such as telephone or audio equipment wiring. The modern square look in switches and sockets is complemented by the mini-trunking which is very easy to install (see Fig. 6.13).

Skirting trunking

A trunking manufactured from PVC or steel and in the shape of a skirting board is frequently used in commercial buildings such as hospitals, laboratories and offices. The trunking is fitted around the walls of a room and contains the wiring for socket outlets and telephone points which are mounted on the lid, as shown in Fig. 6.13.

Where any trunking passes through walls, partitions, ceilings or floors, short lengths of lid should be fitted so that the remainder of the lid may be removed

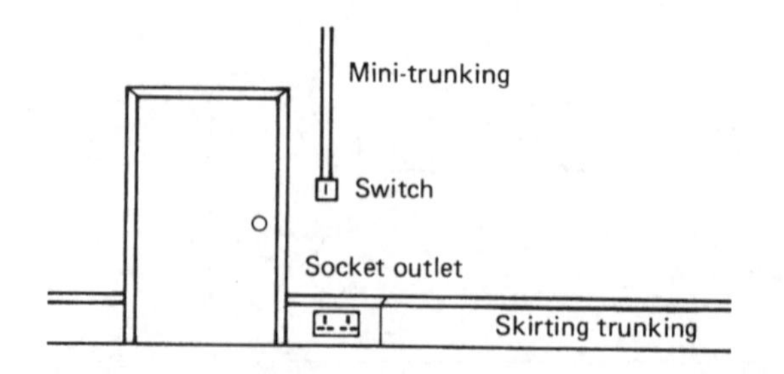

Fig. 6.13 Typical installation of skirting trunking and mini-trunking.

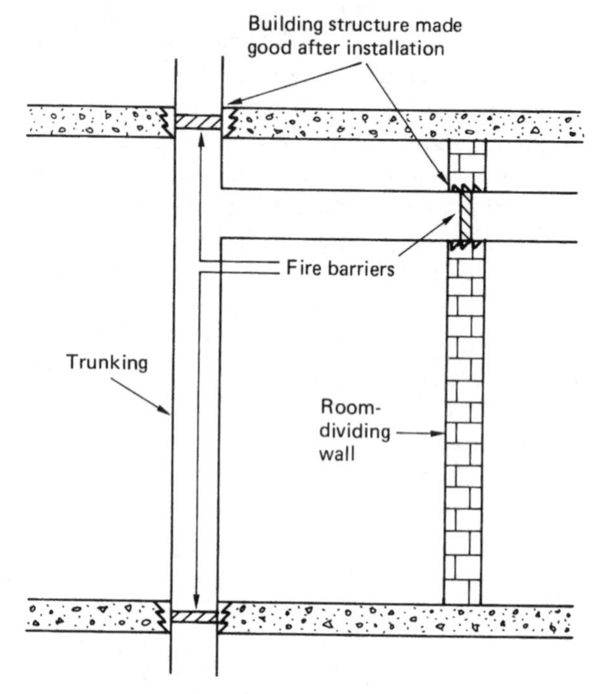

(a) Fire barriers in trunking

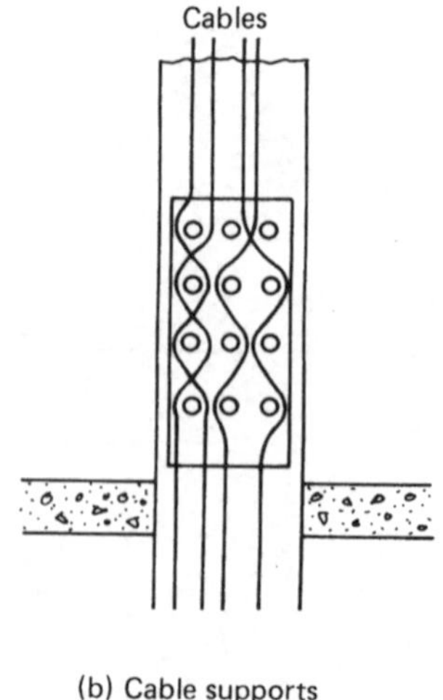

(b) Cable supports in vertical trunking

Fig. 6.14 Installation of trunking.

later without difficulty. Any damage to the structure of the buildings must be made good with mortar, plaster or concrete in order to prevent the spread of fire. Fire barriers must be fitted inside the trunking every 5 m, or at every floor level or room dividing wall, if this is a shorter distance, as shown in Fig. 6.14(a).

Where trunking is installed vertically, the installed conductors must be supported so that the maximum unsupported length of non-sheathed cable does not exceed 5 m. Figure 6.14(b) shows cables woven through insulated pin supports, which is one method of supporting vertical cables.

PVC insulated cables are usually drawn into an erected conduit installation or laid into an erected trunking installation. Table 5D of the *On Site Guide* only gives factors for conduits up to 32 mm in diameter, which would indicate that conduits larger than this are not in frequent or common use. Where a cable enclosure greater than 32 mm is required because of the number or size of the conductors, it is generally more economical and convenient to use trunking.

Trunking capacities

The ratio of the space occupied by all the cables in a conduit or trunking to the whole space enclosed by the conduit or trunking is known as the *space factor*. Where sizes and types of cable and trunking are not covered by the tables in Appendix 5 of the *On Site Guide* a space factor of 45% must not be exceeded. This means that the cables must not fill more than 45% of the space enclosed by the trunking. The tables of Appendix 5 take this factor into account.

To calculate the size of trunking required to enclose a number of cables:

- Identify the cable factor for the particular size of conductor (Table 5E).
- Multiply the cable factor by the number of conductors to give the sum of the cable factors.
- Consider the factors for trunking (Table 5F). The correct size of trunking to accommodate the cables is that trunking which has a factor equal to or greater than the sum of the cable factors.

EXAMPLE

Calculate the minimum size of trunking required to accommodate the following single-core PVC cables:

20 × 1.5 mm	solid conductors
20 × 2.5 mm	solid conductors
21 × 4.0 mm	stranded conductors
16 × 6.0 mm	stranded conductors

From Table 5E, the cable factors are:

for 1.5 mm solid cable	–	8.0
for 2.5 mm solid cable	–	11.9
for 4.0 mm stranded cable	–	16.6
for 6.0 mm stranded cable	–	21.2

The sum of the cable terms is:

$(20 \times 8.0) + (20 \times 11.9) + (21 \times 16.6) + (16 \times 21.2) = 1085.8$

From Table 5F, 75 mm × 38 mm trunking has a factor of 1146 and, therefore, the minimum size of trunking to accommodate these cables is 75 mm × 38 mm, although a larger size, say 75 mm × 50 mm would be equally acceptable if this was more readily available as a standard stock item.

SEGREGATION OF CIRCUITS

Where an installation comprises a mixture of low-voltage and very low-voltage circuits such as mains lighting and power, fire alarm and telecommunication circuits, they must be separated or *segregated* to prevent electrical contact (IEE Regulation 528–01–01).

For the purpose of these regulations various circuits are identified by one of two bands and defined by Part 2 of the Regulations as follows:

Band I	telephone, radio, bell, call and intruder alarm circuits, emergency circuits for fire alarm and emergency lighting.
Band II	mains voltage circuits

When Band I circuits are insulated to the same voltage as Band II circuits, they may be drawn into the same compartment.

When trunking contains rigidly fixed metal barriers along its length, the same trunking may be used to enclose cables of the separate Bands without further precautions, provided that each Band is separated by a barrier, as shown in Fig. 6.15.

Multi-compartment PVC trunking cannot provide band segregations since there is no metal screen between the Bands. This can only be provided in

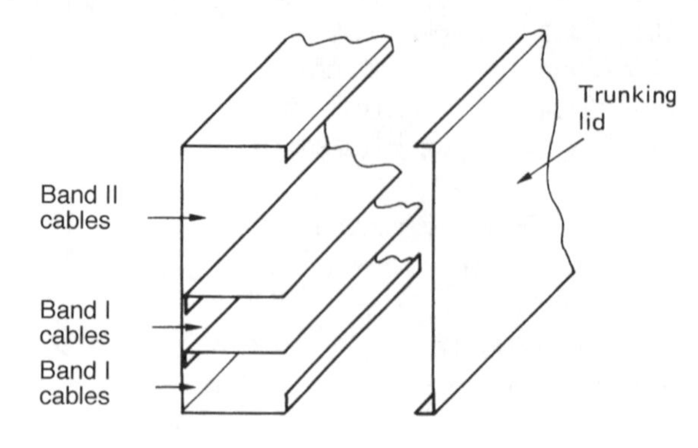

Fig. 6.15 Segregation of cables in trunking.

PVC trunking if screened cables are drawn into the trunking.

CABLE TRAY INSTALLATIONS

Cable tray is a sheet-steel channel with multiple holes. The most common finish is hot-dipped galvanized but PVC-coated tray is also available. It is used extensively on large industrial and commercial installations for supporting MI and SWA cables which are laid on the cable tray and secured with cable ties through the tray holes.

Cable tray should be adequately supported during installation by brackets which are appropriate for the particular installation. The tray should be bolted to the brackets with round-headed bolts and nuts, with the round head inside the tray so that cables drawn along the tray are not damaged.

The tray is supplied in standard widths from 50 mm to 900 mm, and a wide range of bends, tees and reducers are available. Figure 6.16 shows a factory-made 90° bend at B. The tray can also be bent using a cable tray bending machine to create bends such as that shown at A in Fig. 6.16. The installed tray should be securely bolted with round-headed bolts where lengths or accessories are attached, so that there is a continuous earth path which may be bonded to an electrical earth. The whole tray should provide a firm support for the cables and therefore the tray fixings must be capable of supporting the weight of both the tray and cables.

PVC/SWA CABLE INSTALLATIONS

Steel wire armoured PVC insulated cables are now extensively used on industrial installations and often laid on cable tray. This type of installation has the advantage of flexibility, allowing modifications to be made speedily as the need arises. The cable has a steel wire armouring giving mechanical protection and

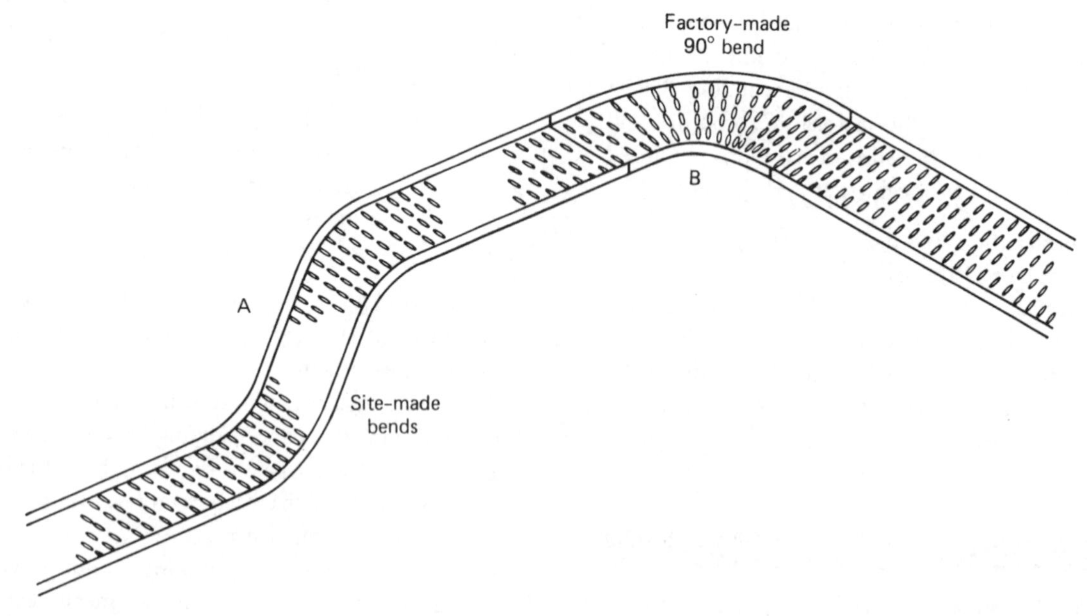

Fig. 6.16 Cable tray with bends.

permitting it to be laid directly in the ground or in ducts, or it may be fixed directly or laid on a cable tray.

It should be remembered that when several cables are grouped together the current rating will be reduced according to the correction factors given in Table 4B1 of the IEE Regulations and Table 6C of the *On Site Guide*.

The cable is easy to handle during installation, is pliable and may be bent to a radius of eight times the cable diameter. The PVC insulation would be damaged if installed in ambient temperatures over 70°C or below 0°C, but once installed the cable can operate at low temperatures.

The cable is terminated with a simple gland which compresses a compression ring on to the steel wire armouring to provide the earth continuity between the switchgear and the cable.

MI CABLE INSTALLATIONS

Mineral insulated cables are available for general wiring as:

- light-duty MI cables for voltages up to 600 V and sizes from 1.0 mm^2 to 10 mm^2 and
- heavy-duty MI cables for voltages up to 1000 V and sizes from 1.0 mm^2 to 150 mm^2.

The cables are available with bare sheaths or with a PVC oversheath. The cable sheath provides sufficient mechanical protection for all but the most severe situations, where it may be necessary to fit a steel sheath or conduit over the cable to give extra protection, particularly near floor level in some industrial situations.

The cable may be laid directly in the ground, in ducts, on cable tray or clipped directly to a structure. It is not affected by water, oil or the cutting fluids used in engineering and can withstand very high temperature or even fire. The cable diameter is small in relation to its current carrying capacity and it should last indefinitely if correctly installed because it is made from inorganic materials. These characteristics make the cable ideal for category 3 emergency circuits, boiler-houses, furnaces, petrol stations and chemical plant installations.

The cable is supplied in coils and should be run off during installation and not spiralled off, as described in Fig. 6.11 for conduit. The cable can be work-hardened if over-handled or over-manipulated. This makes the copper outer sheath stiff and may result in fracture. The outer sheath of the cable must not be penetrated, otherwise moisture will enter the magnesium oxide insulation and lower its resistance. To reduce the risk of damage to the outer sheath during installation, cables should be straightened and formed by hammering with a hide hammer or a block of wood and a steel hammer. When bending MI cables the radius of the bend should not cause the cable to become damaged and clips should provide adequate support (Regulations 522–08–03 and 04 and Tables 4A and 4E of the *On Site Guide*).

The cable must be prepared for termination by removing the outer copper sheath to reveal the copper conductors. This can be achieved by using a rotary stripper tool or, if only a few cables are to be terminated, the outer sheath can be removed with side cutters, peeling off the cable in a similar way to peeling the skin from a piece of fruit with a knife. When enough conductor has been revealed, the outer sheath must be cut off square to facilitate the fitting of the sealing pot, and this can be done with a ringing tool. All excess magnesium oxide powder must be wiped from the conductors with a clean cloth. This is to prevent moisture from penetrating the seal by capillary action.

Cable ends must be terminated with a special seal to prevent the entry of moisture. Figure 6.3 shows a brass screw-on seal and gland assembly, which allows termination of the MI cables to standard switchgear and conduit fittings. The sealing pot is filled with a sealing compound, which is pressed in from one side only to prevent air pockets forming, and the pot closed by crimping home the sealing disc. Such an assembly is suitable for working temperatures up to 105°C. Other compounds or powdered glass can increase the working temperature up to 250°C.

The conductors are not identified during the manufacturing process and so it is necessary to identify them after the ends have been sealed. A simple continuity or polarity test, as described in Chapter 7, can identify the conductors which are then sleeved or identified with coloured markers.

Connection of MI cables can be made directly to motors, but to absorb the vibrations a 360° loop should be made in the cable just before the termination. If excessive vibration is to be expected the MI cable should be terminated in a conduit through-box and the final connection made by flexible conduit.

Copper MI cables may develop a green incrustation or patina on the surface, even when exposed to normal atmospheres. This is not harmful and should not be removed. However, if the cable is exposed to an environment which might encourage corrosion, an MI cable with an overall PVC sheath should be used.

CABLE SELECTION

The size of a cable to be used for an installation depends upon:

- the current rating of the cable under defined installation conditions and
- the maximum permitted drop in voltage as defined by Regulation 525–01.

The factors which influence the current rating are:

1. the design current – the cable must carry the full load current;
2. the type of cable – PVC, MICC, copper conductors or aluminium conductors;
3. the installed conditions – clipped to a surface or installed with other cables in a trunking;
4. the surrounding temperature – cable resistance increases as temperature increases and insulation may melt if the temperature is too high;
5. the type of protection – for how long will the cable have to carry a fault current?

Regulation 525–01 states that the drop in voltage from the supply terminals to the fixed current-using equipment must not exceed 4% of the mains voltage. The volt drop for a particular cable may be found from

$$VD = \text{Factor} \times \text{Design current} \times \text{Length of run}$$

The factor is given in the tables of Appendix 4 of the IEE Regulations and Appendix 6 of the *On Site Guide*.

The cable rating, denoted I_t, may be determined as follows:

$$I_t = \frac{\text{Current rating of protective device}}{\text{Any applicable correction factors}}$$

The cable rating must be chosen to comply with Regulation 433–02–01. The correction factors which may need applying are given below as:

Ca the ambient or surrounding temperature correction factor, which is given in Tables 4C1 and 4C2 of Appendix 4 of the IEE Regulations and 6A1 and 6A2 of the *On Site Guide*.

Cg the grouping correction factor given in Tables 4B1, 4B2 and 4B3, of the IEE Regulations and 6C of the *On Site Guide*.

Cr the 0.725 correction factor to be applied when semi-enclosed fuses protect the circuit as described in item 6.2 of the preface to Appendix 4 of the IEE Regulations.

Ci the correction factor to be used when cables are enclosed in thermal insulation. Regulation 523–04 gives us three possible correction values:

- Where one side of the cable is in contact with thermal insulation we must read the current rating from the column in the table which relates to reference method 4.
- Where the cable is *totally* surrounded over a length greater than 0.5 m we must apply a factor of 0.5.
- Where the cable is *totally* surrounded over a short length, the appropriate factor given in Table 52A of the IEE Regulations or Table 6B of the *On Site Guide* should be applied.

Having calculated the cable rating, the smallest cable should be chosen from the appropriate table which will carry that current. This cable must also meet the voltage drop Regulation 525–01 and this should be calculated as described earlier. When the calculated value is less than 4% of the mains voltage the cable may be considered suitable. If the calculated value is greater than the 4% value, the next larger cable size must be tested until a cable is found which meets both the current rating and voltage drop criteria.

EXAMPLE

A house extension has a total load of 6 kW installed some 18 m away from the mains consumer unit. A PVC insulated and sheathed twin and earth cable will provide a submain to this load and be clipped to the side of the ceiling joists over much of its length in a roof space which is anticipated to reach 35°C in the summer and where insulation is installed up to the top of the joists. Calculate the minimum cable size if the circuit is to be protected (a) by a semi-enclosed fuse to BS 3036 and (b) by a type 2 MCB to BS 3871. Assume a TN-S supply, that is, a supply having a separate neutral and protective conductor throughout.

Let us solve this question using only the tables given in the *On Site Guide*. The tables in the Regulations will give the same values, but this will simplify the problem.

$$\text{Design current } I_b = \frac{\text{Power}}{\text{Volts}} = \frac{6000\ \text{W}}{240\ \text{V}} = 26.09\ \text{A}.$$

Nominal current setting of the protection for this load $I_n = 30$ A.

For (a) the correction factors to be included in this calculation are:

Ca ambient temperature; from Table 6A2 of Appendix 6 the correction factor for 35°C is 0.97.

Cg the grouping correction factor is not applied since the cable is to be clipped direct to a surface and not in contact with other cables.

Cr the protection is by a semi-enclosed fuse and, therefore, a factor of 0.725 must be applied.

Ci thermal insulation is in contact with one side of the cable and we must therefore assume installed method 4.

The cable rating, I_t is given by

$$I_t = \frac{\text{Current rating of protective device}}{\text{The product of the correction factors}}$$

$$= \frac{30\ \text{A}}{0.97 \times 0.725} = 42.66\ \text{A}$$

From column 2 of table 6E1, a 10 mm cable having a rating of 43 A is required to carry this current.

Now test for volt drop: The maximum permissible volt drop is 4% × 230 V = 9.2 V. From Table 6E2 the volt drop per ampere per metre for a 10 mm cable is 4.4 mV.

Therefore, the volt drop for this cable length and load is equal to

$$4.4 \times 10^{-3}\ \text{V/(A m)} \times 26.09\ \text{A} \times 18\ \text{m} = 2.07\ \text{V}.$$

Since this is less than the maximum permissible value of 9.2 V, a 10 mm cable satisfies the current and drop in voltage requirements and is therefore the chosen cable when semi-enclosed fuse protection is used.

For (b) the correction factors to be included in this calculation are:

Ca ambient temperature; from Table 6A1 the correction factor for 35°C is 0.94.

Cg grouping factors need not be applied.

Cr since protection is by MCB no factor need be applied.

Ci thermal insulation once more demands that we assume installed method 4.

The design current is still 26.09 A and we will therefore choose a 30 A MCB for the nominal current setting of the protective device, I_n.

$$\text{Cable rating} = I_t = \frac{30}{0.94} = 31.9\ \text{A}$$

From column 2 of Table 6E1 a 6 mm cable, having a rating of 32 A, is required to carry this current.

Now test for volt drop: from Table 6E2 the volt drop per ampere per metre for a 6 mm cable is 7.3 mV. So the volt drop for this cable length and load is equal to

$$7.3 \times 10^{-3}\ \text{V/(A m)} \times 26.09\ \text{A} \times 18\ \text{m} = 3.43\ \text{V}.$$

Since this is less than the maximum permissible value of 9.2 V, a 6 mm cable satisfies the current and drop in voltage requirements when the circuit is protected by an MCB. From the above calculations it is clear that better protection can reduce the cable size. Even though an MCB is more expensive than a semi-enclosed fuse, the installation of a 6 mm cable with an MCB may be less expensive than 10 mm cable protected by a semi-enclosed fuse. These are some of the decisions which the electrical contractor must make when designing an installation which meets the requirements of the customer and the IEE Regulations.

If you are unsure of the standard fuse and MCB rating of protective devices, you can refer to Tables 2A, 2B, 2C and 2D of the *On Site Guide*.

The *On Site Guide*, Section 7, gives us a simplified method of cable calculation for smaller installations, that is, a domestic, industrial or commercial installation which is fed from a consumer's unit or distribution fuse board fixed at the origin of the supply and protected by a 100 A fuse or less.

Consider part (a) of the above example once more. Protection is provided by a 30 A semi-enclosed fuse to BS 3036. The length of run is 18 m from a TN-S supply. The disconnection time will be 5 seconds.

Looking at Table 7.1 of the *On Site Guide*, we see that for a 30 A radial circuit protected by a BS 3036 fuse, 6.0 mm cable containing a 2.5 mm CPC is only suitable if it is installed according to method 1, that is, clipped direct to a surface according to Appendix 6 of the *On Site Guide*. However, the question tells us that this cable is covered over much of its length with insulating material which is classified as installed by method 4. The 6.0 mm cable is therefore unsuitable for the installed conditions and we must look to the next larger size cable. A 10.0 mm cable having a 4.0 mm CPC and protected by a BS 3036 fuse is suitable for cable runs up to 74 m in length. The length of cable run in this example is only 18 m and therefore the 10.0 mm cable is suitable and confirms the earlier calculation.

Consider part (b) of the example. Protection is provided this time by a 30 A type 2 MCB. The length of run remains 18 m from the TN-S supply. The disconnection time will be 5 seconds.

Table 7.1 of the *On Site Guide* tells us that a 6.0 mm cable with a 2.5 mm CPC and protected by a 30 A type 2 MCB will supply a load 27 m distant from the supply. The cable run is only 18 m and therefore 6.0 mm cable can be used with the better protection offered by the MCB, which also confirms the earlier calculation.

Protection from excess current

Excessive current may flow in a circuit as a result of an overload or a short circuit. An overload or overcurrent is defined as a current which exceeds the rated value in an otherwise healthy circuit. A short circuit is an overcurrent resulting from a fault of negligible impedance between live conductors having a difference in potential under normal operating conditions. Overload currents usually occur in a circuit because it is abused by the consumer or because it has been badly designed or modified by the installer. Short circuits usually occur as a result of an accident which could not have been predicted before the event.

An overload may result in currents of two or three times the rated current flowing in the circuit, while short-circuit currents may be hundreds of times greater than the rated current. In both cases the basic requirement for protection is that the circuit should be interrupted before the fault causes a temperature rise which might damage the insulation, terminations, joints or the surroundings of the conductors. If the device used for overload protection is also capable of breaking a prospective short-circuit current safely, then one device may be used to give protection from both faults (Regulation 432–02–01). Devices which offer protection from overcurrent are:

- semi-enclosed fuses manufactured to BS 3036;
- cartridge fuses manufactured to BS 1361 and BS 1362;
- high breaking capacity fuses (HBC fuses) manufactured to BS 88;
- miniature circuit breakers (MCBs) manufactured to BS 3871.

SEMI-ENCLOSED FUSES (BS 3036)

The semi-enclosed fuse consists of a fuse wire, called the fuse element, secured between two screw terminals in a fuse carrier. The fuse element is connected in series with the load and the thickness of the element is sufficient to carry the normal rated circuit current. When a fault occurs an overcurrent flows and the fuse element becomes hot and melts or 'blows'.

The designs of the fuse carrier and base are also important. They must not allow the heat generated from an overcurrent to dissipate too quickly from the element, otherwise a larger current would be required to 'blow' the fuse. Also if over-enclosed, heat will not escape and the fuse will 'blow' at a lower current. This type of fuse is illustrated in Fig. 6.17. The fuse element should consist of a single strand of plain or tinned copper wire having a diameter appropriate to the current rating as given in Table 6.1.

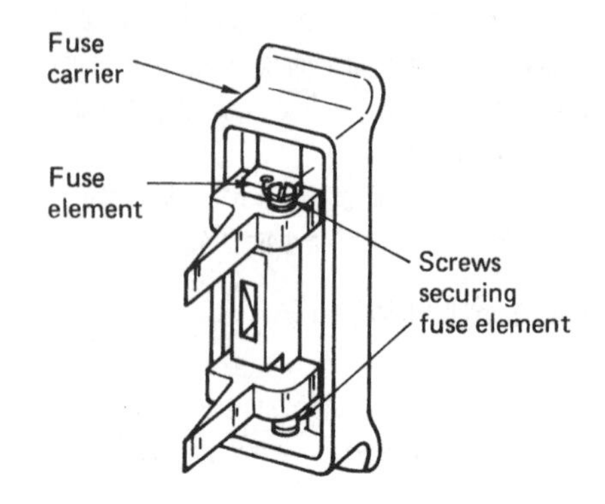

Fig. 6.17 A semi-enclosed fuse.

Table 6.1 Size of fuse element

Current rating (A)	Wire diameter (mm)
5	0.20
10	0.35
15	0.50
20	0.60
30	0.85

Advantages of semi-enclosed fuses

- They are very cheap compared with other protective devices both to install and to replace.
- There are no mechanical moving parts.
- It is easy to identify a 'blown fuse'.

Disadvantages of semi-enclosed fuses

- The fuse element may be replaced with wire of the wrong size either deliberately or by accident.
- The fuse element weakens with age due to oxidization, which may result in a failure under normal operating conditions.
- The circuit cannot be restored quickly since the fuse element requires screw fixing.
- They have low breaking capacity since, in the event of a severe fault, the fault current may vaporize the

fuse element and continue to flow in the form of an arc across the fuse terminals.

- There is a danger from scattering hot metal if the fuse carrier is inserted into the base when the circuit is faulty.

CARTRIDGE FUSES (BS 1361)

The cartridge fuse breaks a faulty circuit in the same way as a semi-enclosed fuse, but its construction eliminates some of the disadvantages experienced with an open-fuse element.

The fuse element is encased in a glass or ceramic tube and secured to end-caps which are firmly attached to the body of the fuse so that they do not blow off when the fuse operates. Cartridge fuse construction is illustrated in Fig. 6.18. With larger-size cartridge fuses, lugs or tags are sometimes brazed on to the end-caps to fix the fuse cartridge mechanically to the carrier. They may also be filled with quartz sand to absorb and extinguish the energy of the arc when the cartridge is brought into operation.

Advantages of cartridge fuses

- They have no mechanical moving parts.
- The declared rating is accurate.
- The element does not weaken with age.
- They have small physical size and no external arcing which permits their use in plug tops and small fuse carriers.
- Their operation is more rapid than rewirable fuses. Operating time is inversely proportional to the fault current.

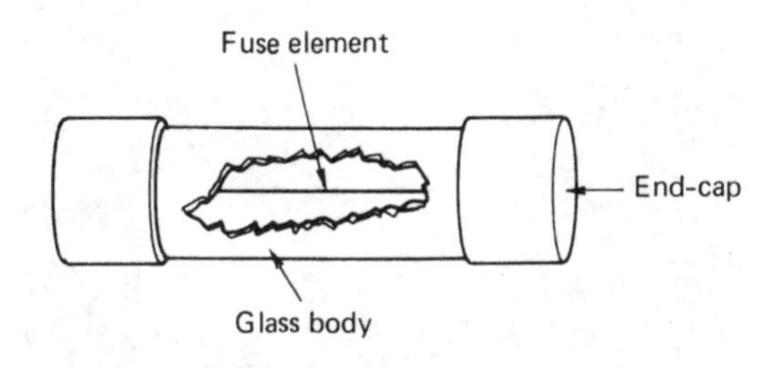

Fig. 6.18 A cartridge fuse.

Disadvantages of cartridge fuses

- They are more expensive to replace than rewirable fuse elements.
- They can be replaced with an incorrect cartridge.
- The cartridge may be shorted out by wire or silver foil in extreme cases of bad practice.
- They are not suitable where extremely high fault currents may develop.

HIGH BREAKING CAPACITY FUSES (BS 88)

As the name might imply, these cartridge fuses are for protecting circuits where extremely high fault currents may develop such as on industrial installations or distribution systems.

The fuse element consists of several parallel strips of pure silver encased in a substantial ceramic cylinder, the ends of which are sealed with tinned brass end-caps incorporating fixing lugs. The cartridge is filled with silica sand to ensure quick arc extraction. Incorporated on the body is an indicating device to show when the fuse has blown. HBC fuse construction is shown in Fig. 6.19.

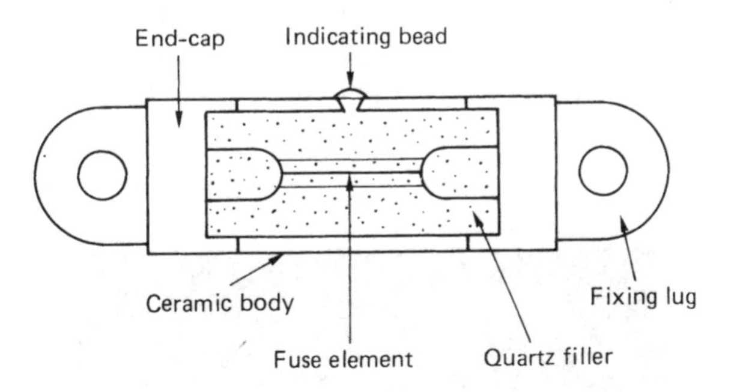

Fig. 6.19 HBC fuse.

Advantages of HBC fuses

- They have no mechanical moving parts.
- The declared rating is accurate.
- The element does not weaken with age.
- Their operation is very rapid under fault conditions.
- They are capable of breaking very heavy fault currents safely.
- They are capable of discriminating between a

persistent fault and a transient fault such as the large starting current taken by motors.

- It is difficult to confuse cartridges since different ratings are made to different physical sizes.

Disadvantages of HBC fuses

- They are very expensive compared to semi-enclosed fuses.

MINIATURE CIRCUIT BREAKERS (BS 3871)

The disadvantage of all fuses is that when they have operated they must be replaced. An MCB overcomes this problem since it is an automatic switch which opens in the event of an excessive current flowing in the circuit and can be closed when the circuit returns to normal.

An MCB of the type shown in Fig. 6.20 incorporates a thermal and magnetic tripping device. The load current flows through the thermal and the electromagnetic mechanisms. In normal operation the current is insufficient to operate either device, but when an overload occurs, the bimetal strip heats up, bends and trips the mechanism. The time taken for this action to occur provides an MCB with the ability to discriminate between an overload which persists for a very short time, for example the starting current of a motor, and an overload due to a fault. The device only trips when a fault current occurs. This slow operating time is ideal for overloads but when a short circuit occurs it is important to break the faulty circuit very quickly. This is achieved by the coil electromagnetic device.

When a large fault current (above about eight times the rated current) flows through the coil a strong magnetic flux is set up which trips the mechanisms almost instantly. The circuit can be restored when the fault is removed by pressing the ON toggle. This latches the various mechanisms within the MCB and 'makes' the switch contact. The toggle switch can also be used to disconnect the circuit for maintenance or isolation or to test the MCB for satisfactory operation. The simplified diagram in Fig. 6.21 shows the various parts within an MCB.

Advantages of MCBs

- Tripping characteristics and therefore circuit protection are set by installer.
- The circuit protection is difficult to interfere with.
- The circuit is provided with discrimination.
- A faulty circuit may be easily and quickly restored.
- The supply may be safely restored by an unskilled operator.

(a)

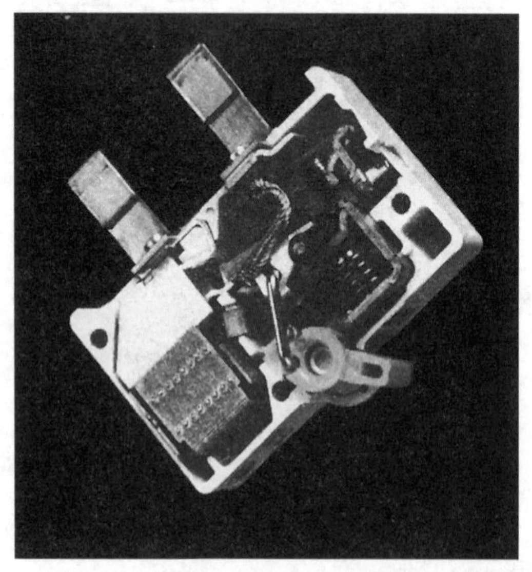

(b)

Fig. 6.20 (a) Interior view of Wylex 'plug-in' MCB; (b) 'plug-in' MCB fits any standard Wylex consumer's unit.

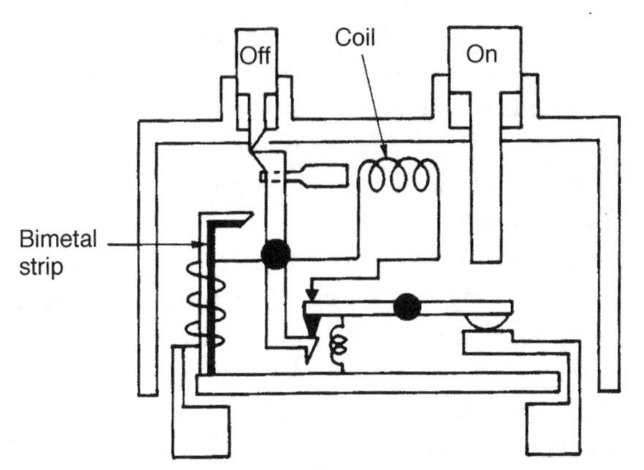

Fig. 6.21 A simplified diagram of an MCB.

Disadvantages of MCBs

- They are very expensive compared to rewirable fuses.
- They contain mechanical moving parts and therefore require regular testing to ensure satisfactory operation under fault conditions.

FUSING FACTOR

The speed with which a protective device will operate under fault conditions gives an indication of the level of protection being offered by that device. This level of protection or fusing performance is given by the fusing factor of the device:

$$\text{Fusing factor} = \frac{\text{Minimum fusing current}}{\text{Current rating}}$$

The minimum fusing current of a device is the current which will cause the fuse or MCB to blow or trip in a given time (BS 88 gives this operating time as 4 hours). The current rating of a device is the current which it will carry continuously without deteriorating.

Thus, a 10 A fuse which operates when 15 A flows will have a fusing factor of 15 ÷ 10 = 1.5.

Since the protective device must carry the rated current it follows that the fusing factor must always be greater than one. The closer the fusing factor is to one, the better is the protection offered by that device.

The fusing factors of the protective devices previously considered are:

- semi-enclosed fuses: between 1.5 and 2
- cartridge fuses: between 1.25 and 1.75
- HBC fuses: less than 1.25
- MCBs: less than 1.5.

In order to give protection to the conductors of an installation:

- the current rating of the protective device must be equal to or less than the current carrying capacity of the conductor;
- the current causing the protective device to operate must not be greater than 1.45 times the current carrying capacity of the conductor to be protected.

The current carrying capacities of cables given in the tables of Appendix 4 of the IEE Regulations assume that the circuit will comply with these requirements and that the circuit protective device will have a fusing factor of 1.45 or less. Cartridge fuses, HBC fuses and MCBs do have a fusing factor less than 1.45 and therefore when this type of protection is afforded the current carrying capacities of cables may be read directly from the tables.

However, semi-enclosed fuses can have a fusing factor of 2. The wiring regulations require that the rated current of a rewirable fuse must not exceed 0.725 times the current carrying capacity of the conductor it is to protect. This factor is derived as follows:

The maximum fusing factor of a rewirable fuse is 2.

Now, if I_n = current rating of the protective device

I_z = current carrying capacity of conductor

I_2 = current causing the protective device to operate.

Then $I_2 = 2\,I_n \leq 1.45\,I_z$

therefore $I_n \leq \dfrac{1.45\,I_z}{2}$

or $I_n \leq 0.725\,I_z$

When rewirable fuses are used, the current carrying capacity of the cables given in the tables is reduced by a factor of 0.725, as detailed in Appendix 4 item 5 of the Regulations.

Earth leakage protection

When it is required to provide the very best protection from electric shock and fire risk, earth fault protection devices are incorporated in the installation. The object of the Regulations concerning these devices, 413–02–15, 471–08–01 and 471–16–02, is to remove an earth fault current in less than 5 seconds, and limit the voltage which might appear on any exposed metal parts under fault conditions to 50 V. They will continue to provide adequate protection throughout the life of the installation even if the earthing conditions deteriorate. This is in direct contrast to the protection provided by overcurrent devices, which require a low resistance earth loop impedance path.

THE RESIDUAL CURRENT DEVICE (RCD)

The basic circuit for a single-phase RCD is shown in Fig. 6.22. The load current is fed through two equal and opposing coils wound on to a common transformer core. The phase and neutral currents in a healthy circuit produce equal and opposing fluxes in the transformer core, which induces no voltage in the tripping coil. However, if more current flows in the phase conductor than in the neutral conductor as a result of a fault between live and earth, an out-of-balance flux will result in an emf being induced in the trip coil which will open the double pole switch and isolate the load. Modern RCDs have tripping sensitivities between 10 and 30 mA, and therefore a faulty circuit can be isolated before the lower lethal limit to

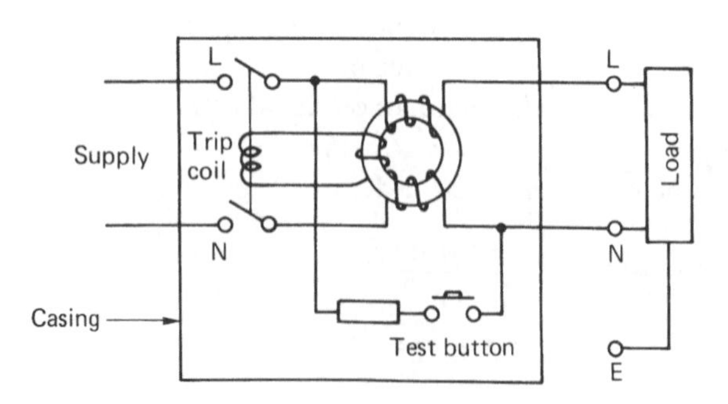

Fig. 6.22 Construction of a residual current device.

human beings (about 50 mA) is reached.

Consumer units can now be supplied which incorporate an RCD, so that any equipment supplied by the consumer unit outside the zone created by the main equipotential bonding, such as a garage or greenhouse, can have the special protection required by Regulation 471–08–01.

Finally, it should perhaps be said that a foolproof method of giving protection to people or animals who simultaneously touch both live and neutral has yet to be devised. The ultimate safety of an installation depends upon the skill and experience of the electrical contractor and the good sense of the user.

Electrical installation circuits

LIGHTING CIRCUITS

Table 1A in Appendix 1 of the *On Site Guide* deals with the assumed current demand of points, and states that for lighting outlets we should assume a current equivalent to a minimum of 100 W per lampholder. This means that for a domestic lighting circuit rated at 5 A, a maximum of 11 lighting outlets could be connected to each circuit. In practice, it is usual to divide the fixed lighting outlets into two or more circuits of seven or eight outlets each. In this way the whole installation is not plunged into darkness if one lighting circuit fuses.

Lighting circuits are usually wired in 1.0 mm or 1.5 mm cable using either a loop-in or joint-box method of installation. The loop-in method is universally employed with conduit installations or when access from above or below is prohibited after installation, as is the case with some industrial installations or blocks of flats. In this method the only joints are at the switches or lighting points, the live conductors being looped from switch to switch and the neutrals from one lighting point to another.

The use of junction boxes with fixed brass terminals is the method often adopted in domestic installations, since the joint boxes can be made accessible but are out of site in the loft area and under floorboards.

All switches and ceiling roses must contain an earth connection (Regulation 471–09–02) and the live conductors must be broken at the switch position in order to comply with the polarity regulations (713–09–01). A ceiling rose may only be connected to installations operating at 250 V maximum and must only accommodate one flexible cord unless it is specially designed to take more than one (553–04–02). Lampholders must comply with Regulation 553–03–02 and be suspended from flexible cords capable of suspending the mass of the luminaire fixed to the lampholder (554–01–01).

The type of circuit used will depend upon the installation conditions and the customer's requirements. One light controlled by one switch is called one-way switch control (see Fig. 6.23). A room with two access doors might benefit from a two-way switch control (see Fig. 6.24), so that the lights may be switched on or off at either position. A long staircase with more than two switches controlling the same lights would require intermediate switching (see Fig. 6.25).

One-way, two-way or intermediate switches can be obtained as plate switches for wall mounting or ceiling mounted cord switches. Cord switches can provide a convenient method of control in bedrooms or bathrooms and for independently controlling an office luminaire.

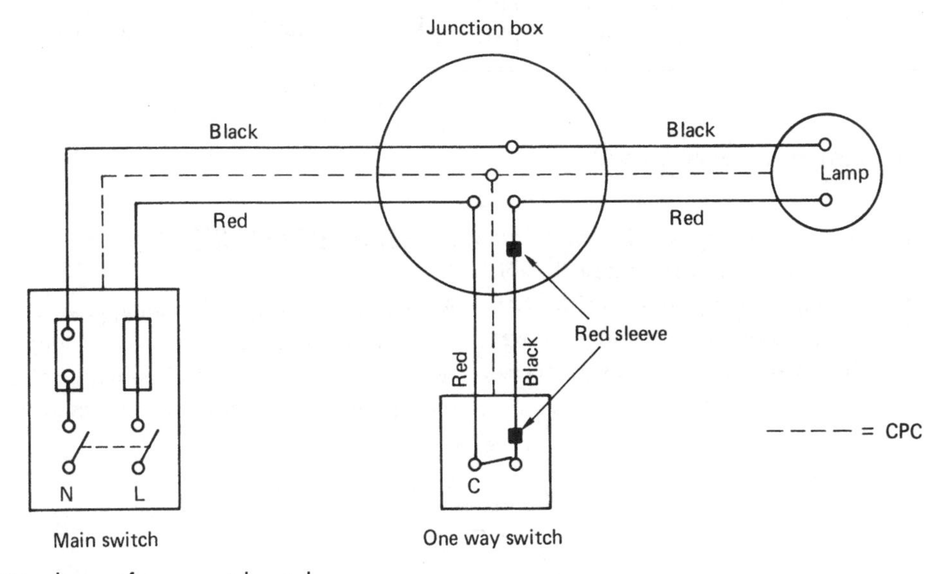

Fig. 6.23 Wiring diagram of one-way switch control.

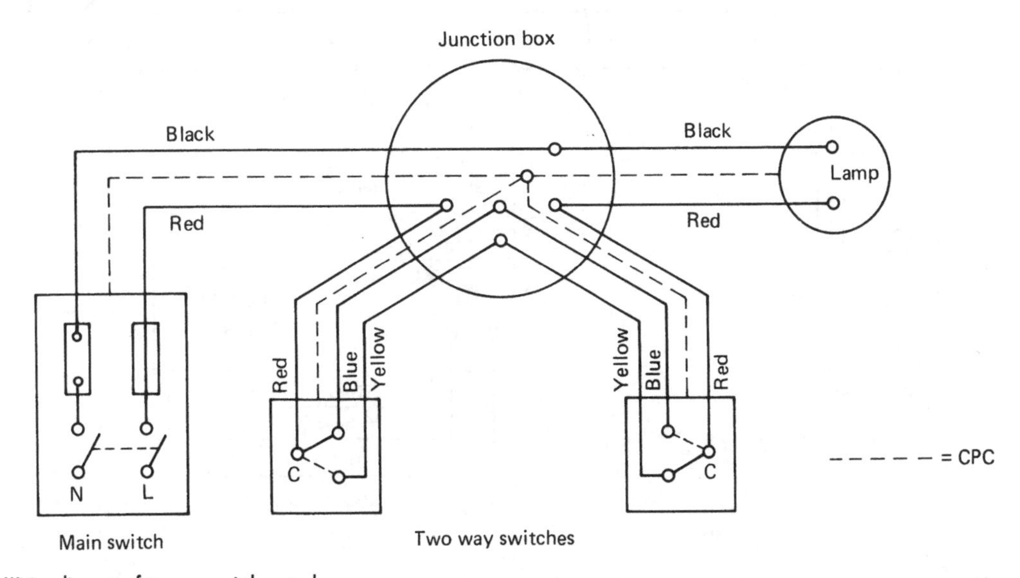

Fig. 6.24 Wiring diagram of two-way switch control.

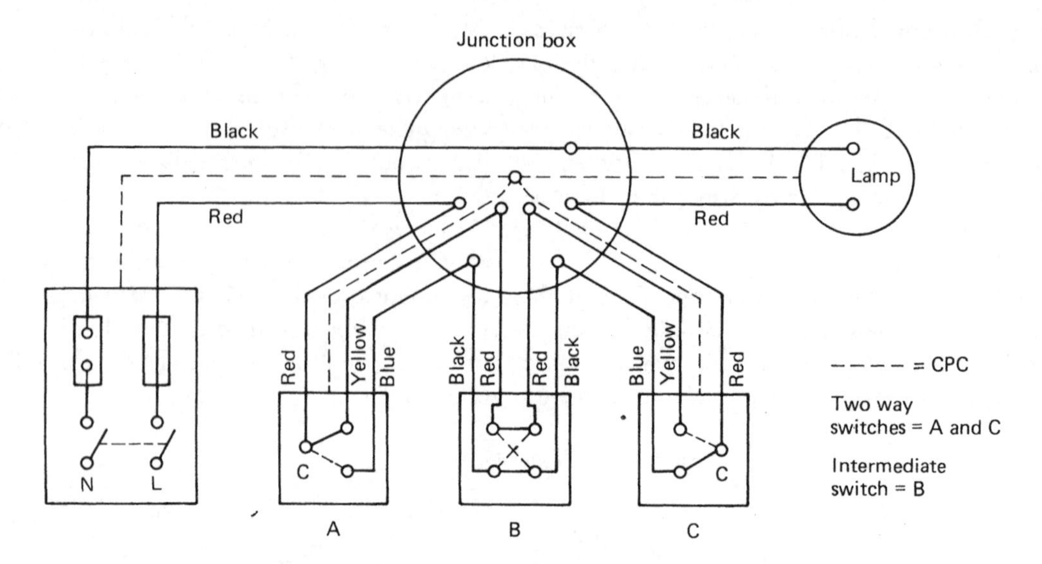

Fig. 6.25 Wiring diagram of intermediate switch control.

To convert an existing one-way switch control into a two-way switch control, a three-core and earth cable is run from the existing switch position to the proposed second switch position. The existing one-way switch is replaced by a two-way switch and connected as shown in Fig. 6.26.

SOCKET OUTLET CIRCUITS

A plug top is connected to an appliance by a flexible cord which should normally be no longer than 2 m (Regulation 553–01–07). Pressing the plug top into a socket outlet connects the appliance to the source of

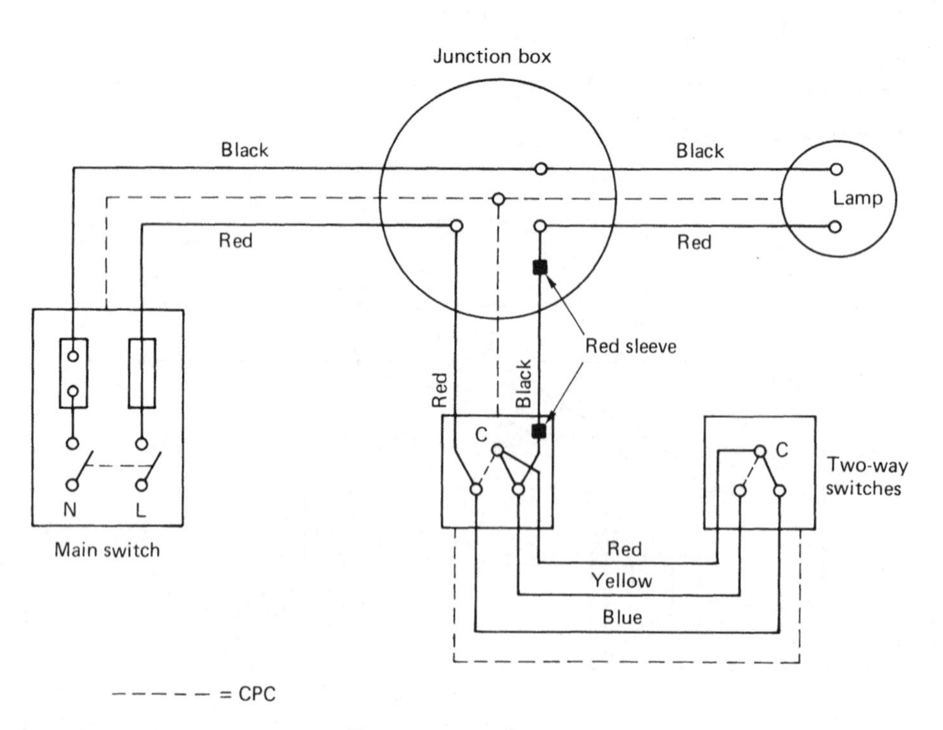

Fig. 6.26 Wiring diagram of one-way to two-way switch control.

supply. Socket outlets therefore provide an easy and convenient method of connecting portable electrical appliances to a source of supply.

Socket outlets can be obtained in 15, 13, 5 and 2 A ratings, but the 13 A flat pin type complying with BS 1363 is the most popular for domestic installations in the United Kingdom. Each 13 A plug top contains a cartridge fuse to give maximum potential protection to the flexible cord and the appliances which it serves.

Socket outlets may be wired on a ring or radial circuit and in order that every appliance can be fed from an adjacent and convenient socket outlet, the number of sockets is unlimited provided that the floor area covered by the circuit does not exceed that given in Table 8A, Appendix 8 of the *On Site Guide*.

RADIAL CIRCUITS

In a radial circuit each socket outlet is fed from the previous one. Live is connected to live, neutral to neutral and earth to earth at each socket outlet. The fuse and cable sizes are given in Table 8A of Appendix 8 but circuits may also be expressed with a block diagram, as shown in Fig. 6.27.

Where two or more circuits are installed in the same premises, the socket outlets and permanently connected equipment should be reasonably shared out among the circuits, so that the total load is balanced.

When designing ring or radial circuits special consideration should be given to the loading in kitchens which may require separate circuits. This is because the maximum demand of current-using equipment in kitchens may exceed the rating of the circuit cable and protection devices.

Ring and radial circuits may be used for domestic or other premises where the maximum demand of the current using equipment is estimated not to exceed the rating of the protective devices for the chosen circuit.

RING CIRCUITS

Ring circuits are very similar to radial circuits in that each socket outlet is fed from the previous one, but in ring circuits the last socket is wired back to the source of supply. Each ring final circuit conductor must be looped into every socket outlet or joint box which forms the ring and must be electrically continuous throughout its length.

The circuit details are given in Table 8A, Appendix 8 of the *On Site Guide* but may also be expressed by the block diagram given in Fig. 6.28.

Spurs to ring circuits

A spur is defined in Part 2 of the Regulations as a branch cable from a ring final circuit.

Non-fused spurs

The total number of non-fused spurs must not exceed the total number of socket outlets and pieces of stationary equipment connected directly in the circuit. The cable used for non-fused spurs must not be less than that of the ring circuit. The requirements concerning spurs are given in Appendix 8 of the *On Site Guide* but the various circuit arrangements may be expressed by the block diagrams of Fig. 6.29.

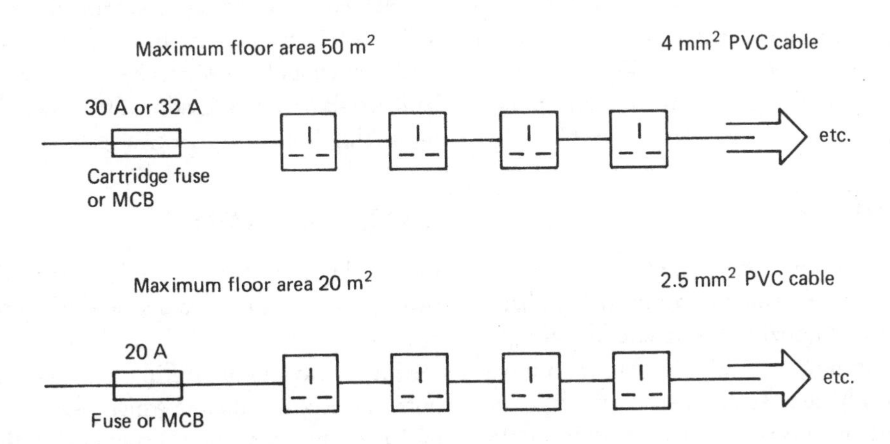

Fig. 6.27 Block diagram of radial circuits.

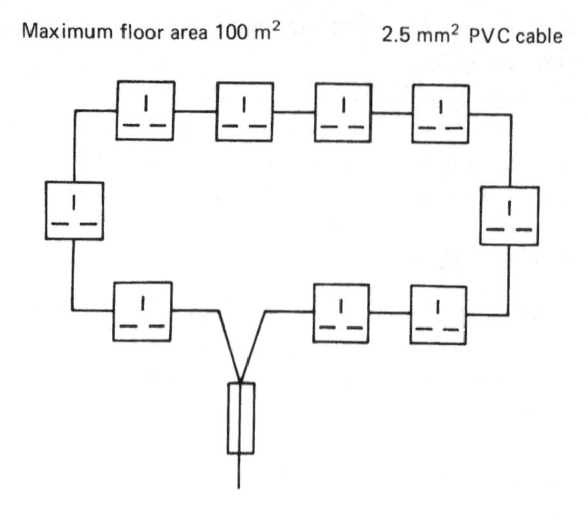

Fig. 6.28 Block diagram of ring circuits.

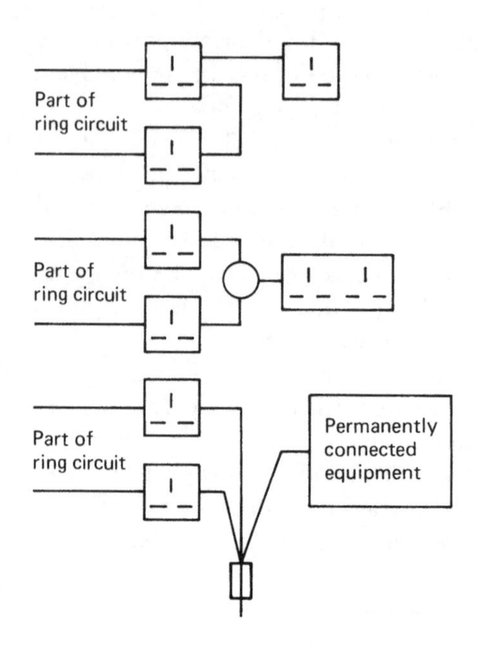

Fig. 6.29 Connection of non-fused spurs.

A non-fused spur may only feed one single or one twin socket or one permanently connected piece of equipment.

Non-fused spurs may be connected into the ring circuit at the terminals of socket outlets or at joint boxes or at the origin of the circuit.

Fused spurs

The total number of fused spurs is unlimited. A fused spur is connected to the circuit through a fused connection unit, the rating of which should be suitable for the conductor forming the spur but should not exceed 13 A. The requirements for fused spurs are also given in Appendix 8 but the various circuit arrangements may be expressed by the block diagrams of Fig. 6.30.

The general arrangement shown in Fig. 6.31 shows 11 socket outlets connected to the ring, three non-fused spur connections and two fused spur connections.

WATER HEATING

A small, single-point over-sink type water heater may be considered as a permanently connected appliance and so may be connected to a ring circuit through a fused connection unit. A water heater of the immersion type is usually rated at a maximum of 3 kW, and could be considered as a permanently connected appliance, fed from a fused connection unit. However, many immersion heating systems are connected into storage vessels of about 150 litres in domestic installations, and Appendix 8 of the *On Site Guide* states that immersion heaters fitted to vessels in excess of 15 litres should be supplied by their own circuit.

Therefore, immersion heaters must be wired on a separate radial circuit when they are connected to water vessels which hold more than 15 litres. Figure 6.32 shows the wiring arrangements for an immersion heater. The hot and cold water connections must be connected to an earth connection in order to meet the supplementary bonding requirements of Regulation 413–05–02. Every switch must be a double-pole (DP) switch and out of reach of anyone using a fixed bath or shower (Regulation 601–08–01) when the immersion heater is fitted to a vessel in a bathroom.

ELECTRIC SPACE HEATING

Electrical heating systems can be broadly divided into two categories: unrestricted local heating and off-peak heating.

Unrestricted local heating may be provided by portable electric radiators which plug into the socket outlets of the installation. Fixed heaters that are wall-mounted or inset must be connected through a fused

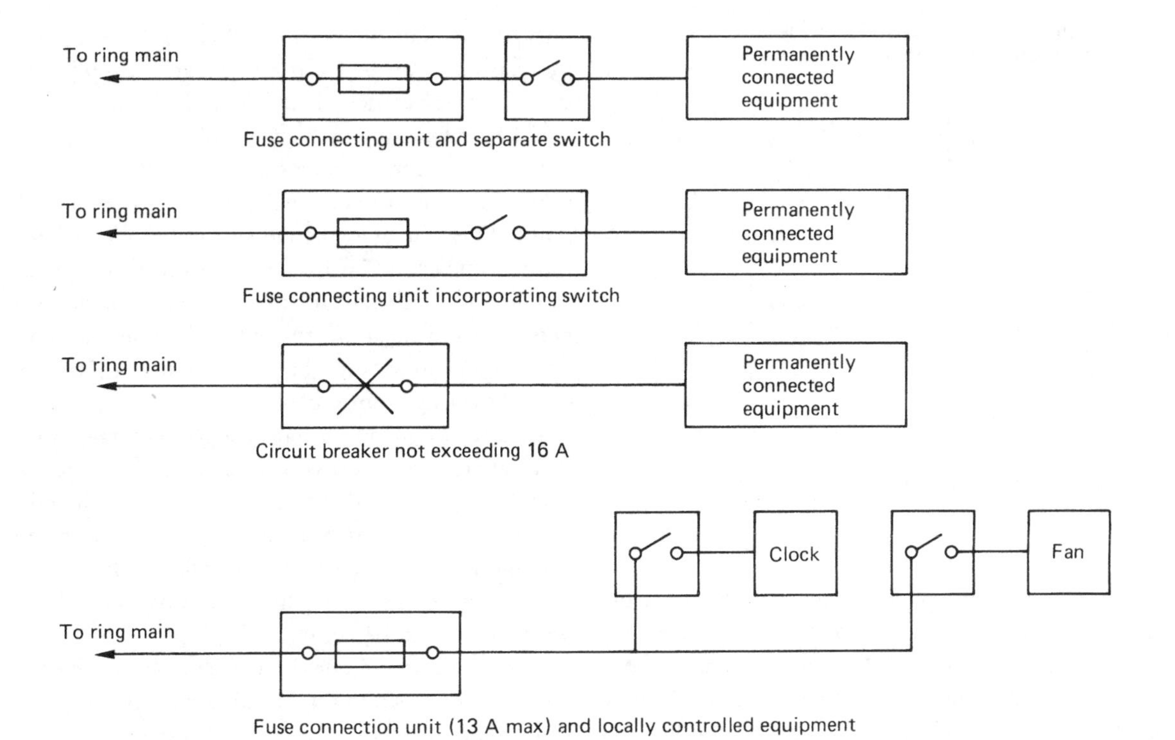

Fig. 6.30 Connection of fused spurs.

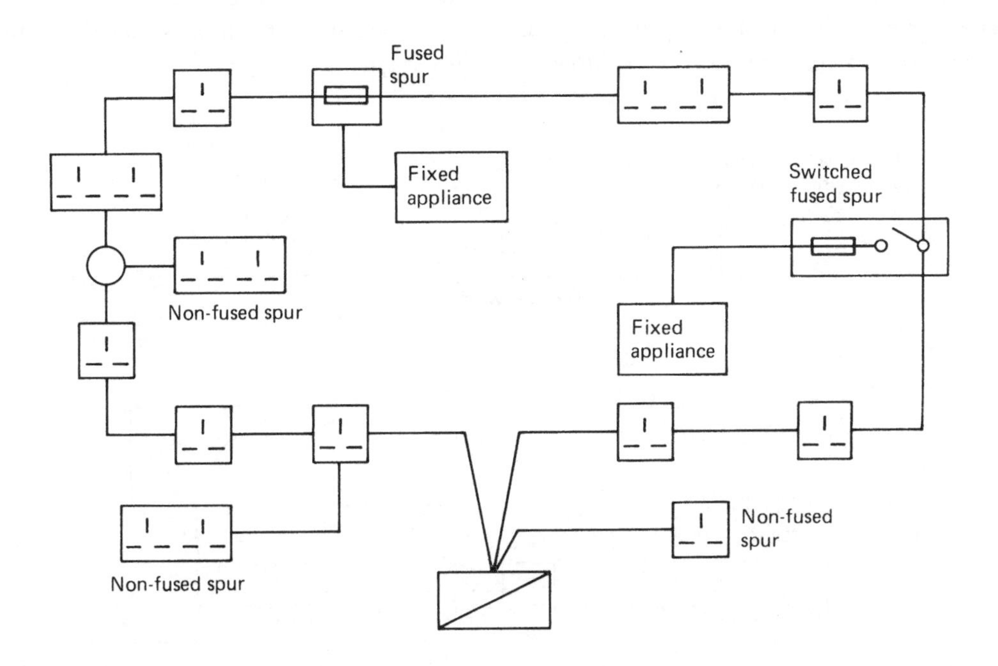

Fig. 6.31 Typical ring circuit with spurs.

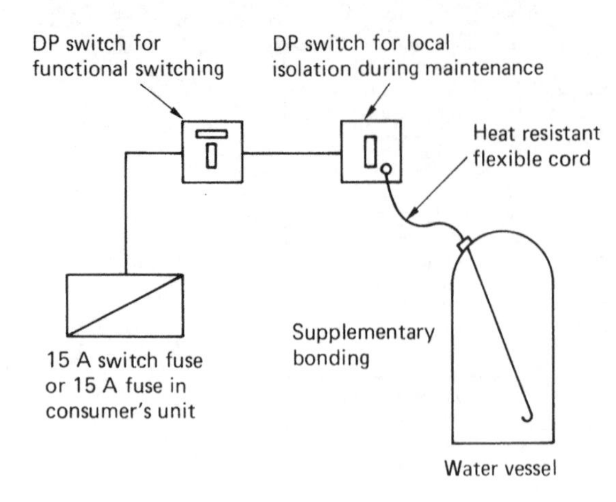

Fig. 6.32 Immersion heater wiring.

connection and incorporate a local switch, either on the heater itself or as a part of the fuse connecting unit, as shown in Fig. 6.30. Heating appliances where the heating element can be touched must have a double-pole switch which disconnects all conductors. This requirement includes radiators which have an element inside a silica-glass sheath (601–12–01).

Off-peak heating systems may provide central heating from storage radiators, ducted warm air or underfloor heating elements. All three systems use the thermal storage principle, whereby a large mass of heat-retaining material is heated during the off-peak period and allowed to emit the stored heat throughout the day. The final circuits of all off-peak heating installations must be fed from a separate supply controlled by an electricity board time clock.

When calculating the size of cable required to supply a single storage radiator, it is good practice to assume a current demand equal to 3.4 kW at each point. This will allow the radiator to be changed at a future time with the minimum disturbance to the installation. Each radiator must have a 20 A double-pole means of isolation adjacent to the heater and the final connection should be via a flex outlet. See Fig. 6.33 for wiring arrangements.

Ducted warm air systems have a centrally sited thermal storage heater with a high storage capacity. The unit is charged during the off-peak period, and a fan drives the stored heat in the form of warm air through large air ducts to outlet grilles in the various rooms. The wiring arrangements for this type of heating are shown in Fig. 6.34.

The single storage heater is heated by an electric element embedded in bricks and rated between 6 kW and 15 kW depending upon its thermal capacity. A radiator of this capacity must be supplied on its own circuit, in cable capable of carrying the maximum current demand and protected by a fuse or MCB of 30 A, 45 A or 60 A as appropriate. At the heater position, a double-pole switch must be installed to terminate the fixed heater wiring. The flexible cables used for the final connection to the heaters must be of the heat-resistant type.

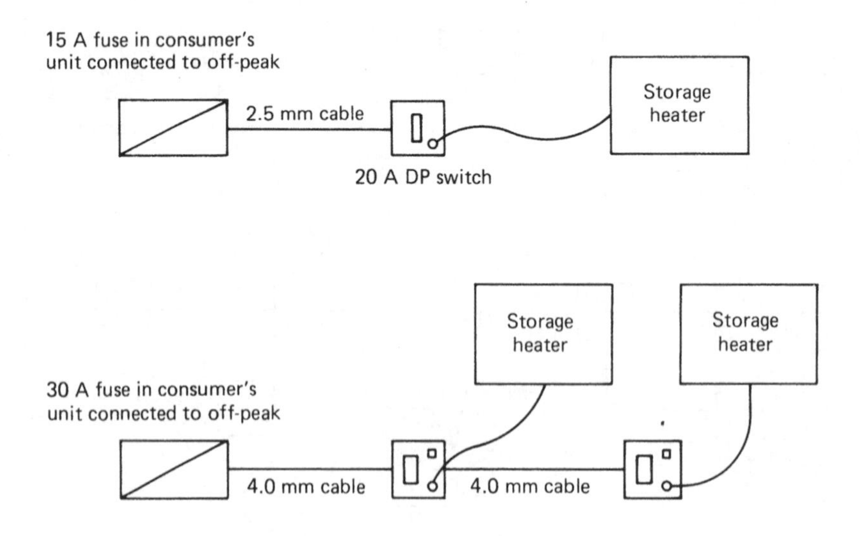

Fig. 6.33 Possible wiring arrangements for storage heaters.

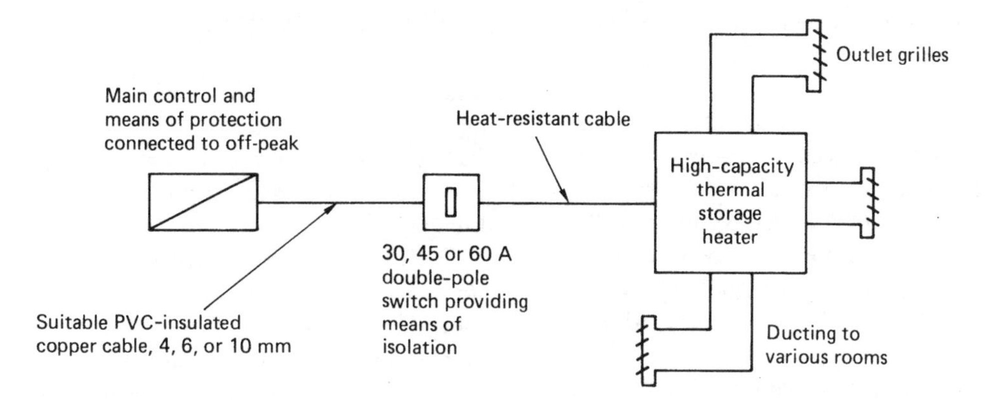

Fig. 6.34 Ducted warm air heating system.

Floor warming installations use the thermal storage properties of concrete. Special cables are embedded in the concrete floor screed during construction. When current is passed through the cables they become heated, the concrete absorbs this heat and radiates it into the room. The wiring arrangements are shown in Fig. 6.35. Once heated, the concrete will give off heat for a long time after the supply is switched off and is, therefore, suitable for connection to an off-peak supply.

Underfloor heating cables installed in bathrooms or shower rooms must incorporate an earthed metallic sheath or be covered by an earthed metallic grid connected to the supplementary bonding (Regulation 601–12–02).

BATHROOM INSTALLATIONS

In rooms containing a fixed bath or shower, additional regulations are specified. This is to reduce the risk of electric shock to people in circumstances where body resistance is lowered because of contact with

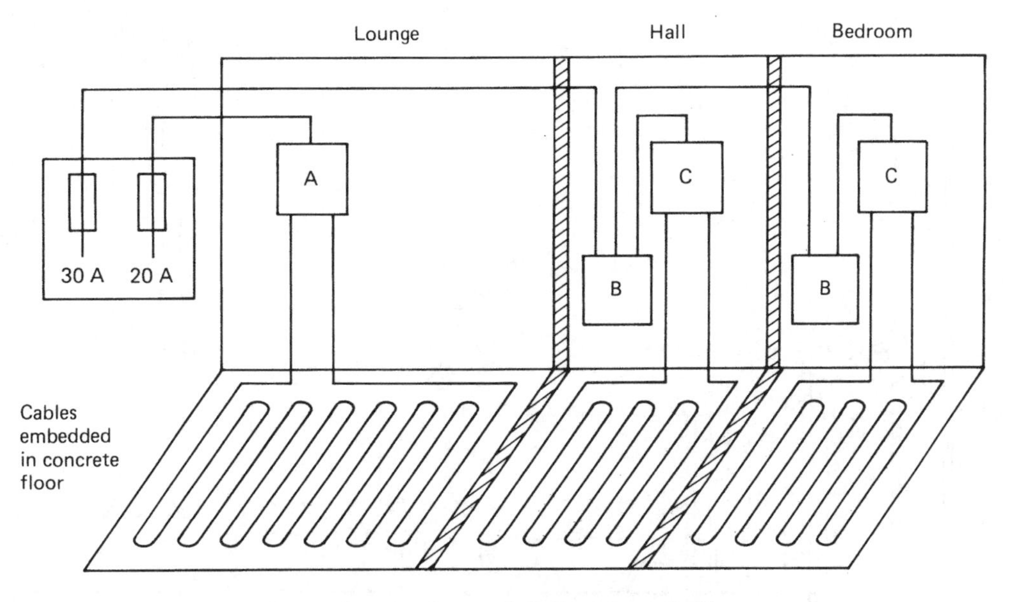

Fig. 6.35 Floor warming installations.

water. The regulations may be found in Section 601 and can be summarized as follows:

- Socket outlets must not be installed and no provision is made for connection of portable appliances.
- Only shaver sockets which comply with BS EN 61184 or BS EN 60238, that is, those which contain an isolating transformer, may be installed.
- Every switch must be inaccessible to anyone using the bath or shower unless it is of the cord-operated type.
- Lampholders which are within 2.5 m of a bath or shower should be shrouded in insulating material or totally enclosed luminaires may be used.
- A supplementary bonding conductor must be provided, in addition to the main equipotential bonding discussed in Chapter 1.

SUPPLEMENTARY BONDING

Modern plumbing methods make considerable use of non-metals (PTFE tape on joints, for example). Therefore, the metalwork of water and gas installations cannot be relied upon to be continuous throughout.

The IEE Regulations describe the need to consider additional or supplementary bonding in situations where there is a high risk of electric shock (for example, in kitchens and bathrooms).

In kitchens, supplementary bonding of hot and cold taps, sink tops and exposed water and gas pipes *is only required* if an earth continuity test proves that they are not already effectively and reliably connected to the main equipotential bonding, having negligible impedance, by the soldered pipe fittings of the

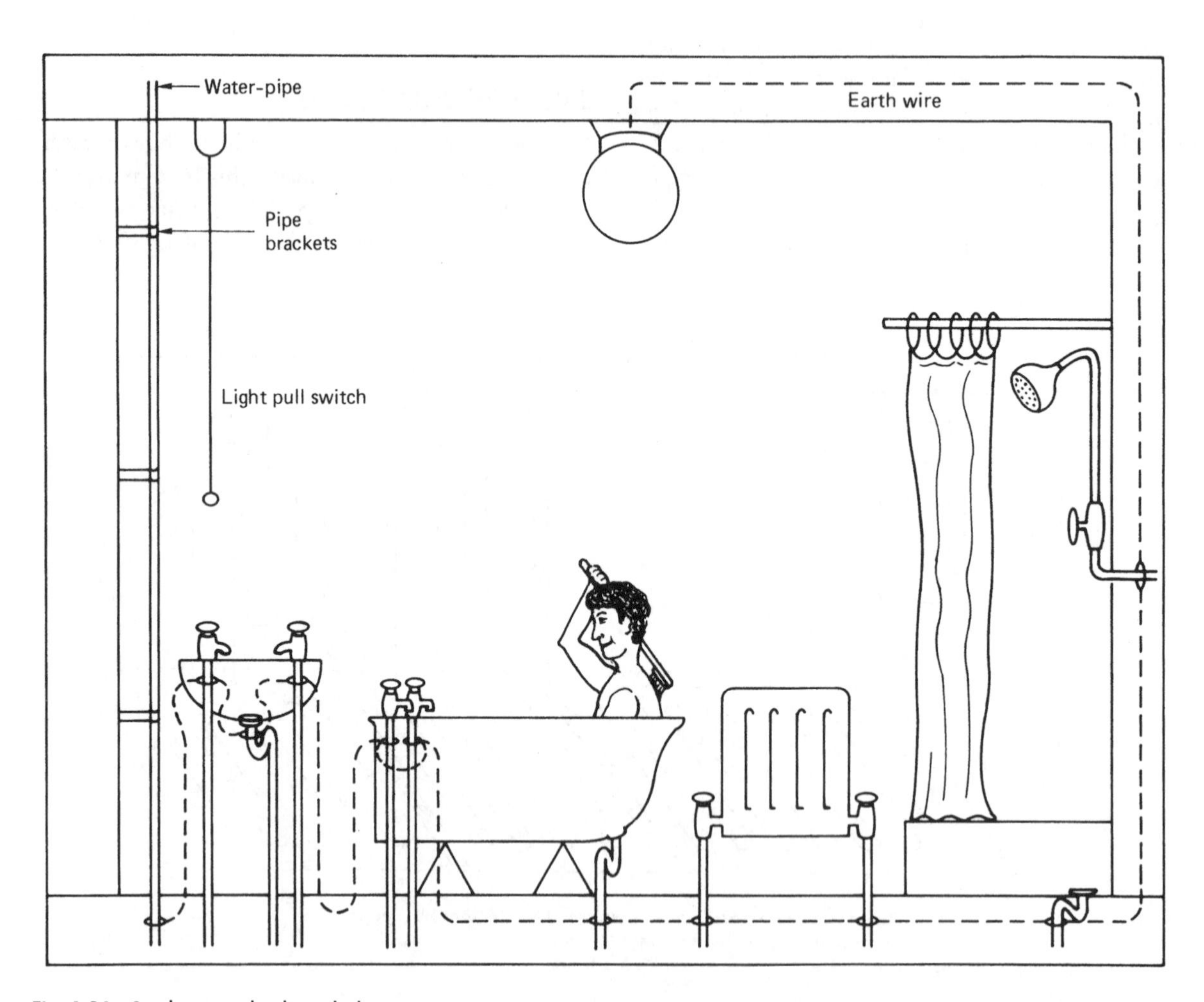

Fig. 6.36 Supplementary bonding in bathrooms.

installation. If the test proves unsatisfactory, the metalwork must be bonded using a single core copper conductor with PVC green/yellow insulation, which will normally be 4 mm² for domestic installations but must comply with Regulations 547–03–01 to 03.

In rooms containing a fixed bath or shower, supplementary bonding conductors *must* be installed to reduce to a minimum the risk of an electric shock (Regulation 601–04–02). Bonding conductors in domestic premises will normally be of 4 mm² copper with PVC insulation to comply with Regulations 547–03–01 to 03 and must be connected between all exposed metalwork (for example, between metal baths, bath and sink taps, shower fittings, metal waste pipes and radiators, as shown in Fig. 6.36.

The bonding connection must be made to a cleaned pipe, using a suitable bonding clip. Fixed at or near the connection must be a permanent label saying 'Safety electrical connection – do not remove' (Regulation 514–13–01) as shown in Fig. 6.37.

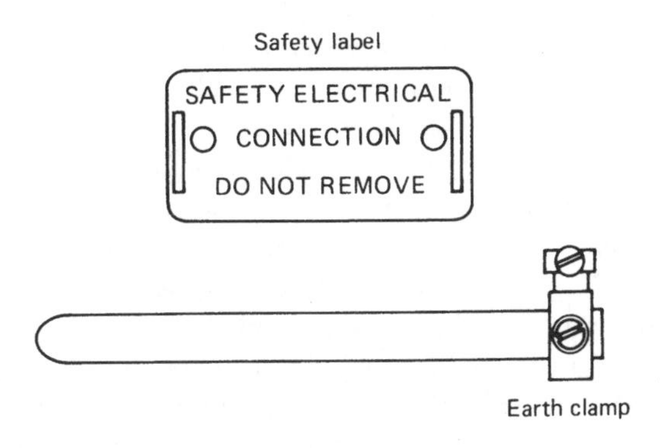

Fig. 6.37 Typical earth bonding clamp.

COOKER CIRCUIT

A cooker with a rating above 3 kW must be supplied on its own circuit but since it is unlikely that in normal use every heating element will be switched on at the same time, a diversity factor may be applied in calculating the cable size, as detailed in Table 1A in Appendix 1 of the *On Site Guide*.

Consider, as an example, a cooker with the following elements fed from a cooker control unit incorporating a 13 A socket:

4×2 kW fast boiling rings = 8000 W
1×2 kW grill = 2000 W
1×2 kW oven = 2000 W
Total loading = 12 000 W

When connected to 250 V

$$\text{Current rating} = \frac{12\ 000}{250} = 48 \text{ A.}$$

Applying the diversity factor of Table 1A,
Total current rating = 48 A.
First 10 amperes = 10 A
30% of 38 A = 11.4 A
Socket outlet = 5 A
Assessed current demand = 10 + 11.4 + 5 = 26.4 A

Therefore, a cable capable of carrying 26.4 A may be used safely rather than a 48 A cable.

A cooking appliance must be controlled by a switch separate from the cooker but in a readily accessible position (Regulation 476–03–04). Where two cooking appliances are installed in one room, such as split-level cookers, one switch may be used to control both appliances provided that neither appliance is more than 2 m from the switch (*On Site Guide*, Appendix 8).

Light sources

The two most common artificial sources of illumination to be found in a domestic installation are the general lighting service lamp (GLS lamp) and the fluorescent lamp. Both may be connected to the circuits discussed previously and shown in Figs 6.23 to 6.26.

GLS LAMPS

General lighting service incandescent filament lamps produce light as a result of the heating effect of an electric current. The modern lamp consists of a fine tungsten wire which is first coiled, and then coiled again. This coiled coil arrangement reduces the cooling of the filament, and increases the light output by operating the filament at a high temperature (about 2900°C). The filament is supported on a metal spider which is held by a glass stem and the whole arrangement is enclosed in a glass envelope, or bulb, as shown in Fig. 6.38.

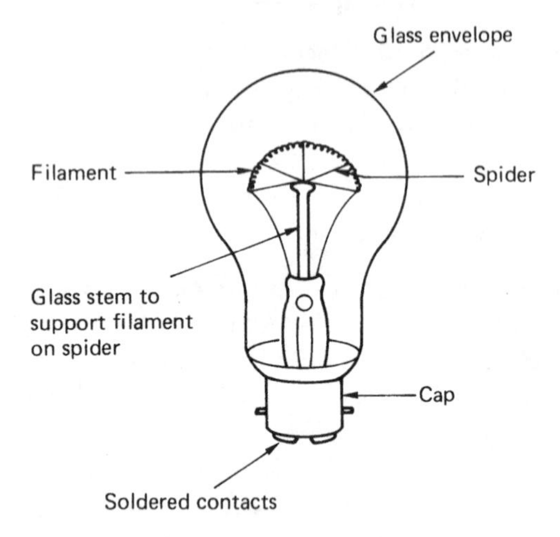

Fig. 6.38 The construction of a GLS lamp.

The light output covers the visible spectrum, giving a warm white to yellow light with good colour rendering. This means that the colours of articles viewed in the light of a GLS lamp are faithfully reproduced. The efficacy of a modern GLS lamp is about 14 lumens per watt, over its intended life of 1000 hours. There are no special regulations relating to GLS lamps but, because the glass envelope becomes very hot during operation, it must not come into contact with combustible materials.

The GLS lamp is unchallenged as the domestic light source. Despite the availability of more efficient lamps, more GLS lamps are produced each year than any other type.

FLUORESCENT LAMPS

Fluorescent lamps are low-pressure mercury vapour discharge lamps. They do not produce light by means of an incandescent filament but by the excitation of a metallic vapour within the tube. Fluorescent tubes contain mercury vapour at low pressure which emits blue and ultraviolet light when ionized; this is converted to white light by the fluorescent powder coated on the inside of the tube. Figure 6.39 shows a switch-start fluorescent lamp circuit in which a glow-type starter switch is now standard. A glow-type starter switch consists of two bimetallic strip electrodes, encased in a glass bulb containing an inert gas. The starter switch is initially open-circuit. When the supply is switched on the full mains voltage is developed across these contacts and a glow discharge takes place between them. This warms the switch electrodes and they bend toward each other until the switch makes contact. The current flows through the lamp electrodes which become heated so that a cloud of electrons is formed at each end of the tube, which glows. When the contacts in the starter switch are made, the glow discharge between the contacts is extinguished since no voltage is developed across the switch. The starter switch contacts cool and, after a few seconds, spring apart. Since there is a choke in series with the lamp, the breaking of this inductive circuit causes a voltage surge across the lamp electrodes, which is sufficient to strike the main arc in the tube. If the lamp does not strike first time the process is repeated.

When the main arc has been struck in the low-pressure mercury vapour, the current is limited by the choke. The capacitor across the mains supply provides power factor correction, and the capacitor across the starter switch contact is for radio interference suppression.

Fluorescent tubes have an efficacy of between 30 and 70 lumens per watt depending upon the tube colour and usually operate trouble-free for the rated life of the lamp, which is 7500 hours. A flickering effect, caused by the failure to establish the main arc in the tube, can be experienced but is usually eliminated by replacing the starter lamp or the tube if it is blackening at the tube ends.

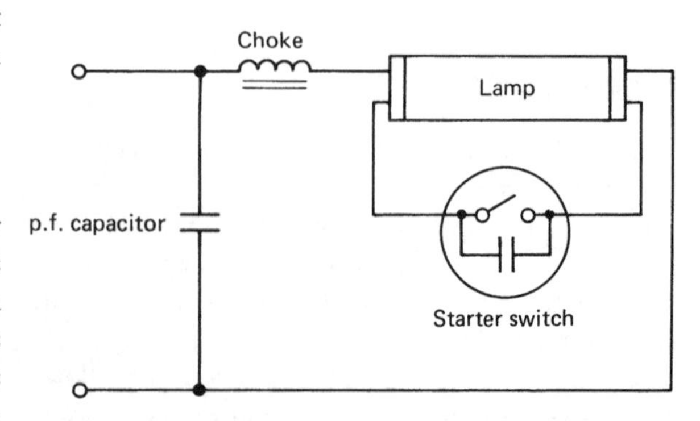

Fig. 6.39 Fluorescent lamp and circuit.

When calculating the assumed rating of discharge circuits it is necessary to consider the lamp and any associated control gear, such as the choke and capacitor. Appendix 1 of the *On Site Guide* tells us that where more exact information is not available, the rating of the final circuits for discharge lamps can be taken as the rated lamp wattage multiplied by 1.8. Therefore, an 80 W fluorescent lamp luminaire will have an assumed demand of $80 \times 1.8 = 144$ W.

Discharge lighting circuits are inductive and can cause excessive wear on the functional switch contacts. Where discharge lighting circuits are to be switched, the rating of such a switch must be suitable for an inductive load.

ENERGY-EFFICIENT LAMPS

Energy-efficient lamps are miniature fluorescent lamps designed to replace ordinary GLS lamps. They are available in a variety of shapes and sizes so that they can be fitted into existing light fittings. Figure 6.40 shows three typical shapes. Lamps of the 'stick' type give most of their light output radially, while those of the flat 'double D' type give most of their light output above and below.

Energy-efficient lamps use electricity much more efficiently than equivalent GLS lamps. For example, a 20 W energy-efficient lamp will give the same light output as a 100 W GLS lamp. An 11 W energy efficient lamp is equivalent to a 60 W GLS lamp. Energy-efficient lamps also have a lifespan about eight times longer than GLS lamps and so they do use energy very efficiently.

However, energy-efficient lamps are very expensive to purchase and they do take a few minutes to attain full brilliance after switching on. They cannot be controlled by a dimmer switch and are unsuitable for incorporating in an automatic presence detector because these are usually switched on only for short periods, but energy-efficient lamps are excellent for outside security lighting which is left on for several hours each night.

The electrical contractor, in discussion with a customer, must balance the advantages and disadvantages of energy-efficient lamps compared to other sources of illumination for each individual installation.

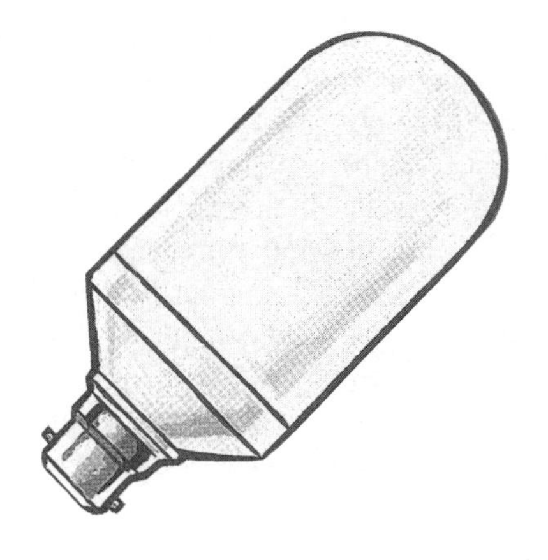

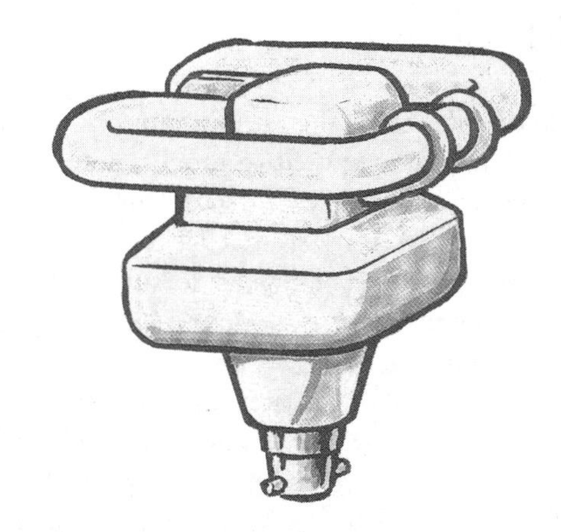

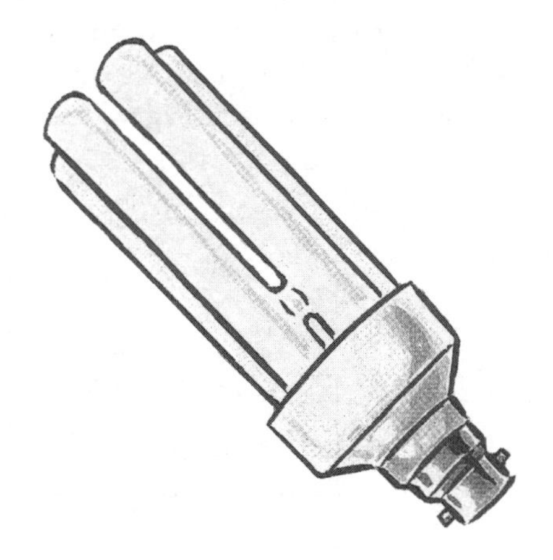

Fig. 6.40 Energy-efficient lamps.

STROBOSCOPIC EFFECT

The flicker effect of any light source can lead to the risk of a stroboscopic effect. This causes rotating or reciprocating machinery to appear to be running at speeds different than their actual speed, and in extreme cases a circular saw or lathe chuck may appear stationary when rotating. A stroboscopic light is used to good effect when electronically 'timing' a car, by making the crank shaft appear stationary when the engine is running so that the top dead centre position (TDC) may be found.

All discharge lamps used on a.c. circuits flicker, often unobtrusively, due to the arc being extinguished every half cycle as the lamp current passes through zero. The elimination of this flicker is desirable in all commercial installations and particularly those which use rotating machinery. The elimination of stroboscopic effect is discussed in detail in Chapter 10 of *Advanced Electrical Installation Work*.

Fire alarm circuits (BS 5839: 1980)

Through one or more of the various statutory Acts, all public buildings are required to provide an effective means of giving a warning of fire so that life and property may be protected. An effective system is one which gives a warning of fire while sufficient time remains for the fire to be put out and any occupants to leave the building.

Fire alarm circuits are wired as either normally open or normally closed. In a *normally open circuit*, the alarm call points are connected in parallel with each other so that when any alarm point is initiated the circuit is completed and the sounder gives a warning of fire. The arrangement is shown in Fig. 6.41. It is essential for some parts of the wiring system to continue operating even when attacked by fire. For this reason the master control and sounders should be wired in MI or FP200 cable. The alarm call points of a normally open system must also be wired in MI or FP200 cable, unless a monitored system is used. In its simplest form this system requires a high-value resistor to be connected across the call-point contacts, which permits a small current to circulate and operate an indicator, declaring the circuit healthy. With a monitored system, PVC insulated cables may be used to wire the alarm call points.

In a *normally closed circuit*, the alarm call points are connected in series to normally closed contacts as shown in Fig. 6.42. When the alarm is initiated, or if a break occurs in the wiring, the alarm is activated. The sounders and master control unit must be wired in MI or FP200 cable, but the call points may be wired in PVC insulated cable since this circuit will always 'fail safe'.

ALARM CALL POINTS

Manually operated alarm call points should be provided in all parts of a building where people may be present, and should be located so that no one need walk more than 30 m from any position within the premises in order to give an alarm. A breakglass manual call point is shown in Fig. 6.43. They should be located on exit routes and, in particular, on the floor landings of staircases and exits to the street. They should be fixed at a height of 1.4 m above the floor at easily accessible, well-illuminated and conspicuous positions.

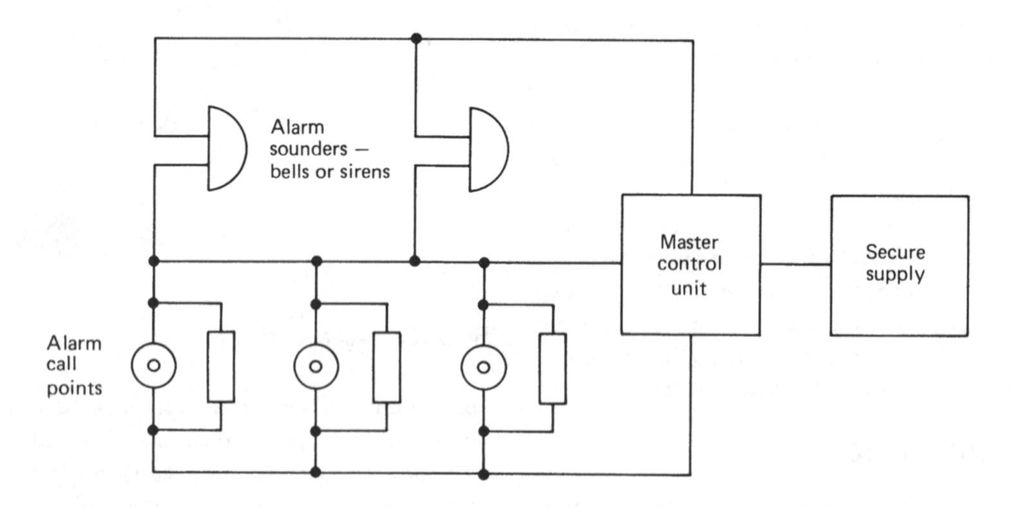

Fig. 6.41 A simple normally open fire alarm circuit.

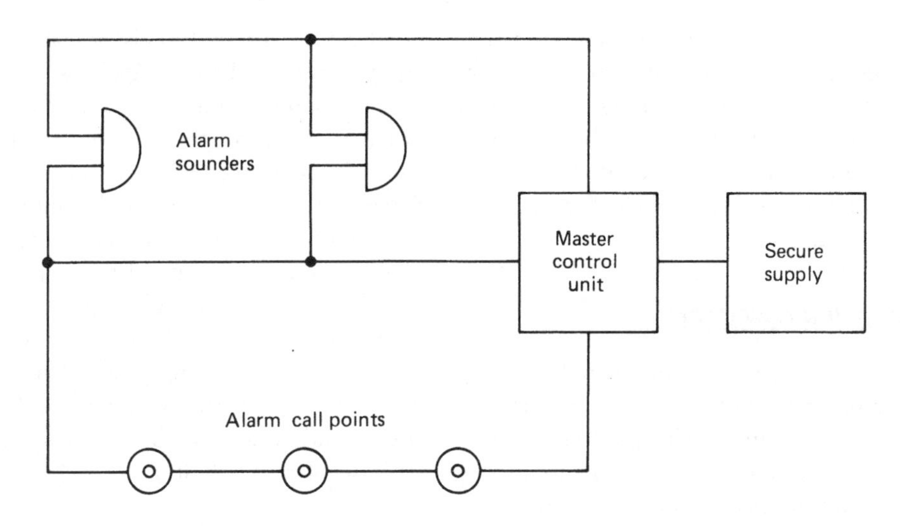

Fig. 6.42 A simple normally closed fire alarm circuit.

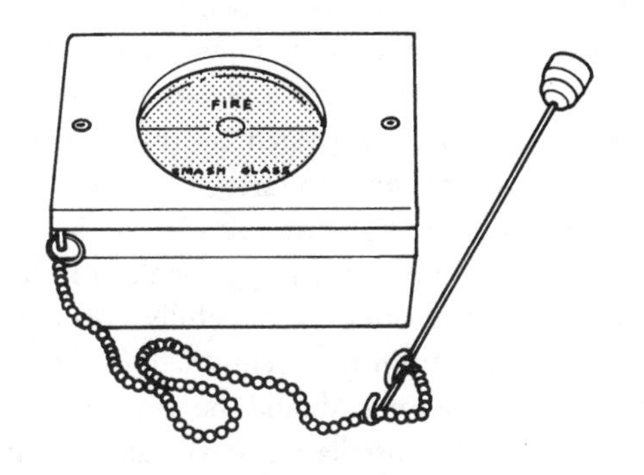

Fig. 6.43 Breakglass manual call point.

Automatic detection of fire is possible with heat and smoke detectors. These are usually installed on the ceilings and at the top of stair wells of buildings because heat and smoke rise. Smoke detectors tend to give a faster response than heat detectors, but whether manual or automatic call points are used should be determined by their suitability for the particular installation. They should be able to discriminate between a fire and the normal environment in which they are to be installed.

SOUNDERS

The position and numbers of sounders should be such that the alarm can be distinctly heard above the

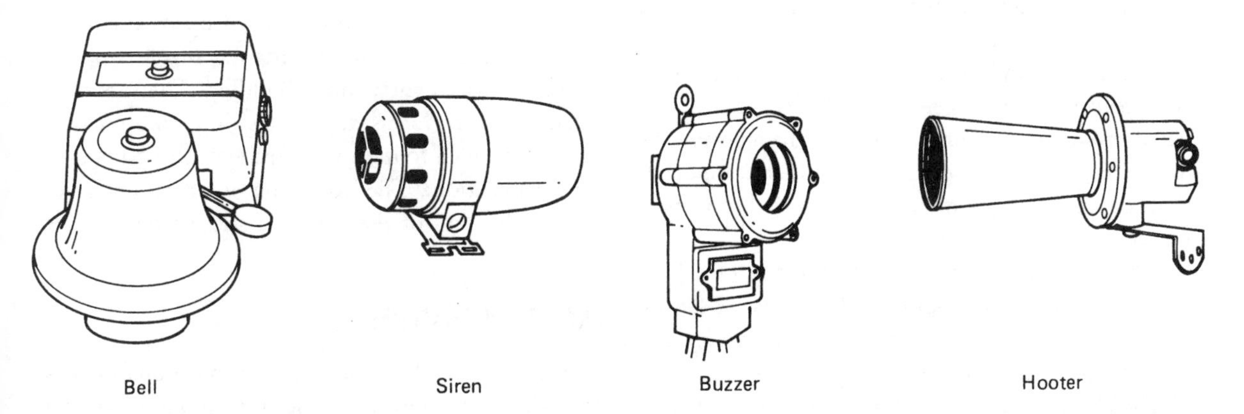

Fig. 6.44 Typical fire alarm sounders.

background noise in every part of the premises. The sounders should produce a minimum of 65 decibels, or 5 decibels above any ambient sound which might persist, for more than 30 seconds. Bells, hooters or sirens may be used but in any one installation they must all be of the same type. Examples of sounders are shown in Fig. 6.44.

FIRE ALARM DESIGN CONSIDERATIONS

Since all fire alarm installations must comply with the relevant statutory regulations, good practice recommends that contact be made with the local fire prevention officer at the design stage in order to identify any particular local regulations and obtain the necessary certification.

Larger buildings must be divided into zones so that the location of the fire can be quickly identified by the emergency services. The zones can be indicated on an indicator board situated in, for example, a supervisor's office or the main reception area.

In selecting the zones, the following rules must be considered:

1. Each zone should not have a floor area in excess of 2000 m^2.
2. Each zone should be confined to one storey, except where the total floor area of the building does not exceed 300 m^2.
3. Staircases and very small buildings should be treated as one zone.
4. Each zone should be a single fire compartment. This means that the walls, ceilings and floors are capable of containing the smoke and fire.

At least one fire alarm sounder will be required in each zone, but all sounders in the building must operate when the alarm is activated.

The main sounders may be silenced by an authorized person, once the general public have been evacuated from the building, but the current must be diverted to a supervisory buzzer which cannot be silenced until the system has been restored to its normal operational state.

A fire alarm installation may be linked to the local fire brigade's control room by the British Telecom network, if the permission of the fire authority and local BT office is obtained.

The electricity supply to the fire alarm installation must be secure in the most serious conditions. In practice the most reliable supply is the mains supply, backed up by a 'standby' battery supply in case of mains failure. The supply should be exclusive to the fire alarm installation, fed from a separate switch fuse, painted red and labelled, 'Fire Alarm – Do Not Switch Off'. Standby battery supplies should be capable of maintaining the system in full normal operation for at least 24 hours and, at the end of that time, be capable of sounding the alarm for at least 30 minutes.

Fire alarm circuits are category 3 circuits and consequently cables forming part of a fire alarm installation must be physically segregated from all other circuits and from each other unless wired in MI cables (IEE Regulation 528–01).

Intruder alarm circuits

The circuitry of intruder alarms is very similar to that of fire alarms, the installation being wired as an open or closed circuit system as shown in Figs 6.41 and 6.42. Instead of the alarm call points, the intruder alarm installation is activated by pressure switches, proximity switches or detectors.

It is usual to connect two sounders on an intruder alarm installation, one inside to make the intruder apprehensive and anxious, hopefully encouraging a rapid departure from the premises, and one outside. The outside sounder should be displayed prominently, since the installation of an alarm system is thought to deter the casual intruder and a ringing alarm encourages neighbours and officials to investigate a possible criminal act.

The supply must be secure, and this is usually achieved by a mains supply backed up by a battery standby supply. The alarm initiating switches detect an unauthorized entry into a building. They are provided in many forms and almost any combination may be installed to meet the requirements of a particular installation. The following information gives some indication of the characteristics of the various types.

PROXIMITY SWITCHES

They are designed for the discreet protection of doors and windows. They are made from moulded plastic and are about the size of a chewing gum packet. One moulding contains a reed switch, the other a magnet, and when they are placed close together the magnet maintains the contacts of the reed switch in either an open or closed position. Opening the door or window separates the two mouldings, and the switch is activated, triggering the alarm.

PRESSURE PADS

Pressure pad switches are placed under the carpet close to a door. Treading on the carpet activates the switch and the alarm system.

PASSIVE INFRARED DETECTORS

These are activated by a moving body which is warmer than the surroundings. They have a range of 15 m and a detection zone of 90° but can be accidentally triggered by domestic pets.

INFRARED BEAM DETECTORS

These consist of a transmitter and receiver placed at opposite ends of a room up to 20 m apart. Breaking the infrared beam activates the alarm system.

ULTRASONIC DETECTORS

Any movement detected by the sensor triggers the alarm system. The detection range is variable, between 2 and 8 metres.

The type and extent of the intruder alarm installation and, therefore, the cost, will depend upon many factors including the type and position of the building, the contents of the building, the insurance risk involved and the peace of mind offered by an alarm system to the owner or occupier of the building.

Intruder alarm circuits are category 2 circuits and should be segregated from mains supply cables or insulated to the highest voltage present if run in a common enclosure with category 1 cables.

Emergency lighting (BS 5266: 1975)

Emergency lighting should be planned, installed and maintained to the highest standards of reliability and integrity, so that it will operate satisfactorily when called into action, no matter how infrequently this may be.

Emergency lighting is not required in private homes because the occupants are familiar with their surroundings, but in public buildings people are in unfamiliar surroundings. In an emergency people do not always act rationally, but well-illuminated and easily identified exit routes can help to reduce panic.

Emergency lighting is provided for two reasons; to illuminate escape routes, called 'escape' lighting; and to enable a process or activity to continue after a normal lights failure, called 'standby' lighting.

Escape lighting is usually required by local and national statutory authorities under legislative powers. The escape lighting scheme should be planned so that identifiable features and obstructions are visible in the lower levels of illumination which may prevail during an emergency. Exit routes should be clearly indicated by signs and illuminated to a uniform level, avoiding bright and dark areas.

Standby lighting is required in hospital operating theatres and in industry, where an operation or process once started must continue, even if the mains lighting fails. Standby lighting may also be required for security reasons. The cash points in public buildings may need to be illuminated at all times to discourage acts of theft occurring during a mains lighting failure.

EMERGENCY SUPPLIES

Since an emergency occurring in a building may cause the mains supply to fail, the emergency lighting should be supplied from a source which is independent from the main supply. In most premises the alternative power supply would be from batteries, but generators may also be used. Generators can have a large capacity and duration, but a major disadvantage is the delay time while the generator runs up to speed and takes over the load. In some premises a delay of more than 5 seconds is considered unacceptable, and in these cases a battery supply is required to supply the load until the generator can take over.

The emergency lighting supply must have an adequate capacity and rating for the specified duration of time (IEE Regulation 313–02). BS 5266 states that after a battery is discharged by being called into operation for its specified duration of time, it should be capable of once again operating for the specified duration of time following a recharge period of not longer than 24 hours. The duration of time for which the emergency lighting should operate will be specified by a statutory authority but is normally 1–3 hours. BS 5266 states that escape lighting should operate for a minimum of 1 hour. Standby lighting operation time will depend upon financial considerations and the importance of continuing the process or activity.

There are two possible modes of operation for emergency lighting installations: maintained and non-maintained.

Maintained emergency lighting

The emergency lamps are continuously lit using the normal supply when this is available, and change over to an alternative supply when the mains supply fails. The advantage of this system is that the lamps are continuously proven healthy and any failure is immediately obvious. It is a wise precaution to fit a supervisory buzzer in the emergency supply to prevent accidental discharge of the batteries, since it is not otherwise obvious which supply is being used.

Maintained emergency lighting is normally installed in theatres, cinemas, discotheques and places of entertainment where the normal lighting may be dimmed or extinguished while the building is occupied. The emergency supply for this type of installation is often supplied from a central battery, the emergency lamps being wired in parallel from the low-voltage supply as shown in Fig. 6.45.

Non-maintained emergency lighting

The emergency lamps are only illuminated if the normal mains supply fails. Failure of the main supply de-energizes a solenoid and a relay connects the emergency lamps to a battery supply, which is maintained in a state of readiness by a trickle charge from the normal mains supply. When the normal supply is restored, the relay solenoid is energized, breaking the relay contacts, which disconnects the emergency lamps, and the charger recharges the battery. Figure 6.46 illustrates this arrangement.

The disadvantage with this type of installation is that broken lamps are not detected until they are called into operation in an emergency, unless regularly maintained. The emergency supply is usually provided by a battery contained within the luminaire, together with the charger and relay, making the unit self-contained. Self-contained units are cheaper and easier to install than a central battery system, but the central battery can have a greater capacity and duration, and permit a range of emergency lighting luminaires to be installed.

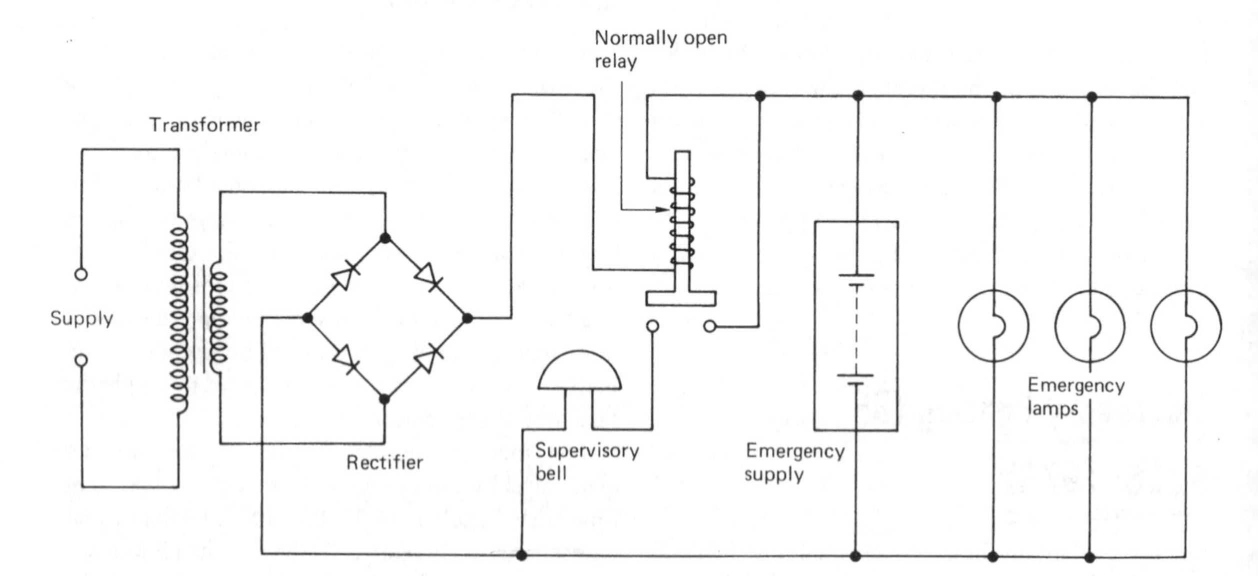

Fig. 6.45 Maintained emergency lighting.

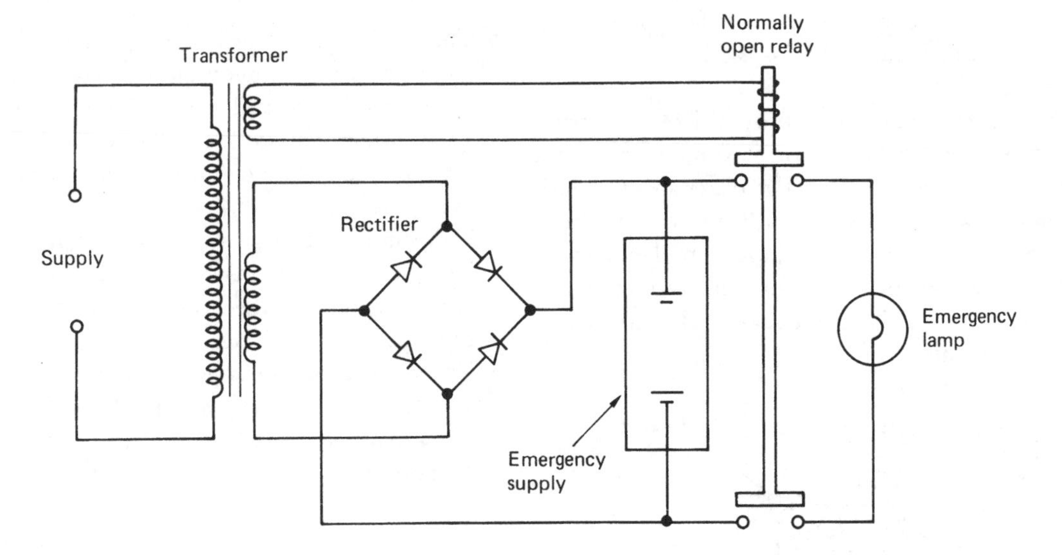

Fig. 6.46 Non-maintained emergency lighting.

MAINTENANCE

The contractor installing the emergency lighting should provide a test facility which is simple to operate and secure against unauthorized interference. The emergency lighting installation must be segregated completely from any other wiring, so that a fault on the main electrical installation cannot damage the emergency lighting installation (IEE Regulation 528–01). Figure 6.15 shows a trunking which provides for segregation of circuits.

The batteries used for the emergency supply should be suitable for this purpose. Motor vehicle batteries are not suitable for emergency lighting applications, except in the starter system of motor-driven generators. The fuel supply to a motor-driven generator should be checked. The battery room of a central battery system must be well ventilated and, in the case of a motor-driven generator, adequately heated to ensure rapid starting in cold weather.

BS 5266 recommends that the full load should be carried by the emergency supply for at least 1 hour in every 6 months. After testing, the emergency system must be carefully restored to its normal operative state. A record should be kept of each item of equipment and the date of each test by a qualified or responsible person. It may be necessary to produce the record as evidence of satisfactory compliance with statutory legislation to a duly authorized person.

Self-contained units are suitable for small installations of up to about 12 units. The batteries contained within these units should be replaced about every 5 years, or as recommended by the manufacturer.

PRIMARY CELLS

A primary cell cannot be recharged. Once the active chemicals are exhausted, the cell must be discarded.

Primary cells, in the form of Leclanche cells, are used extensively as portable power sources for radios and torches and have an emf of 1.5 V. Larger voltages are achieved by connecting cells in series. Thus, a 6 V supply can be provided by connecting four cells in series.

Mercury primary cells have an emf of 1.35 V, and can have a very large capacity in a small physical size. They have a long shelf life and leakproof construction, and are used in watches and hearing aids.

SECONDARY CELLS

A secondary cell has the advantage of being rechargeable. If the cell is connected to a suitable electrical supply, electrical energy is stored on the plates of the cell as chemical energy. When the cell is connected to a load, the chemical energy is converted to electrical energy.

A lead-acid cell is a secondary cell. Each cell delivers about 2 V, and when six cells are connected in series a 12 V battery is formed of the type used on motor vehicles. Figure 6.47 shows the construction of a lead-acid battery.

A lead-acid battery is constructed of lead plates which are deeply ribbed to give maximum surface area for a given weight of plate. The plates are assembled in groups, with insulating separators between them. The separators are made of a porous insulating material, such as wood or ebonite, and the whole assembly is immersed in a dilute sulphuric acid solution in a plastic container.

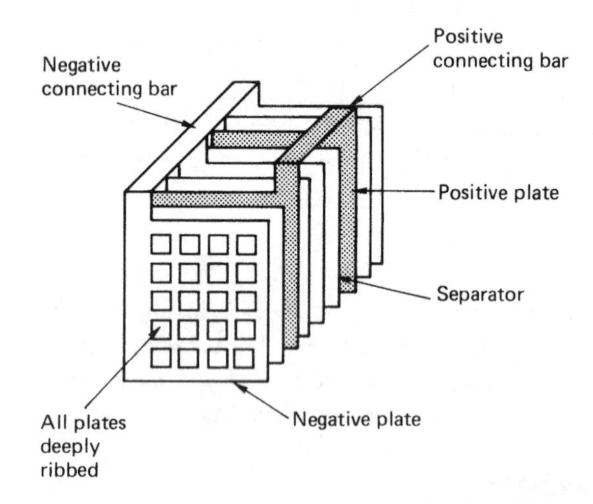

Fig. 6.47 The construction of a lead-acid battery.

BATTERY RATING

The capacity of a cell to store charge is a measure of the total quantity of electricity which it can cause to be displaced around a circuit after being fully charged. It is stated in ampere-hours, abbreviation Ah, and calculated at the 10 hour rate which is the steady load current which would completely discharge the battery in 10 hours. Therefore, a 50 Ah battery will provide a steady current of 5 A for ten hours.

MAINTENANCE OF LEAD-ACID BATTERIES

- The plates of the battery must always be covered by the dilute sulphuric acid. If the level falls, it must be topped up with distilled water.
- Battery connections must always be tight and should be covered with a thin coat of petroleum jelly.
- The specific gravity or relative density of the battery gives the best indication of its state of charge. A discharged cell will have a specific gravity of 1.150, which will rise to 1.280 when fully charged. The specific gravity of a cell can be tested with a hydrometer.
- To maintain a battery in good condition it should be regularly trickle-charged. A rapid charge or discharge encourages the plates to buckle, and may cause permanent damage.
- The room used to charge a battery must be well ventilated because the charged cell gives off hydrogen and oxygen, which are explosive in the correct proportions.

Simple communication systems

The purpose of any communication system is to convey information between two physically remote points. Any communication system must include a transmitter for sending the information, a transmission circuit or medium, and a receiver. To transmit signals comprehensible to the human ear the system must produce an audible note. The simplest system is a morse code or buzzer which produces a sound whenever a key activates the circuit. This sound may be of one frequency only. For the transmission of speech, a complex mixture of frequencies is required. Therefore, the communication system must be capable of accommodating a wide range of frequencies. The range of frequencies required by the communication system is called the bandwidth; the simpler the system, the smaller the bandwidth.

The bandwidth of speech is from 30 Hz to 5 kHz. Music requires an even greater bandwidth, from 20 Hz to 20 kHz. The telephone system is designed to transmit on a limited bandwidth from 300 Hz to 3.4 kHz. This is suitable for acceptable voice reproduction, but, because the upper and lower frequencies are cut off, the quality of reproduction is not perfect.

SIMPLE TELEPHONE CIRCUIT

A telephone handset contains a receiver, the ear piece and a transmitter for speaking into. The transmitter converts the pressure waves from the spoken word into electrical signals, which are transmitted along conductors to the receiver of another handset which converts the electrical signals into sound waves. This provides the basis of a simple two-way communication system, and a suitable circuit is shown in Fig. 6.48.

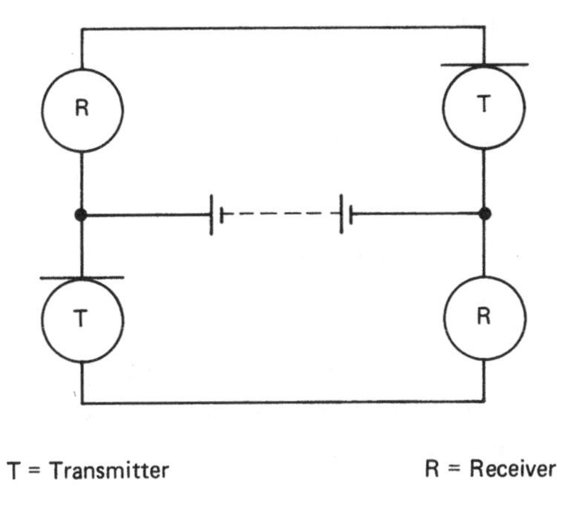

Fig. 6.48 A simple telephone circuit.

TELEPHONE SOCKET OUTLETS

The installation of telecommunications equipment could, for many years, only be undertaken by British Telecom engineers, but as a result of the recent liberalization of BT the electrical contractor may now supply and install telecommunications equipment.

On new premises the electrical contractor may install sockets and the associated wiring to the point of intended line entry, but the connection of the incoming line to the installed master socket must only be made by the telephone company's engineer.

On existing installations, additional secondary sockets may be installed to provide an extended plug-in facility as shown in Fig. 6.49. Any number of secondary sockets may be connected in parallel, but the number of telephones which may be connected at any one time is restricted.

Each telephone or extension bell is marked with a ringing equivalence number (REN) on the underside. Each exchange line has a maximum capacity of REN 4 and therefore the total REN values of all the connected telephones must not exceed four if they are to work correctly.

An extension bell may be connected to the installation by connecting the two bell wires to terminals 3 and 5 of a telephone socket. The extension bell must be of the high impedance type having a REN rating. All equipment connected to a BT exchange line must display the green circle of approval.

The multicore cable used for wiring extension socket outlets should be of a type intended for use with telephone circuits, which will normally be between 0.4 mm and 0.68 mm in cross-section.

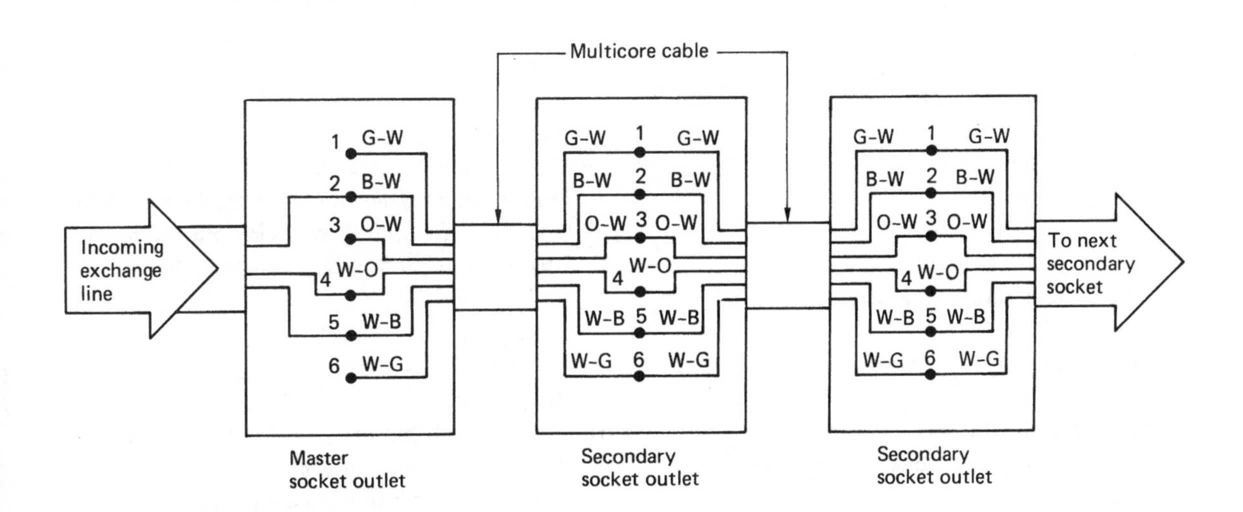

Fig. 6.49 Telephone circuit outlet connection diagram.

Telephone cable conductors are identified in Table 6.2 and the individual terminals in Table 6.3. The conductors should be connected as shown in Fig. 6.49. Telecommunications cables are Band I circuits and must be segregated from Band II circuits containing mains cables (IEE Regulation 528–01).

Table 6.2 Telephone cable identification

Code	Base colour	Stripe
G–W	Green	White
B–W	Blue	White
O–W	Orange	White
W–O	White	Orange
W–B	White	Blue
W–G	White	Green

Table 6.3 Telephone socket terminal identification. Terminals 1 and 6 are frequently unused, and therefore 4-core cable may normally be installed. Terminal 4, on the incoming exchange line, is only used on a PBX line for earth recall.

Socket terminal	Circuit
1	Spare
2	Speech circuit
3	Bell circuit
4	Earth recall
5	Speech circuit
6	Spare

Exercises

1 Overload or overcurrent protection is offered by a:
(a) transistor
(b) transformer
(c) functional switch
(d) circuit breaker.

2 The current rating of a protective device is the current which:
(a) it will carry continuously without deterioration
(b) will cause the device to operate
(c) will cause the device to operate within 4 hours
(d) is equal to the fusing factor.

3 On a domestic installation the lighting circuit comprises 15 outlets. Good practice would suggest:
(a) one final circuit
(b) two final circuits
(c) five final circuits
(d) 15 final circuits.

4 Domestic lighting circuits are usually wired in:
(a) 1.0 mm cable
(b) 2.5 mm cable
(c) 4.0 mm cable
(d) 6.0 mm cable

5 Domestic ring final circuits are usually wired in:
(a) 1.0 mm cable
(b) 2.5 mm cable
(c) 4.0 mm cable
(d) 6.0 mm cable.

6 The regulations recommend that the flexible cable connecting an appliance to a socket outlet should not exceed:
(a) 1 m
(b) 2 m
(c) 3 m
(d) 4 m.

7 A 13 A plug top always:
(a) has round pins
(b) has copper pins
(c) contains a thermal overload
(d) contains a cartridge fuse.

8 A radial circuit, wired in 2.5 mm PVC insulated and sheathed cable and protected by a 20 A fuse, may feed any number of socket outlets, provided that the total floor area does not exceed:
(a) 10 m^2
(b) 20 m^2
(c) 50 m^2
(d) 100 m^2.

9 A ring circuit wired in 2.5 mm PVC insulated and sheathed cable and protected by a 30 A fuse, may feed any number of socket outlets, provided that the total floor area does not exceed:
(a) 10 m^2
(b) 20 m^2
(c) 50 m^2
(d) 100 m^2.

10 A non-fused spur to a ring circuit may feed:
(a) any number of socket outlets
(b) any number of fixed appliances
(c) only one single- or twin-socket outlet
(d) only two fixed appliances.

11 Block storage radiators connected to an off-peak supply have the advantage of:
(a) a higher thermal output than other radiators
(b) using half-price electricity
(c) a slim, lightweight and portable construction
(d) controllable radiant heat.

12 In a room containing a fixed bath or shower:
(a) no electrical equipment may be installed
(b) only switched socket outlets may be installed
(c) every socket outlet must be inaccessible to anyone using the bath or shower
(d) every switch must be inaccessible to anyone using the bath or shower.

13 A cooking appliance must be controlled by a switch which:
(a) is incorporated in the appliance
(b) is separate from the cooker but within 2 m of it
(c) is cord-operated
(d) incorporates a pilot light.

14 Most domestic lighting is provided by:
(a) fluorescent lamps
(b) GLS lamps
(c) discharge lamps
(d) series-connected lamps.

15 The stroboscopic effect may be dangerous in:
(a) a domestic environment
(b) an industrial environment
(c) a commercial environment
(d) a business environment.

16 Fluorescent lamps, when compared with GLS lamps:
(a) give a whiter light
(b) operate at a higher temperature
(c) have a longer useful life
(d) require no additional circuitry.

17 The choke in a fluorescent light fitting:
(a) provides power-factor correction
(b) provides radio-interference suppression
(c) provides a voltage surge to strike the arc
(d) prevents the stroboscopic effect.

18 The assumed rating of a discharge circuit is taken as the rated lamp wattage multiplied by a factor of:
(a) 1.25
(b) 1.5
(c) 1.8
(d) 2.

19 Five 125 V discharge luminaires are connected to one final circuit. The assumed rating of the circuit will be:
(a) 125 W
(b) 250 W
(c) 625 W
(d) 1125 W.

20 The capacity of a battery to store charge is measured in:
(a) watts
(b) volt-amperes
(c) ampere-ohms
(d) ampere-hours.

21 With the aid of a sketch, describe the construction of a semi-enclosed fuse and a cartridge fuse. State the advantages and disadvantages of each type of fuse and identify one typical application for each device.

22 With the aid of a sketch, describe the construction of a cartridge fuse and an HBC fuse. State the advantages of each type of fuse and identify one typical application for each device.

23 Describe the operation of a MCB:
(a) when carrying the rated current
(b) during overcurrent conditions
(c) during short-circuit conditions.
State the advantages of an MCB compared with a fuse.

24 Describe the operation of an RCD and state one application for this device.

25 Describe, with the aid of a diagram, the supplementary bonding requirements of the IEE Regulations for a room containing a fixed bath or shower.

26 Sketch the circuit diagram for a normally open fire alarm circuit, with four call points and three sounders. Describe the appearance and installation position of a breakglass call point.

27 Sketch the circuit diagram for a normally closed fire alarm circuit, with six call points and three sounders. Describe two typical fire alarm sounders.

28 Describe a simple intruder alarm installation suitable for a domestic dwelling. Indicate the type and position of detectors, sounders, supply and wiring.

29 Sketch the circuit diagram for a maintained emergency lighting installation. Describe some of the factors to be considered when choosing the position of the luminaires.

30 Sketch the circuit diagram for a non-maintained emergency lighting installation. State the advantages and disadvantages of a maintained system and a non-maintained system.

31 Describe the construction of a motor vehicle battery. Describe how to test and maintain a battery in good condition.

32 One advantage of a steel conduit installation, compared with a PVC conduit installation, is that it:
(a) may be easily rewired
(b) may be installed more quickly
(c) offers greater mechanical protection
(d) may hold more conductors for a given conduit size.

33 The earth continuity of a metallic conduit installation will be improved if:
(a) black enamel conduit is replaced by galvanized conduit
(b) the installation is painted with galvanized paint
(c) the installation is painted with bright orange paint
(d) all connections are made tight and secure during installation.

34 The earth continuity of a metallic trunking installation may be improved if:
(a) copper earth straps are fitted across all joints
(b) galvanized trunking is used
(c) all joints are painted with galvanized paint
(d) a space factor of 45% is not exceeded.

35 Circuits of categories 1 and 2 can
(a) never be installed in the same trunking
(b) only be installed in the same trunking if they are segregated by metal enclosures
(c) only be installed in the same trunking if a space factor of 45% is not exceeded
(d) only be wired in MI cables.

36 An industrial installation of PVC/SWA cables laid on cable tray offers the advantage over other types of installation of:
(a) greater mechanical protection
(b) greater flexibility in response to changing requirements
(c) higher resistance to corrosion in an industrial atmosphere
(d) flameproof installation suitable for hazardous areas.

37 The cables which can best withstand high temperatures are:
(a) MI cables
(b) PVC cables with asbestos oversleeves
(c) PVC/SWA cables
(d) PVC cables in galvanized conduit.

38 Four 1 mm cables and four 2.5 mm cables are to be run in a metal conduit which contains one right-angle bend and one double set. The distance between the boxes is 8 m. Find the size of conduit required to enclose these cables.

39 Determine the minimum size of trunking required to contain the following stranded cables:
(a) 20×1.5 mm cables
(b) 16×2.5 mm cables
(c) 10×4.0 mm cables
(d) 20×6.0 mm cables.

40 Calculate the number of 1.0 mm cables which may be drawn into a 5 m straight run of 20 mm conduit.

41 Calculate the number of 2.5 mm cables which may be drawn into a 20 mm plastic conduit along with a 4.0 mm CPC if the distance between the boxes is 10 m and contains one right-angled bend.

42 Determine the size of galvanized steel conduit required to contain PVC insulated conductors if the distance between two boxes is 5 m and the conduit has two bends of 90°. The conduit must contain ten 1.5 mm cables and four 2.5 mm cables.

43 Calculate the number of PVC insulated 4.0 mm cables which may be installed in a 75×75 mm trunking.

44 (a) Calculate the minimum size of vertical trunking required to contain 20×10 mm PVC insulated cables.
(b) Explain why fire barriers are fitted in vertical trunking.
(c) Explain how and why cables are supported in vertical trunking.

7

INSPECTION AND TESTING

Inspection and testing techniques

The testing of an installation implies the use of instruments to obtain readings. However, a test is unlikely to identify a cracked socket outlet, a chipped or loose switch plate, a missing conduit-box lid or saddle, so it is also necessary to make a visual inspection of the installation.

All new installations must be inspected and tested before connection to the mains, and all existing installations should be periodically inspected and tested to ensure that they are safe and meet the regulations of the IEE (Regulations 711 to 744).

The method used to test an installation may inject a current into the system. This current must not cause danger to any person or equipment in contact with the installation, even if the circuit being tested is faulty. The test procedures must be followed carefully and in the correct sequence, as indicated by Regulation 713–01–01. This ensures that the protective conductors are correctly connected and secure before the circuit is energized.

The installation must be visually inspected before testing begins. The aim of the visual inspection is to confirm that all equipment and accessories are undamaged and comply with the relevant British and European Standards, and also that the installation has been securely and correctly erected. Regulation 712–01–03 gives a check-list for the initial visual inspection of an installation, including:

- connection of conductors
- identification of conductors
- routeing of cables in safe zones
- selection of conductors for current carrying capacity and volt drop
- connection of single-pole devices for protection or switching in phase conductors only
- correct connection of socket outlets, lampholders, accessories and equipment
- presence of fire barriers, suitable seals and protection against thermal effects
- methods of protection against electric shock, including the insulation of live parts and placement of live parts out of reach by fitting appropriate barriers and enclosures
- prevention of detrimental influences (e.g. corrosion)
- presence of appropriate devices for isolation and switching
- presence of undervoltage protection devices
- choice and setting of protective devices
- labelling of circuits, fuses, switches and terminals
- selection of equipment and protective measures appropriate to external influences
- adequate access to switchgear and equipment
- presence of danger notices and other warning notices
- presence of diagrams, instruction and similar information
- appropriate erection method.

The check-list is a guide, it is not exhaustive or detailed, and should be used to identify relevant items for inspection, which can then be expanded upon. For example, the first item on the check-list, connection of conductors, might be further expanded to include the following:

- Are connections secure?
- Are connections correct? (conductor identification)

- Is the cable adequately supported so that no strain is placed on the connections?
- Does the outer sheath enter the accessory?
- Is the insulation undamaged?
- Does the insulation proceed up to but not *into* the connection?

This is repeated for each appropriate item on the checklist.

Those tests which are relevant to the installation must then be carried out in the sequence given in Regulation 713–01–01 and Sections 9 and 10 of the *On Site Guide* for reasons of safety and accuracy. These tests are as follows:

Before the supply is connected

1 Test for continuity of protective conductors, including main and supplementary bonding.
2 Test the continuity of all ring final circuit conductors.
3 Test for insulation resistance.
4 Test for polarity using the continuity method.
5 Test the earth electrode resistance.

With the supply connected

6 Recheck polarity using a voltmeter or approved test lamp.
7 Test the earth fault loop impedance.
8 Carry out functional testing (e.g. operation of RCDs).

If any test fails to comply with the Regulations, then *all* the preceding tests must be repeated after the fault has been rectified. This is because the earlier test results may have been influenced by the fault (Regulation 713–01–01).

There is an increased use of electronic devices in electrical installation work, for example, in dimmer switches and ignitor circuits of discharge lamps. These devices should temporarily be disconnected so that they are not damaged by the test voltage of, for example, the insulation resistance test (Regulation 713–04).

TEST INSTRUMENTS

The test instruments and test leads used by the electrician for testing an electrical installation must meet all the requirements of the relevant regulations. The Health and Safety Executive has published Guidance Notes GS 38 for test equipment used by electricians. The IEE Regulations (BS 7671) also specify the test voltage or current required to carry out particular tests satisfactorily. All testing must, therefore, be carried out using an 'approved' test instrument if the test results are to be valid. The test instrument must also carry a calibration certificate, otherwise the recorded results may be void. Calibration certificates usually last for a year. Test instruments must, therefore, be tested and recalibrated each year by an approved supplier. This will maintain the accuracy of the instrument to an acceptable level, usually within 2% of the true value.

Modern digital test instruments are reasonably robust, but to maintain them in good working order they must be treated with care. An approved test instrument costs equally as much as a good-quality camera; it should, therefore, receive the same care and consideration.

Continuity tester

To measure accurately the resistance of the conductors in an electrical installation we must use an instrument which is capable of producing an open circuit voltage of between 4 V and 24 V a.c. or d.c., and deliver a short-circuit current of not less than 200 mA (Regulation 713–02). The functions of continuity testing and insulation resistance testing are usually combined in one test instrument.

Insulation resistance tester

The test instrument must be capable of detecting insulation leakage between live conductors and between live conductors and earth. To do this and comply with Regulation 713–04 the test instrument must be capable of producing a test voltage of 250 V, 500 V or 1000 V and deliver an output current of not less than 1 mA at its normal voltage.

Earth fault loop impedance tester

The test instrument must be capable of delivering fault currents as high as 25 A for up to 40 ms using the supply voltage. During the test, the instrument does an Ohm's law calculation and displays the test result as a resistance reading.

RCD tester

Where circuits are protected by a residual current device we must carry out a test to ensure that the device will operate very quickly under fault conditions and within the time limits set by the IEE Regulations. The instrument must, therefore, simulate a fault and measure the time taken for the RCD to operate. The instrument is, therefore, calibrated to give a reading measured in milliseconds to an in-service accuracy of 10%.

If you purchase good-quality 'approved' test instruments and leads from specialist manufacturers they will meet all the Regulations and Standards and therefore give valid test results. However, to carry out all the tests required by the IEE Regulations will require a number of test instruments and this will represent a major capital investment in the region of £1000.

Let us now consider the individual tests.

1 CONTINUITY OF PROTECTIVE CONDUCTORS (713–02)

The object of the test is to ensure that the circuit protective conductor (CPC) is correctly connected, is electrically sound and has a total resistance which is low enough to permit the overcurrent protective device to operate within the disconnection time requirements of Regulation 413–02–08, should an earth fault occur. Every protective conductor must be separately tested from the consumer's earthing terminal to verify that it is electrically sound and correctly connected, including any main and supplementary bonding conductors.

A d.c. test using an ohmmeter continuity tester is suitable where the protective conductors are of copper or aluminium up to 35 mm^2. The test is made with the supply disconnected, measuring from the consumer's earthing terminal to the far end of each CPC, as shown in Fig. 7.1. The resistance of the long test lead is subtracted from these readings to give the resistance value of the CPC. The result is recorded on an installation schedule such as that given in Appendix 7 of the *On Site Guide*.

Where steel conduit or trunking forms the protective conductor, the standard test described above may be used, but additionally the enclosure must be visually checked along its length to verify the integrity of all the joints.

If the inspecting engineer has grounds to question the soundness and quality of these joints then the phase-earth loop impedance test described later in this chapter should be carried out.

If, after carrying out this further test, the inspecting engineer still questions the quality and soundness of the protective conductor formed by the metallic conduit or trunking then a further test can be done using an a.c. voltage not greater than 50 V at the frequency of the installation and a current approaching 1.5 times the design current of the circuit, but not greater than 25 A.

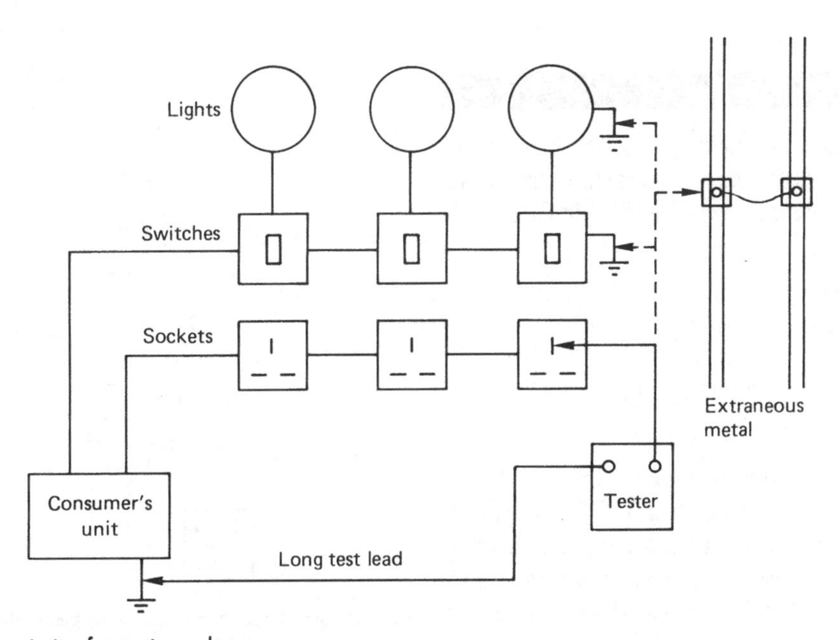

Fig. 7.1 Testing continuity of protective conductors.

This test can be done using a low-voltage transformer and suitably connected ammeters and voltmeters, but a number of commercial instruments are available such as the Clare tester, which give a direct reading in ohms.

Because fault currents will flow around the earth fault loop path, the measured resistance values must be low enough to allow the overcurrent protective device to operate quickly. For a satisfactory test result, the resistance of the protective conductor should be consistent with those values calculated for a phase conductor of similar length and cross-sectional area. Values of resistance per metre for copper and aluminium conductors are given in Table 9A of the *On Site Guide*. The resistances of some other metallic containers are given in Table 7.1.

Table 7.1 Resistance values of some metallic containers

Metallic sheath	Size (mm)	Resistance at 20°C (mΩ/m)
Conduit	20	1.25
	25	1.14
	32	0.85
Trunking	50 × 50	0.949
	75 × 75	0.526
	100 × 100	0.337

EXAMPLE

The CPC for a ring final circuit is formed by a 1.5 mm² copper conductor of 50 m approximate length. Determine a satisfactory continuity test value for the CPC using the value given in Table 9A of the *On Site Guide*. From Table 9A:

Resistance/metre for a
1.5 mm² copper conductor = 12.10 mΩ/m
Therefore,
the resistance of 50 m $= 50 \times 12.10 \times 10^{-3}$
$= 0.605\ \Omega$

The protective conductor resistance values calculated by this method can only be an approximation since the length of the CPC can only be estimated. Therefore, in this case, a satisfactory test result would be obtained if the resistance of the protective conductor was about 0.6 Ω. A more precise result is indicated by the earth fault loop impedance test which is carried out later in the sequence of tests.

2 TESTING FOR CONTINUITY OF RING FINAL CIRCUIT CONDUCTORS (713–03)

The object of the test is to ensure that all ring circuit cables are continuous around the ring, that is, that there are no breaks and no interconnections in the ring and that all connections are electrically and mechanically sound. This test also verifies the polarity of each socket outlet.

The test is made with the supply disconnected, using an ohmmeter as follows:

Disconnect and separate the conductors of both legs of the ring at the main fuse. There are three steps to this test:

Step 1

Measure the resistance of the phase conductors (L_1 and L_2), the neutral conductors (N_1 and N_2) and the protective conductors (E_1 and E_2) at the mains position as shown in Fig. 7.2. End-to-end live and neutral con-

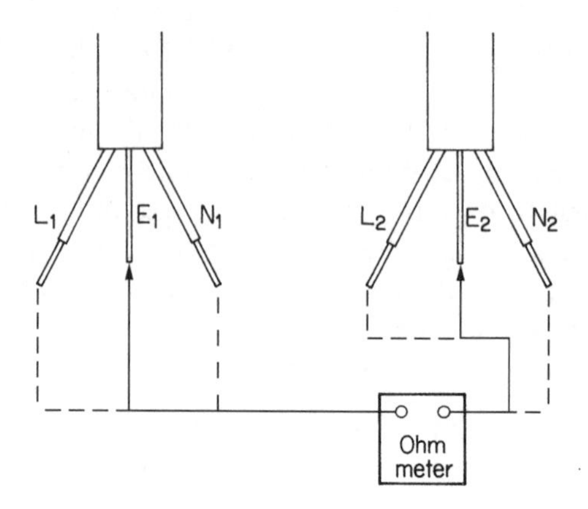

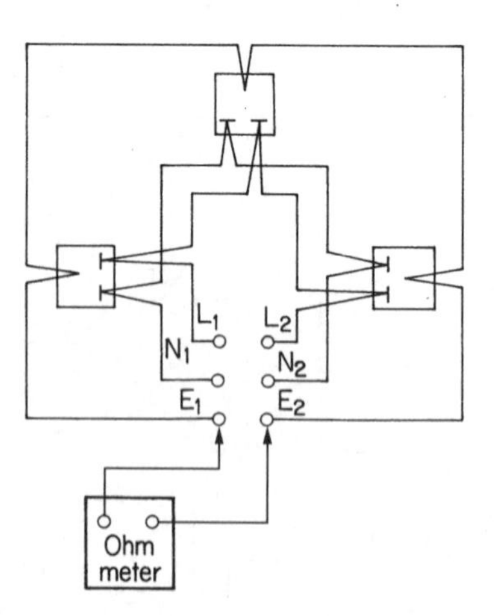

Fig. 7.2 Step 1 test: measuring the resistance of phase, neutral and protective conductors.

ductor readings should be approximately the same (i.e. within 0.05 Ω) if the ring is continuous. The protective conductor reading will be 1.67 times as great as these readings if 2.5/1.5 mm cable is used. Record the results on a table such as that shown in Table 7.2.

Step 2

The live and neutral conductors should now be temporarily joined together as shown in Fig. 7.3. An ohmmeter reading should then be taken between live and neutral at *every* socket outlet on the ring circuit. The readings obtained should be substantially the same, provided that there are no breaks or multiple loops in the ring. Each reading should have a value of approximately half the live and neutral ohmmeter readings measured in step 1 of this test. Sockets connected as a spur will have a slightly higher value of resistance because they are fed by only one cable, while each socket on the ring is fed by two cables. Record the results on a table such as that shown in Table 7.2.

Table 7.2 Table which may be used to record the readings taken when carrying out the continuity of ring final circuit conductors tests according to IEE Regulation 713–02

Test	Ohmmeter connected to:	Ohmmeter readings	This gives a value for
Step 1	L_1 and L_2		r_1
	N_1 and N_2		
	E_1 and E_2		r_2
Step 2	Live and neutral at each socket		
Step 3	Live and earth at each socket		$R_1 + R_2$
As a check $(R_1 + R_2)$ value should equal			$\frac{r_1 + r_2}{4}$

Step 3

Where the circuit protective conductor is wired as a ring, for example where twin and earth cables or plastic conduit is used to wire the ring, temporarily join

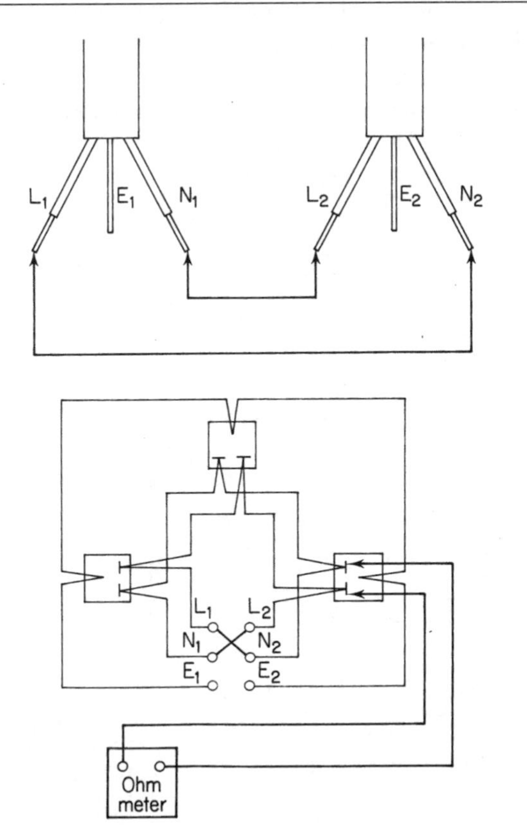

Fig. 7.3 Step 2 test: connection of mains conductors and test circuit conditions.

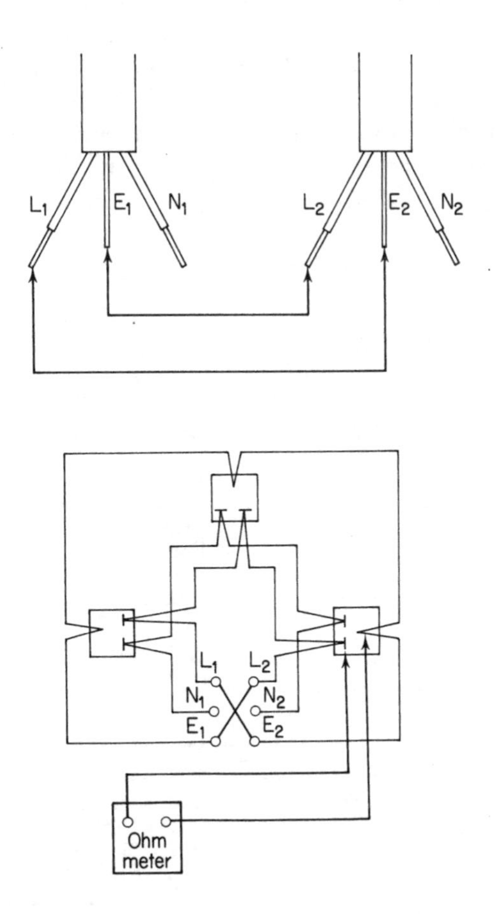

Fig. 7.4 Step 3 test: connection of mains conductors and test circuit conditions.

the live and circuit protective conductors together as shown in Fig. 7.4. An ohmmeter reading should then be taken between live and earth at *every* socket outlet on the ring. The readings obtained should be substantially the same provided that there are no breaks or multiple loops in the ring. This value is equal to $R_1 + R_2$ for the circuit. Record the results on an installation schedule such as that given in Appendix 7 of the *On Site Guide* or a table such as that shown in Table 7.2. The step 3 value of $R_1 + R_2$ should be equal to $(r_1 + r_2)/4$, where r_1 and r_2 are the ohmmeter readings from step 1 of this test (see Table 7.2).

3 INSULATION RESISTANCE (713–04)

The object of the test is to verify that the quality of the insulation is satisfactory and has not deteriorated or short-circuited. The test should be made at the consumer's unit with the mains switch off, all fuses in place and all switches closed. Neon lamps, capacitors and electronic circuits should be disconnected, since they will respectively glow, charge up or be damaged by the test.

There are two tests to be carried out using an insulation resistance tester which must supply a voltage of 500 V d.c. for 230 V and 400 V installations. These are phase and neutral conductors to earth and between phase conductors. The procedures are:

Phase and neutral conductors to earth

1 Remove all lamps.
2 Close all switches and circuit breakers.
3 Disconnect appliances.
4 Test separately between the phase conductor and earth, *and* between the neutral conductor and earth, for *every* distribution circuit at the consumer's unit as shown in Fig. 7.5(a). Record the results on an installation schedule such as that given in Appendix 7 of the *On Site Guide.*

Between phase conductors

1 Remove all lamps.
2 Close all switches and circuit breakers.
3 Disconnect appliances.
4 Test between phase and neutral conductors of *every* distribution circuit at the consumer's unit as shown in Fig. 7.5(b) and record the result.

The insulation resistance readings for each test must be not less than 0.5 MΩ for a satisfactory result (IEE Regulation 713–04–02).

Where equipment is disconnected for the purpose of the insulation resistance test, the equipment itself

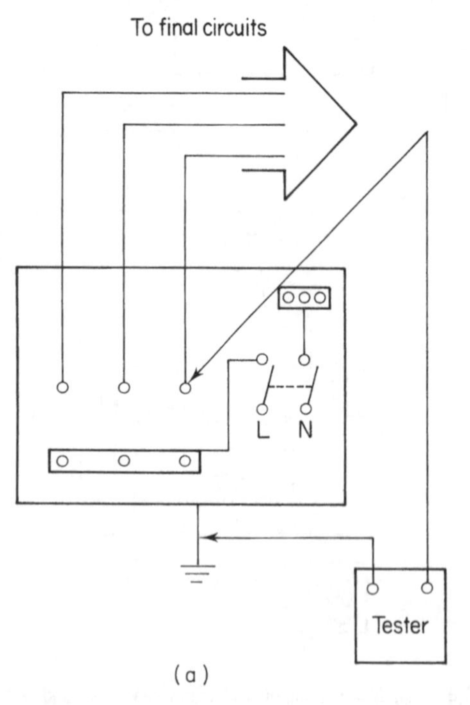

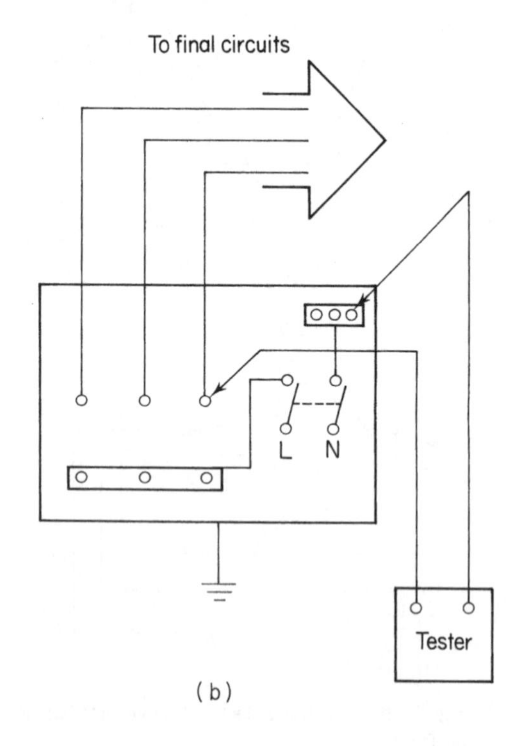

Fig. 7.5 Insulation resistance test.

must be insulation resistance tested between all live parts (that is, live and neutral conductors connected together) and the exposed conductive parts. The insulation resistance of these tests should be not less than 0.5 MΩ (IEE Regulation 713–04–04).

Although an insulation resistance reading of 0.5 MΩ complies with the Regulations, the IEE Guidance Notes tell us that a reading of less than 2 MΩ might indicate a latent but not yet visible fault in the installation. In these cases each circuit should be separately tested to obtain a reading greater than 2 MΩ.

4 POLARITY (713–09)

The object of this test is to verify that all fuses, circuit breakers and switches are connected in the phase or live conductor only, and that all socket outlets are correctly wired and Edison screw-type lampholders have the centre contact connected to the live conductor. It is important to make a polarity test on the installation since a visual inspection will only indicate conductor identification.

The test is done with the supply disconnected using an ohmmeter or continuity tester as follows:

1 Switch off the supply at the main switch.
2 Remove all lamps and appliances.
3 Fix a temporary link between the phase and earth connections on the consumer's side of the main switch.
4 Test between the 'common' terminal and earth at each switch position.
5 Test between the centre pin of any Edison screw lampholders and any convenient earth connection.
6 Test between the live pin (that is, the pin to the right of earth) and earth at each socket outlet as shown in Fig. 7.6.

For a satisfactory test result the ohmmeter or continuity meter should read approximately zero.

Remove the test link and record the results on an installation schedule such as that given in Appendix 7 of the *On Site Guide*.

5 EARTH ELECTRODE RESISTANCE (713–10)

Low-voltage supplies having earthing arrangements which are independent of the supply cable are classified as TT systems. For this type of supply it is necessary to sink an earth electrode into the general mass of earth, which then forms a part of the earth return in conjunction with a residual current device. To verify the resistance of an electrode used in this way, the following test method may be applied:

1 Disconnect the installation equipotential bonding from the earth electrode to ensure that the test current passes only through the earth electrode.

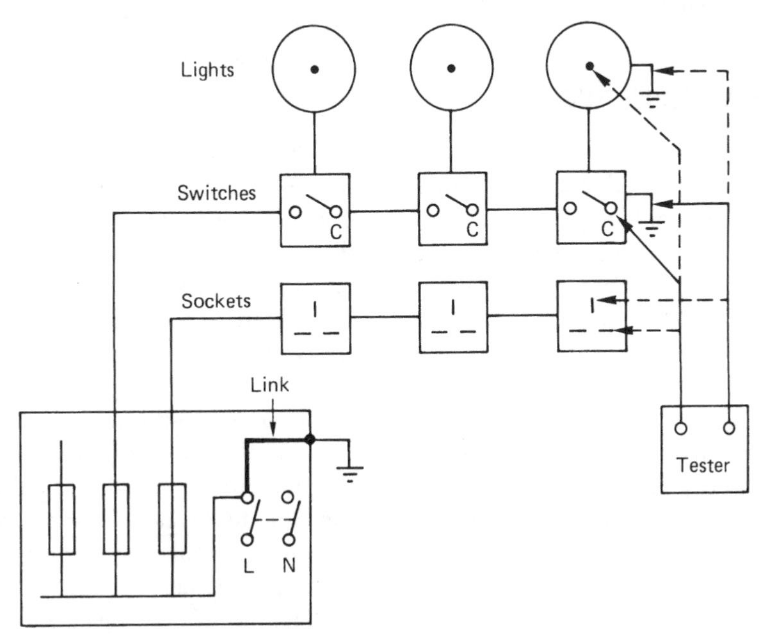

Fig. 7.6 Polarity test.

2 Switch off the consumer's unit to isolate the installation.
3 Using a phase earth loop impedance tester, test between the incoming phase conductor and the earth electrode.

Record the result on an installation schedule such as that given in Appendix 7 of the *On Site Guide*.

Section 10.3.5 of the *On Site Guide* tells us that the recommended maximum value of the earth fault loop impedance for a TT installation is 220 Ω. Since most of the circuit impedance will be made up of the earth electrode resistance, we can say that an acceptable value for the measurement of the earth electrode resistance will be about 200 Ω or less.

Providing the first five tests were satisfactory, the supply may now be switched on and the final tests completed with the supply connected.

6 POLARITY – SUPPLY CONNECTED

Using an approved voltmeter or test lamp and probes which comply with the HSE Guidance Note GS38, again carry out a polarity test to verify that all fuses, circuit breakers and switches are connected in the live conductor. Test from the common terminal of switches to earth, the live pin of each socket outlet to earth and the centre pin of any Edison screw lampholders to earth. In each case the voltmeter or test lamp should indicate the supply voltage for a satisfactory result.

7 EARTH FAULT LOOP IMPEDANCE (SUPPLY CONNECTED) (713–11)

The object of this test is to verify that the impedance of the whole earth fault current loop phase to earth is low enough to allow the overcurrent protective device to operate within the disconnection time requirements of Regulations 413–02–08 and 09, should an earth fault occur.

The whole earth fault current loop examined by this test is comprised of all the installation protective conductors, the earthing terminal and earth conductors, the earthed neutral point and the secondary winding of the supply transformer and the phase conductor from the transformer to the point of the fault in the installation.

The test will, in most cases, be done with a purpose-made phase earth loop impedance tester which circulates a current in excess of 10 A around the loop for a very short time, so reducing the danger of a faulty circuit. The test is made with the supply switched on, from the furthest point of *every* final circuit, including lighting, socket outlets and any fixed appliances. Record the results on an installation schedule.

Purpose-built testers give a readout in ohms and a satisfactory result is obtained when the loop impedance does not exceed the appropriate values given in Tables 2A, 2B and 2C of Appendix 2 of the *On Site Guide* or Table 41B1 and 41B2 or Table 604B2, 605B1 and 605B2 of the IEE Regulations.

8 FUNCTIONAL TESTING OF RCD – SUPPLY CONNECTED (713–12)

The object of the test is to verify the effectiveness of the residual current device, that it is operating with the correct sensitivity and proving the integrity of the electrical and mechanical elements. The test must simulate an appropriate fault condition and be independent of any test facility incorporated in the device.

When carrying out the test, all loads normally supplied through the device are disconnected.

Functional testing of a ring circuit protected by a general-purpose RCD to BS 4293 in a split-board consumer unit is carried out as follows:

1 Using the standard lead supplied with the test instrument, disconnect all other loads and plug in the test lead to the socket at the centre of the ring (that is, the socket at the furthest point from the source of supply).
2 Set the test instrument to the tripping current of the device and at a phase angle of 0°.
3 Press the test button – the RCD should trip and disconnect the supply within 200 ms.
4 Change the phase angle from 0° to 180° and press the test button once again. The RCD should again trip within 200 ms. Record the highest value of these two results on an installation schedule such as that given in Appendix 7 of the *On Site Guide*.
5 Now set the test instrument to 50% of the rated tripping current of the RCD and press the test button. The RCD should *not trip* within 2 seconds. This test is testing the RCD for inconvenience *or* nuisance tripping.
6 Finally, the effective operation of the test button incorporated within the RCD should be tested to

prove the integrity of the mechanical elements in the tripping device. This test should be repeated every 3 months.

If the RCD fails any of the above tests it should be changed for a new one.

Where the residual current device has a rated tripping current not exceeding 30 mA and has been installed to reduce the risk associated with direct contact, as indicated in Regulation 412–06–02, a residual current of 150 mA should cause the circuit breaker to open within 40 ms.

Certification and reporting (Chapter 74)

Following the inspection and testing of an installation, a certificate should be given by the electrical contractor or responsible person to the person ordering the work.

The certificate should be in the form set out in Appendix 6 of the IEE Regulations and Appendix 7 of the *On Site Guide*. It should include the test values which verify that the installation complies with the regulations for electrical installations at the time of testing.

An 'Electrical Installation certificate' should be used for the initial certification of a new electrical installation or for an alteration or addition to an existing installation.

All installations should be tested and inspected periodically and a 'periodic inspection' certificate issued. Suggested periodic inspection intervals are given below:

- domestic installations – 10 years
- industrial installations – 3 years
- agricultural installations – 3 years
- caravan site installations – 1 year
- caravans – 3 years
- temporary installations on construction sites – 3 months

Exercises

1 The test required by the Regulations to ascertain that the circuit protective conductor is correctly connected is:
(a) continuity of ring final circuit conductors
(b) continuity of protective conductors
(c) earth electrode resistance
(d) protection by electrical separation.

2 A visual inspection of a new installation must be carried out:
(a) during the erection period
(b) during testing upon completion
(c) after testing upon completion
(d) before testing upon completion.

3 One objective of the polarity test is to verify that:
(a) lampholders are correctly earthed
(b) final circuits are correctly fused
(c) the CPC is continuous throughout the installation
(d) the protective devices are connected in the live conductor.

4 When testing a 230 V installation an insulation resistance tester must supply a voltage of:
(a) less than 50 V
(b) 500 V
(c) less than 500 V
(d) greater than twice the supply voltage but less than 1000 V.

5 The value of a satisfactory insulation resistance test on each final circuit of a 230 V installation must be:
(a) less than 1 Ω
(b) less than 0.5 mΩ
(c) not less than 0.5 MΩ
(d) not less than 1 MΩ.

6 The value of a satisfactory insulation resistance test on a disconnected piece of equipment is:
(a) less than 1 Ω
(b) less than 0.5 MΩ
(c) not less than 0.5 MΩ
(d) not less than 1 MΩ.

7 The maximum inspection and retest period for a general electrical installation is:
(a) 3 months
(b) 3 years
(c) 5 years
(d) 10 years.

8 The CPC of a lighting final circuit is formed by approximately 70 m of 1.0 mm copper conductor. Calculate a satisfactory value for a continuity test on the CPC given that the resistance per metre of 1.0 mm copper is 18.1 mΩ/m.

9 The CPC of an installation is formed by

approximately 200 m of 50 mm × 50 mm trunking. Determine a satisfactory test result for this CPC, using the information given in Table 7.1. Describe briefly a suitable instrument to carry out this test.

10 Describe how a polarity test should be carried out on a domestic installation comprising eight light positions and ten socket outlets. The final circuits are to be supplied by a consumer unit.

11 Describe how to carry out a continuity test of ring final circuit conductors. State the values to be obtained for a satisfactory test.

12 Describe how to carry out an earth fault loop impedance test. Sketch a circuit diagram and indicate the test circuit path.

13 Describe how to carry out an insulation resistance test on a domestic installation. State the type of instrument to be used and the values of a satisfactory test.

8

ELECTRICAL SCIENCE

Units

Very early units of measurement were based on the things easily available – the length of a stride, the distance from the nose to the outstretched hand, the weight of a stone and the time-lapse of one day. Over the years, new units were introduced and old ones were modified. Different branches of science and engineering were working in isolation, using their own units, and the result was an overwhelming variety of units.

In all branches of science and engineering there is a need for a practical system of units which everyone can use. In 1960 the General Conference of Weights and Measures agreed to an international system called the Système International d'Unités (abbreviated to SI units). SI units are based upon a small number of fundamental units from which all other units may be derived (see Table 8.1).

Table 8.1 SI units

SI unit	Measure of	Symbol
The fundamental units		
metre	length	m
kilogram	mass	kg
second	time	s
ampere	electric current	A
kelvin	thermodynamic temperature	K
candela	luminous intensity	cd
Some derived units		
coulomb	charge	C
joule	energy	J
newton	force	N
ohm	resistance	Ω
volt	potential difference	V
watt	power	W

Like all metric systems, SI units have the advantage that prefixes representing various multiples or sub-multiples may be used to increase or decrease the size of the unit by various powers of 10. Some of the more common prefixes and their symbols are shown in Table 8.2.

Table 8.2 Prefixes for use with SI units

Prefix	Symbol	Multiplication factor		
mega	M	$\times 10^6$	or	$\times 1\,000\,000$
kilo	k	$\times 10^3$	or	$\times 1000$
hecto	h	$\times 10^2$	or	$\times 100$
deca	da	$\times 10$	or	$\times 10$
deci	d	$\times 10^{-1}$	or	$\div 10$
centi	c	$\times 10^{-2}$	or	$\div 100$
milli	m	$\times 10^{-3}$	or	$\div 1000$
micro	μ	$\times 10^{-6}$	or	$\div 1\,000\,000$

Basic circuit theory

All matter is made up of atoms which arrange themselves in a regular framework within the material. The atom is made up of a central, positively charged nucleus, surrounded by negatively charged electrons. The electrical properties of a material depend largely upon how tightly these electrons are bound to the central nucleus.

A *conductor* is a material in which the electrons are loosely bound to the central nucleus and are, therefore, free to drift around the material at random from one atom to another, as shown in Fig. 8.1(a). Materials which are good conductors include copper, brass, aluminium and silver.

An *insulator* is a material in which the outer electrons are tightly bound to the nucleus and so there are

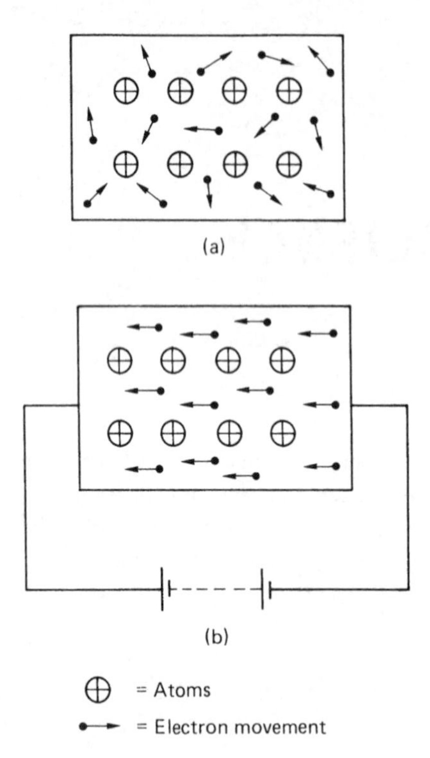

Fig. 8.1 Atoms and electrons on a material.

no free electrons to move around the material. Good insulating materials are PVC, rubber, glass and wood.

If a battery is attached to a conductor as shown in Fig. 8.1(b), the free electrons drift purposefully in one direction only. The free electrons close to the positive plate of the battery are attracted to it since unlike charges attract, and the free electrons near the negative plate will be repelled from it. For each electron entering the positive terminal of the battery, one will be ejected from the negative terminal, so the number of electrons in the conductor remains constant.

This drift of electrons within a conductor is known as an electric *current*, measured in amperes and given the symbol *I*. For a current to continue to flow, there must be a complete circuit for the electrons to move around. If the circuit is broken by opening a switch, for example, the electron flow and therefore the current will stop immediately.

To cause a current to flow continuously around a circuit, a driving force is required, just as a circulating pump is required to drive water around a central heating system. This driving force is the *electromotive force* (abbreviated to emf). Each time an electron passes through the source of emf, more energy is provided to send it on its way around the circuit.

An emf is always associated with energy conversion, such as chemical to electrical in batteries and mechanical to electrical in generators. The energy introduced into the circuit by the emf is transferred to the load terminals by the circuit conductors. The *potential difference* (abbreviated to p.d.) is the change in energy levels measured across the load terminals. This is also called the volt drop or terminal voltage, since emf and p.d. are both measured in volts. Every circuit offers some opposition to current flow, which we call the circuit *resistance*, measured in ohms (symbol Ω), to commemorate the famous German physicist Georg Simon Ohm, who was responsible for the analysis of electrical circuits.

OHM'S LAW

In 1826, Ohm published details of an experiment he had done to investigate the relationship between the current passing through and the potential difference between the ends of a wire. As a result of this experiment, he arrived at a law, now known as Ohm's law, which says that *the current passing through a conductor under constant temperature conditions is proportional to the potential difference across the conductor*. This may be expressed mathematically as

$$V = I \times R \text{ (V)}$$

Transposing this formula, we also have

$$I = \frac{V}{R} \text{ (A)} \qquad \text{and} \qquad R = \frac{V}{I} \text{ }(\Omega)$$

EXAMPLE 1

An electric heater, when connected to a 230 V supply, was found to take a current of 4 A. Calculate the element resistance.

$$R = \frac{V}{I}$$

$$\therefore R = \frac{230 \text{ V}}{4 \text{ A}} = 57.5\ \Omega$$

EXAMPLE 2

The insulation resistance measured between phase conductors on a 400 V supply was found to be 2 MΩ. Calculate the leakage current.

$$I = \frac{V}{R}$$

$$\therefore I = \frac{400\text{ V}}{2 \times 10^6\ \Omega} = 200 \times 10^{-6}\text{ A} = 200\ \mu\text{A}$$

EXAMPLE 3

When a 4 Ω resistor was connected across the terminals of an unknown d.c. supply, a current of 3 A flowed. Calculate the supply voltage.

$$V = I \times R$$

$$\therefore V = 3\text{ A} \times 4\ \Omega = 12\text{ V}$$

RESISTIVITY

The resistance or opposition to current flow varies for different materials, each having a particular constant value. If we know the resistance of, say, 1 metre of a material, then the resistance of 5 metres will be five times the resistance of 1 metre.

The *resistivity* (symbol ρ – the Greek letter 'rho') of a material is defined as the resistance of a sample of unit length and unit cross-section. Typical values are given in Table 8.3. Using the constants for a particular material we can calculate the resistance of any length and thickness of that material from the equation.

$$R = \frac{\rho l}{a}\ (\Omega)$$

where

ρ = the resistivity constant for the material (Ω m)
l = the length of the material (m)
a = the cross-sectional area of the material (m^2).

Table 8.3 gives the resistivity of silver as 16.4×10^{-9} Ω m, which means that a sample of silver 1 metre long and 1 metre in cross-section will have a resistance of 16.4×10^{-9} Ω.

Table 8.3 Resistivity values

Material	Resistivity (Ω m)
Silver	16.4×10^{-9}
Copper	17.5×10^{-9}
Aluminium	28.5×10^{-9}
Brass	75.0×10^{-9}
Iron	100.0×10^{-9}

EXAMPLE 1

Calculate the resistance of 100 metres of copper cable of 1.5 mm² cross-sectional area if the resistivity of copper is taken as 17.5×10^{-9} Ω m.

$$R = \frac{\rho l}{a}\ (\Omega)$$

$$\therefore R = \frac{17.5 \times 10^{-9}\ \Omega \times 100\text{ m}}{1.5 \times 10^{-6}\text{ m}^2} = 1.16\ \Omega$$

EXAMPLE 2

Calculate the resistance of 100 metres of aluminium cable of 1.5 mm² cross-sectional area if the resistivity of aluminium is taken as 28.5×10^{-9} Ω m.

$$R = \frac{\rho l}{a}\ (\Omega)$$

$$\therefore R = \frac{28.5 \times 10^{-9}\ \Omega\text{ m} \times 100\text{ m}}{1.5 \times 10^{-6}\text{ m}^2} = 1.9\ \Omega$$

The above examples show that the resistance of an aluminium cable is some 60% greater than a copper conductor of the same length and cross-section. Therefore, if an aluminium cable is to replace a copper cable, the conductor size must be increased to carry the rated current as given by the tables in Appendix 4 of the IEE Regulations and Appendix 6 of the *On Site Guide.*

The other factor which affects the resistance of a material is the temperature, and we will consider this next.

TEMPERATURE COEFFICIENT

The resistance of most materials changes with temperature. In general, conductors increase their resistance as the temperature increases and insulators decrease their resistance with a temperature increase. Therefore, an increase in temperature has a bad effect upon the electrical properties of a material.

Each material responds to temperature change in a different way, and scientists have calculated constants for each material which are called the *temperature coefficient of resistance* (symbol α – the Greek letter 'alpha'). Table 8.4 gives some typical values.

Table 8.4 Temperature coefficient values

Material	Temperature coefficient (Ω/Ω°C)
Silver	0.004
Copper	0.004
Aluminium	0.004
Brass	0.001
Iron	0.006

Using the constants for a particular material and substituting values into the following formulae the resistance of a material at different temperatures may be calculated. For a temperature increase from 0°C:

$$R_t = R_0(1 + \alpha t)\ (\Omega)$$

where

R_t = the resistance at the new temperature t°C
R_0 = the resistance at 0°C
α = the temperature coefficient for the particular material.

For a temperature increase between two intermediate temperatures above 0°C:

$$\frac{R_1}{R_2} = \frac{(1 + \alpha t_1)}{(1 + \alpha t_2)}$$

where

R_1 = the resistance at the original temperature
R_2 = the resistance at the final temperature
α = the temperature coefficient for the particular material.

If we take a 1 Ω resistor of, say, copper, and raise its temperature by 1°C, the resistance will increase by 0.004 Ω to 1.004 Ω. This increase of 0.004 Ω is the temperature coefficient of the material.

EXAMPLE 1

The field winding of a d.c. motor has a resistance of 100 Ω at 0°C. Determine the resistance of the coil at 20°C if the temperature coefficient is 0.004 Ω/Ω°C.

$$R_t = R_0\ (1 + \alpha t)\ (\Omega)$$
$$\therefore\ R_t = 100\ \Omega\ (1 + 0.004\ \Omega/\Omega°\text{C} \times 20°\text{C})$$
$$R_t = 100\ \Omega\ (1 + 0.08)$$
$$R_t = 108\ \Omega$$

EXAMPLE 2

The field winding of a shunt generator has a resistance of 150 Ω at an ambient temperature of 20°C. After running for some time the mean temperature of the generator rises to 45°C. Calculate the resistance of the winding at the higher temperature if the temperature coefficient of resistance is 0.004 Ω/Ω°C.

$$\frac{R_1}{R_2} = \frac{(1 + \alpha t_1)}{(1 + \alpha t_2)}$$

$$\frac{150\ \Omega}{R_2} = \frac{1 + 0.004\ \Omega/\Omega°\text{C} \times 20°\text{C}}{1 + 0.004\ \Omega/\Omega°\text{C} \times 45°\text{C}}$$

$$\frac{150\ \Omega}{R_2} = \frac{1.08}{1.18}$$

$$\therefore\ R_2 = \frac{150\ \Omega \times 1.18}{1.08} = 164\ \Omega.$$

It is clear from the last two sections that the resistance of a cable is affected by length, thickness, temperature and type of material. Since Ohm's law tells us that current is inversely proportional to resistance, these factors must also influence the current carrying capacity of a cable. The tables of current ratings in Appendix 4 of the IEE Regulations and Appendix 6 of the *On Site Guide* contain correction factors so that current ratings may be accurately determined under defined installation conditions. Cable selection is considered in Chapter 6.

Resistors

In an electrical circuit resistors may be connected in series, in parallel, or in various combinations of series and parallel connections.

SERIES-CONNECTED RESISTORS

In any series circuit a current I will flow through all parts of the circuit as a result of the potential difference supplied by a battery V_T. Therefore, we say that in a series circuit the current is common throughout that circuit.

When the current flows through each resistor in the circuit, R_1, R_2 and R_3 for example in Fig. 8.2, there

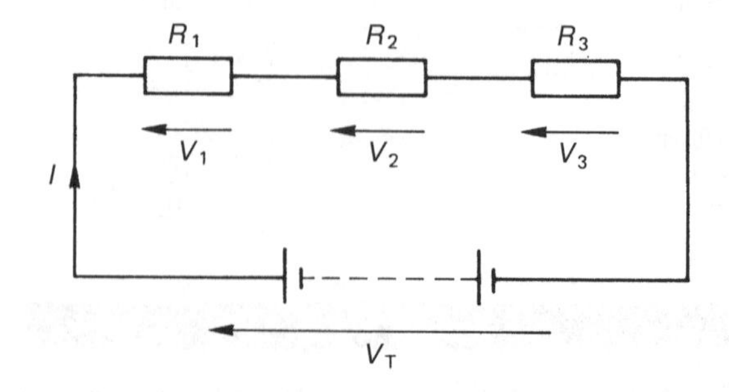

Fig. 8.2 A series circuit.

will be a voltage drop across that resistor whose value will be determined by the values of I and R, since from Ohm's law $V = I \times R$. The sum of the individual voltage drops, V_1, V_2 and V_3 for example in Fig. 8.2, will be equal to the total voltage V_T.

We can summarize these statements as follows. For any series circuit, I is common throughout the circuit and

$$V_T = V_1 + V_2 + V_3 \quad (1)$$

Let us call the total circuit resistance R_T. From Ohm's law we know that $V = I \times R$ and therefore

$$\begin{aligned} &\text{total voltage } V_T = I \times R_T \\ &\text{voltage drop across } R_1 \text{ is } V_1 = I \times R_1 \\ &\text{voltage drop across } R_2 \text{ is } V_2 = I \times R_2 \\ &\text{voltage drop across } R_3 \text{ is } V_3 = I \times R_3 \end{aligned} \quad (2)$$

We are looking for an expression for the total resistance in any series circuit and, if we substitute equations (2) into equation (1) we have:

$$V_T = V_1 + V_2 + V_3$$

$$\therefore I \times R_T = I \times R_1 + I \times R_2 + I \times R_3$$

Now, since I is common to all terms in the equation, we can divide both sides of the equation by I. This will cancel out I to leave us with an expression for the circuit resistance:

$$R_T = R_1 + R_2 + R_3$$

Note that the derivation of this formula is given for information only. Craft students need only state the expression $R_T = R_1 + R_2 + R_3$ for series connections.

PARALLEL-CONNECTED RESISTORS

In any parallel circuit, as shown in Fig. 8.3, the same voltage acts across all branches of the circuit. The total current will divide when it reaches a resistor junction, part of it flowing in each resistor. The sum of the individual currents, I_1, I_2 and I_3 for example in Fig. 8.3, will be equal to the total current I_T.

We can summarize these statements as follows. For any parallel circuit, V is common to all branches of the circuit and

$$I_T = I_1 + I_2 + I_3 \quad (3)$$

Let us call the total resistance R_T.

From Ohm's law we know that $I = \frac{V}{R}$, and therefore

$$\begin{aligned} &\text{the total current } I_T = \frac{V}{R_T} \\ &\text{the current through } R_1 \text{ is } I_1 = \frac{V}{R_1} \\ &\text{the current through } R_2 \text{ is } I_2 = \frac{V}{R_2} \\ &\text{the current through } R_3 \text{ is } I_3 = \frac{V}{R_3} \end{aligned} \quad (4)$$

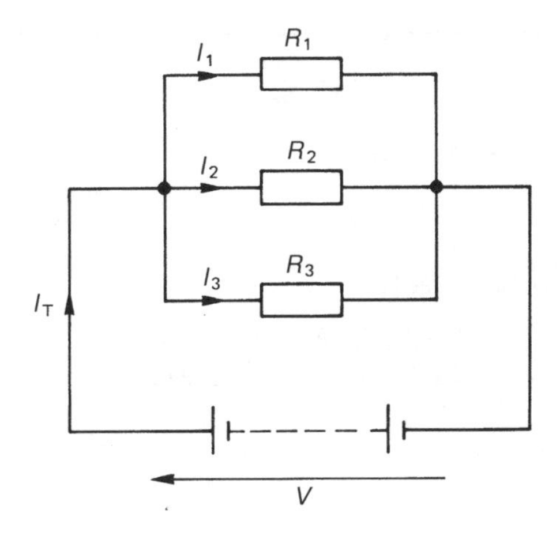

Fig. 8.3 A parallel circuit.

We are looking for an expression for the equivalent resistance R_T in any *parallel* circuit and, if we substitute equations (4) into equation (3) we have:

$$I_T = I_1 + I_2 + I_3$$

$$\therefore \frac{V}{R_T} = \frac{V}{R_1} + \frac{V}{R_2} + \frac{V}{R_3}$$

Now, since V is common to all terms in the equation, we can divide both sides by V, leaving us with an expression for the circuit resistance:

$$\frac{1}{R_T} = \frac{1}{R_1} + \frac{1}{R_2} + \frac{1}{R_3}$$

Note that the derivation of this formula is given for information only. Craft students need only state the expression $1/R_T = 1/R_1 + 1/R_2 + 1/R_3$ for parallel connections.

EXAMPLE

Three 6 Ω resistors are connected (a) in series (see Fig. 8.4), and (b) in parallel (see Fig. 8.5), across a 12 V battery. For each method of connection, find the total resistance and the values of all currents and voltages.

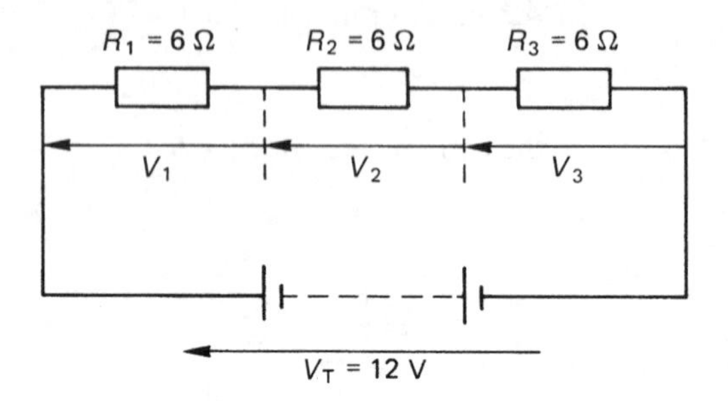

Fig. 8.4 Resistors in series.

For any series connection

$$R_T = R_1 + R_2 + R_3$$
$$\therefore R_T = 6\,\Omega + 6\,\Omega + 6\,\Omega = 18\,\Omega$$

Total current $I_T = \dfrac{V}{R_T}$

$$\therefore I_T = \frac{12\text{ V}}{18\,\Omega} = 0.67\text{ A}$$

The voltage drop across R_1 is

$$V_1 = I \times R_1$$
$$\therefore V_1 = 0.67\text{ A} \times 6\,\Omega = 4\text{ V}$$

The voltage drop across R_2 is

$$V_2 = I \times R_2$$
$$\therefore V_2 = 0.67\text{ A} \times 6\,\Omega = 4\text{ V}$$

The voltage drop across R_3 is

$$V_3 = I \times R_3$$
$$\therefore V_3 = 0.67\text{ A} \times 6\,\Omega = 4\text{ V}$$

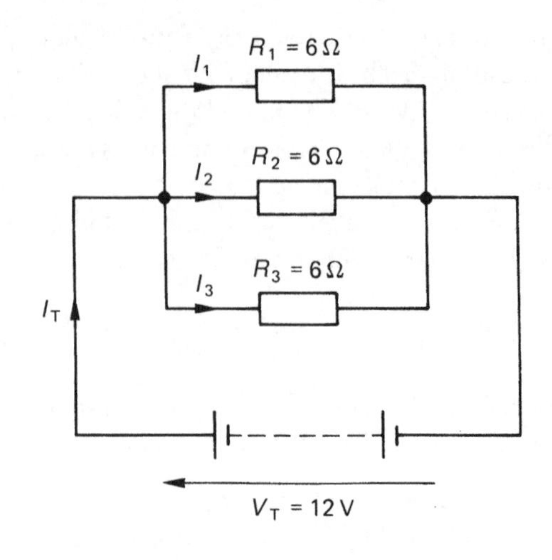

Fig. 8.5 Resistors in parallel.

For any parallel connection,

$$\frac{1}{R_T} = \frac{1}{R_1} + \frac{1}{R_2} + \frac{1}{R_3}$$

$$\therefore \frac{1}{R_T} = \frac{1}{6\,\Omega} + \frac{1}{6\,\Omega} + \frac{1}{6\,\Omega}$$

$$\frac{1}{R_T} = \frac{1+1+1}{6\,\Omega} = \frac{3}{6\,\Omega}$$

$$R_T = \frac{6\,\Omega}{3} = 2\,\Omega$$

Total current $I_T = \dfrac{V}{R_T}$

$$\therefore I_T = \frac{12\text{ V}}{2\,\Omega} = 6\text{ A}$$

The current flowing through R_1 is

$$I_1 = \frac{V}{R_1}$$

$$\therefore I_1 = \frac{12\text{ V}}{6\,\Omega} = 2\text{ A}$$

The current flowing through R_2 is

$$I_2 = \frac{V}{R_2}$$

$$\therefore I_2 = \frac{12\text{ V}}{6\ \Omega} = 2\text{ A}$$

The current flowing through R_3 is

$$I_3 = \frac{V}{R_3}$$

$$\therefore I_3 = \frac{12\text{ V}}{6\ \Omega} = 2\text{ A}$$

SERIES AND PARALLEL COMBINATIONS

The most complex arrangement of series and parallel resistors can be simplified into a single equivalent resistor by combining the separate rules for series and parallel resistors.

EXAMPLE 1

Resolve the circuit shown in Fig. 8.6 into a single resistor and calculate the potential difference across each resistor.

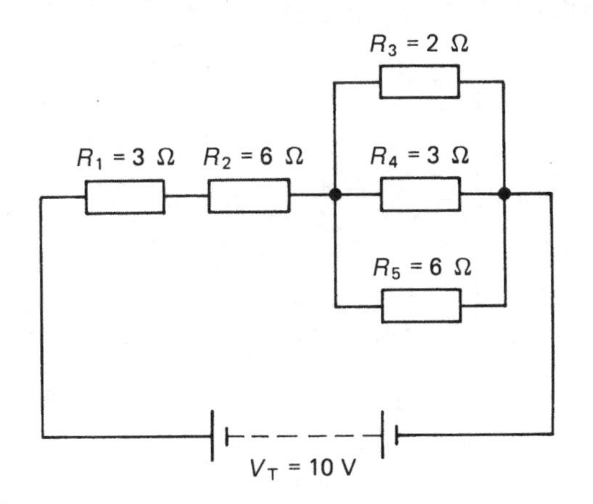

Fig. 8.6 A series/parallel circuit.

By inspection, the circuit contains a parallel group consisting of R_3, R_4 and R_5 and a series group consisting of R_1 and R_2 in series with the equivalent resistor for the parallel branch.

Consider the parallel group. We will label this group R_P. Then

$$\frac{1}{R_P} = \frac{1}{R_3} + \frac{1}{R_4} + \frac{1}{R_5}$$

$$\frac{1}{R_P} = \frac{1}{2\ \Omega} + \frac{1}{3\ \Omega} + \frac{1}{6\ \Omega}$$

$$\frac{1}{R_P} = \frac{3+2+1}{6\ \Omega} = \frac{6}{6\ \Omega}$$

$$R_P = \frac{6\ \Omega}{6} = 1\ \Omega$$

Figure 8.6 may now be represented by the more simple equivalent shown in Fig. 8.7.

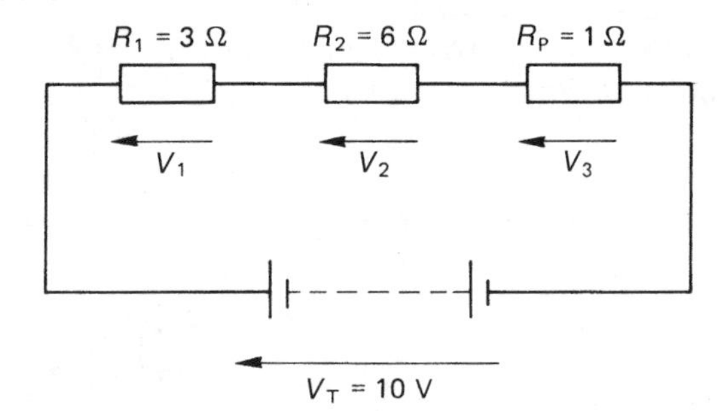

Fig. 8.7 Equivalent series circuit.

Since all resistors are now in series,

$$R_T = R_1 + R_2 + R_P$$

$$\therefore R_T = 3\ \Omega + 6\ \Omega + 1\ \Omega = 10\ \Omega$$

Thus, the circuit may be represented by a single equivalent resistor of value 10 Ω as shown in Fig. 8.8. The total current flowing in the circuit may be found by using Ohm's law:

$$I_T = \frac{V_T}{R_T} + \frac{10\text{ V}}{10\ \Omega} = 1\text{ A}$$

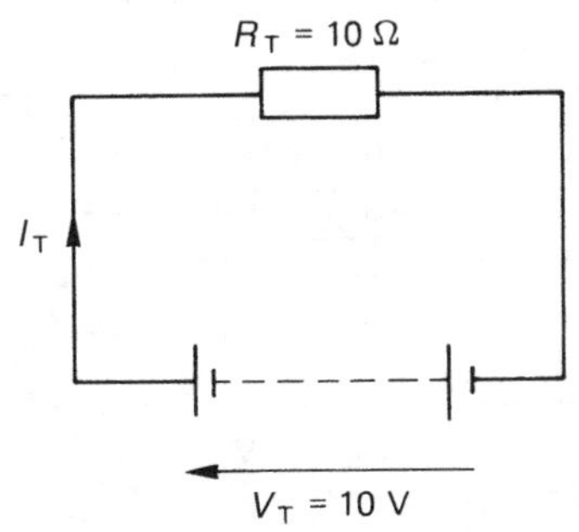

Fig. 8.8 Single equivalent resistor for Fig. 8.6.

The potential differences across the individual resistors are

$$V_1 = I \times R_1 = 1\,\text{A} \times 3\,\Omega = 3\,\text{V}$$
$$V_2 = I \times R_2 = 1\,\text{A} \times 6\,\Omega = 6\,\text{V}$$
$$V_P = I \times R_P = 1\,\text{A} \times 1\,\Omega = 1\,\text{V}$$

Since the same voltage acts across all branches of a parallel circuit the same p.d. of 1 V will exist across each resistor in the parallel branch R_3, R_4 and R_5.

EXAMPLE 2

Determine the total resistance and the current flowing through each resistor for the circuit shown in Fig. 8.9.

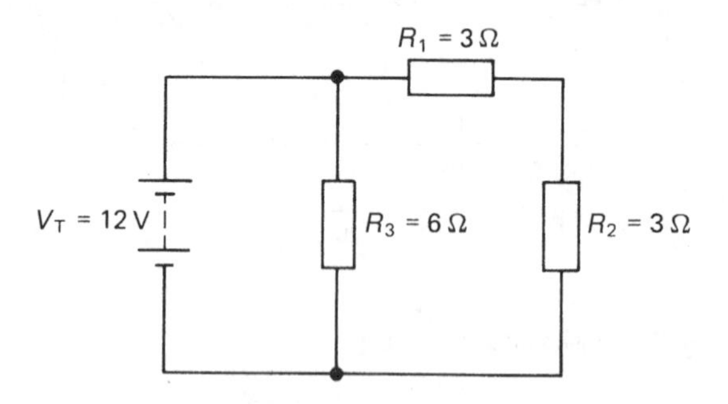

Fig. 8.9 A series/parallel circuit for Example 2.

By inspection, it can be seen that R_1 and R_2 are connected in series while R_3 is connected in parallel across R_1 and R_2. The circuit may be more easily understood if we redraw it as in Fig. 8.10.

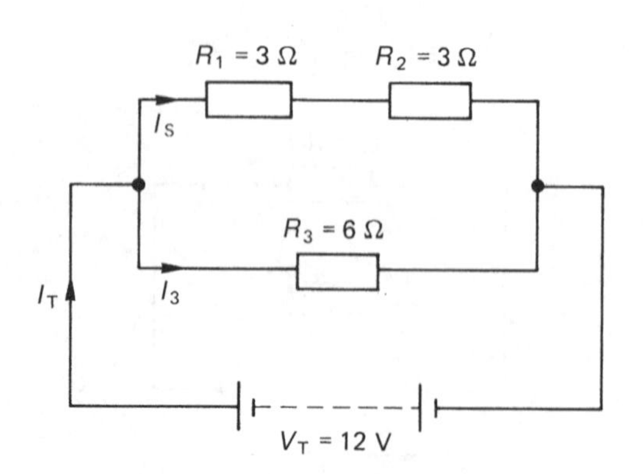

Fig. 8.10 Equivalent circuit for Example 2.

For the series branch, the equivalent resistor can be found from

$$R_S = R_1 + R_2$$
$$\therefore R_S = 3\,\Omega + 3\,\Omega = 6\,\Omega$$

Figure 8.10 may now be represented by a more simple equivalent circuit, as in Fig. 8.11.

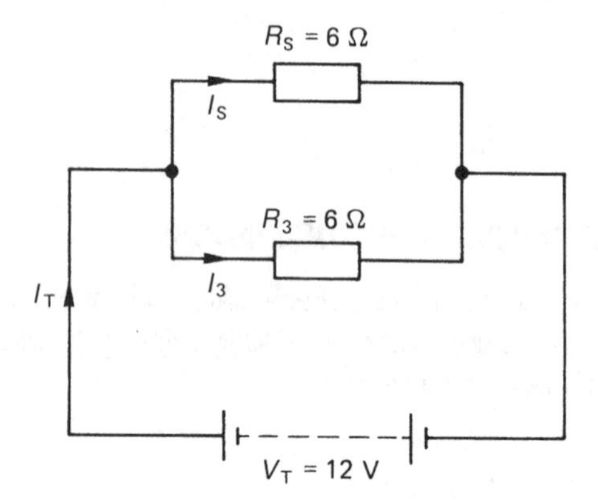

Fig. 8.11 Simplified equivalent circuit for Example 2.

Since the resistors are now in parallel, the equivalent resistance may be found from

$$\frac{1}{R_T} = \frac{1}{R_S} + \frac{1}{R_3}$$

$$\therefore \frac{1}{R_T} = \frac{1}{6\,\Omega} + \frac{1}{6\,\Omega}$$

$$\frac{1}{R_T} = \frac{1+1}{6\,\Omega} = \frac{2}{6\,\Omega}$$

$$R_T = \frac{6\,\Omega}{2} = 3\,\Omega$$

The total current is

$$I_T = \frac{V}{R_T} = \frac{12\,\text{V}}{3\,\Omega} = 4\,\text{A}$$

Let us call the current flowing through resistor R_3 I_3.

$$\therefore I_3 = \frac{V}{R_3} = \frac{12\,\text{V}}{6\,\Omega} = 2\,\text{A}$$

Let us call the current flowing through both resistors R_1 and R_2, as shown in Fig. 8.10, I_S.

$$\therefore I_S = \frac{V}{R_S} = \frac{12\ \text{V}}{6\ \Omega} = 2\ \text{A}$$

Power and energy

POWER

Power is the rate of doing work and is measured in watts:

$$\text{Power} = \frac{\text{Work done}}{\text{Time taken}}\ (\text{W})$$

In an electrical circuit,

$$\text{Power} = \text{Voltage} \times \text{Current}\ (\text{W}) \qquad (5)$$

Now from Ohm's law

$$\text{Voltage} = I \times R\ (\text{V}) \qquad (6)$$

$$\text{Current} = \frac{V}{R}\ (\text{A}) \qquad (7)$$

Substituting equation (6) into equation (5), we have

$$\text{Power} = (I \times R) \times \text{Current} = I^2 \times R\ (\text{W})$$

and substituting equation (7) into equation (5) we have

$$\text{Power} = \text{Voltage} \times \frac{V}{R} = \frac{V^2}{R}\ (\text{W})$$

We can find the power of a circuit by using any of the three formulae

$$P = V \times I, \qquad P = I^2 \times R, \qquad P = \frac{V^2}{R}$$

ENERGY

Energy is a concept which engineers and scientists use to describe the ability to do work in a circuit or system:

$$\text{Energy} = \text{Power} \times \text{Time}$$

but, since Power = Voltage × Current

then Energy = Voltage × Current × Time

The SI unit of energy is the joule, where time is measured in seconds. For practical electrical installation circuits this unit is very small and therefore the kilowatt-hour (kWh) is used for domestic and commercial installations. Electricity Board meters measure 'units' of electrical energy, where each 'unit' is 1 kWh. So,

Energy in joules = Voltage × Current × Time in seconds

Energy in kWh = kW × Time in hours

EXAMPLE 1

A domestic immersion heater is switched on for 40 minutes and takes 15 A from a 200 V supply. Calculate the energy used during this time.

$$\text{Power} = \text{Voltage} \times \text{Current}$$
$$\text{Power} = 200\ \text{V} \times 15\ \text{A} = 3000\ \text{W or } 3\ \text{kW}$$
$$\text{Energy} = \text{kW} \times \text{Time in hours}$$
$$\text{Energy} = 3\ \text{kW} \times \frac{40\ \text{min}}{60\ \text{min/h}} = 2\ \text{kWh}$$

This immersion heater uses 2 kWh in 40 minutes, or 2 'units' of electrical energy every 40 minutes.

EXAMPLE 2

Two 50 Ω resistors may be connected to a 200 V supply. Determine the power dissipated by the resistors when they are connected (a) in series, (b) each resistor separately connected and (c) in parallel.

For (a), the equivalent resistance when resistors are connected in series is given by

$$R_T = R_1 + R_2$$
$$\therefore R_T = 50\ \Omega + 50\ \Omega = 100\ \Omega$$
$$\text{Power} = \frac{V^2}{R_T}\ (\text{W})$$
$$\therefore \text{Power} = \frac{200\ \text{V} \times 200\ \text{V}}{100\ \Omega} = 400\ \text{W}$$

For (b), each resistor separately connected has a resistance of 50 Ω.

$$\text{Power} = \frac{V^2}{R}\ (\text{W})$$
$$\therefore \text{Power} = \frac{200\ \text{V} \times 200\ \text{V}}{50\ \Omega} = 800\ \text{W}$$

For (c), the equivalent resistance when resistors are connected in parallel is given by

$$\frac{1}{R_T} = \frac{1}{R_1} + \frac{1}{R_2}$$

$$\therefore \frac{1}{R_T} = \frac{1}{50\,\Omega} + \frac{1}{50\,\Omega}$$

$$\frac{1}{R_T} = \frac{1+1}{50\,\Omega} = \frac{2}{50\,\Omega}$$

$$R_T = \frac{50\,\Omega}{2} = 25\,\Omega$$

$$\text{Power} = \frac{V^2}{R_T}\ \text{(W)}$$

$$\therefore \text{Power} = \frac{200\text{ V} \times 200\text{ V}}{25\,\Omega} = 1600\text{ W}$$

This example shows that by connecting resistors together in different combinations of series and parallel connections, we can obtain various power outputs: in this example, 400, 800 and 1600 W. This theory finds a practical application in the three heat switch used to control a boiling ring.

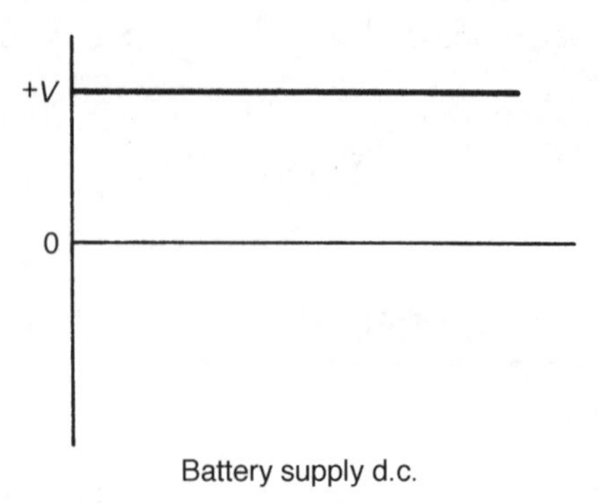

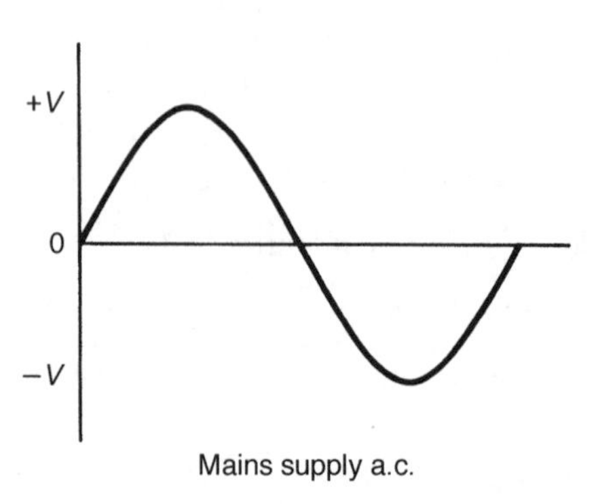

Fig. 8.12 Unidirectional and alternating supply.

Alternating current theory

The supply which we obtain from a car battery is a unidirectional or d.c. supply, whereas the mains electricity supply is alternating or a.c. (see Fig. 8.12).

Most electrical equipment makes use of alternating current supplies, and for this reason a knowledge of alternating waveforms and their effect upon resistive, capacitive and inductive loads is necessary for all practising electricians.

When a coil of wire is rotated inside a magnetic field a voltage is induced in the coil. The induced voltage follows a mathematical law known as the sinusoidal law and, therefore, we can say that a sine wave has been generated. Such a waveform has the characteristics displayed in Fig. 8.13.

In the UK we generate electricity at a frequency of 50 Hz and the time taken to complete each cycle is given by

$$T = \frac{1}{f}$$

$$\therefore T = \frac{1}{50\text{ Hz}} = 0.02\text{ s}$$

An alternating waveform is constantly changing from zero to a maximum, first in one direction, then in the opposite direction, and so the instantaneous values of the generated voltage are always changing. A useful description of the electrical effects of an a.c. waveform can be given by the maximum, average and rms values of the waveform.

The maximum or peak value is the greatest instantaneous value reached by the generated waveform. Cable and equipment insulation levels must be equal to or greater than this value.

The average value is the average over one half-cycle of the instantaneous values as they change from zero to a maximum and can be found from the following formula applied to the sinusoidal waveform shown in Fig. 8.14:

$$V_{av} = \frac{V_1 + V_2 + V_3 + V_4 + V_5 + V_6}{6} = 0.637\,V_{max}$$

For any sinusoidal waveform the average value is equal to 0.637 of the maximum value.

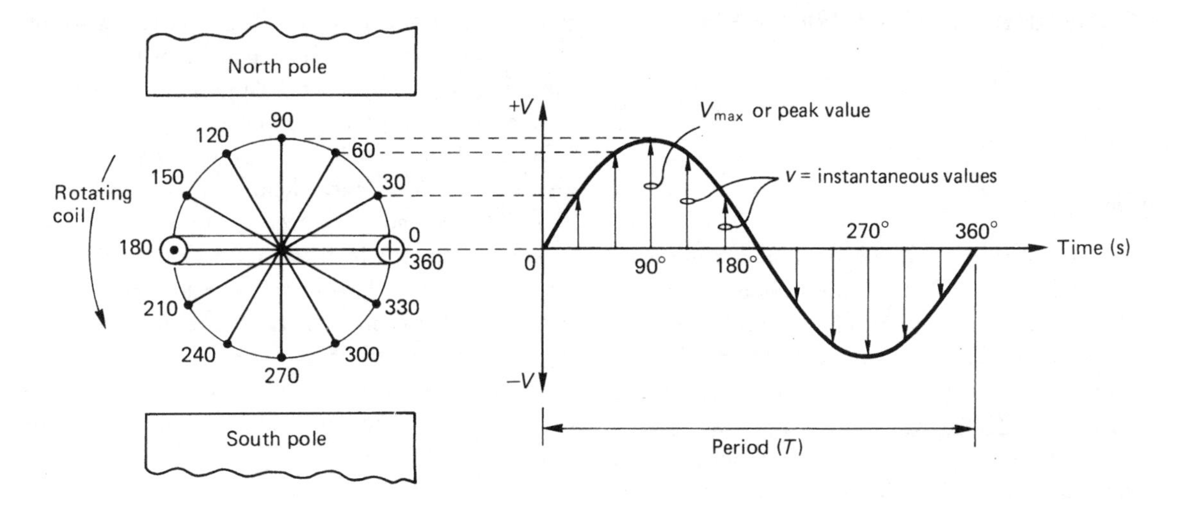

Fig. 8.13 Characteristics of a sine wave.

The rms value is the square root of the mean of the individual squared values and is the value of an a.c. voltage which produces the same heating effect as a d.c. voltage. The value can be found from the following formula applied to the sinusoidal waveform shown in Fig. 8.14.

$$V_{rms} = \sqrt{\frac{V_1^2 + V_2^2 + V_3^2 + V_4^2 + V_5^2 + V_6^2}{6}}$$

$$= 0.7071\ V_{max}$$

For any sinusoidal waveform the rms value is equal to 0.7071 of the maximum value.

EXAMPLE

The sinusoidal waveform applied to a particular circuit has a maximum value of 325.3 V. Calculate the average and rms value of the waveform.

$$\text{average value } V_{av} = 0.637 \times V_{max}$$

$$\therefore\ V_{av} = 0.637 \times 325.3 = 207.2\text{ V}$$

$$\text{rms value } V_{rms} = 0.7071 \times V_{max}$$

$$V_{rms} = 0.7071 \times 325.3 = 230\text{ V}$$

When we say that the main supply to a domestic property is 230 V we really mean 230 V rms. Such a waveform has an average value of about 207.2 V and a maximum value of almost 325.3 V but because the rms value gives the d.c. equivalent value we almost always give the rms value without identifying it as such.

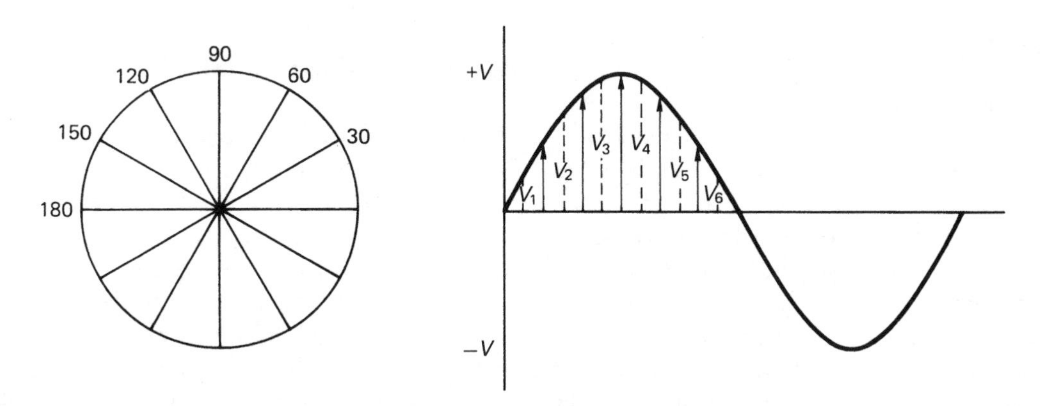

Fig. 8.14 Sinusoidal waveform showing instantaneous values of voltage.

THE THREE EFFECTS OF AN ELECTRIC CURRENT

When an electric current flows in a circuit it can have one or more of the following three effects: *heating*, *magnetic* or *chemical*.

Heating effect

The movement of electrons within a conductor, which is the flow of an electric current, causes an increase in the temperature of the conductor. The amount of heat generated by this current flow depends upon the type and dimensions of the conductor and the quantity of current flowing. By changing these variables, a conductor may be operated hot and used as the heating element of a fire, or be operated cool and used as an electrical installation conductor.

The heating effect of an electric current is also the principle upon which a fuse gives protection to a circuit. The fuse element is made of a metal with a low melting point and forms a part of the electrical circuit. If an excessive current flows, the fuse element overheats and melts, breaking the circuit.

Magnetic effect

Whenever a current flows in a conductor a magnetic field is set up around the conductor like an extension of the insulation. The magnetic field increases with the current and collapses if the current is switched off. A conductor carrying current and wound into a solenoid produces a magnetic field very similar to a permanent magnet, but has the advantage of being switched on and off by any switch which controls the circuit current.

The magnetic effect of an electric current is the principle upon which electric bells, relays, instruments, motors and generators work.

Chemical effect

When an electric current flows through a conducting liquid, the liquid is separated into its chemical parts. The conductors which make contact with the liquid are called the anode and cathode. The liquid itself is called the electrolyte, and the process is called *electrolysis*.

Electrolysis is an industrial process used in the refining of metals and electroplating. It was one of the earliest industrial applications of electric current. Most of the aluminium produced today is extracted from its ore by electrochemical methods. Electroplating serves a double purpose by protecting a base metal from atmospheric erosion and also giving it a more expensive and attractive appearance. Silver and nickel plating has long been used to enhance the appearance of cutlery, candlesticks and sporting trophies.

An anode and cathode of dissimilar metal placed in an electrolyte can react chemically and produce an emf. When a load is connected across the anode and cathode, a current is drawn from this arrangement, which is called a cell. A battery is made up of a number of cells. It has many useful applications in providing portable electrical power, but electrochemical action can also be undesirable since it is the basis of electrochemical corrosion which rots our motor cars, industrial containers and bridges.

Magnetism

The Greeks knew as early as 600 BC that a certain form of iron ore, now known as magnetite or lodestone, had the property of attracting small pieces of iron. Later, during the Middle Ages, navigational compasses were made using the magnetic properties of lodestone. Small pieces of lodestone attached to wooden splints floating in a bowl of water always came to rest pointing in a north–south direction. The word lodestone is derived from an old English word meaning 'the way', and the word magnetism is derived from Magnesia, the place where magnetic ore was first discovered.

Iron, nickel and cobalt are the only elements which are attracted strongly by a magnet. These materials are said to be *ferromagnetic*. Copper, brass, wood, PVC and glass are not attracted by a magnet and are, therefore, described as *non-magnetic*.

SOME BASIC RULES OF MAGNETISM

1. Lines of magnetic flux have no physical existence, but they were introduced by Michael Faraday (1791–1867) as a way of explaining the magnetic energy existing in space or in a material. They help us to visualize and explain the magnetic effects. The symbol used for magnetic flux is the Greek letter Φ (phi) and the unit of magnetic flux is the weber (symbol Wb), pronounced 'veber', to commemorate the work of the German physicist Wilhelm Weber (1804–91).

2 Lines of magnetic flux always form closed loops.
3 Lines of magnetic flux behave like stretched elastic bands, always trying to shorten themselves.
4 Lines of magnetic flux never cross over each other.
5 Lines of magnetic flux travel along a magnetic material and always emerge out of the 'north pole' end of the magnet.
6 Lines of magnetic flux pass through space and non-magnetic materials undisturbed.
7 The region of space through which the influence of a magnet can be detected is called the *magnetic field* of that magnet.
8 The number of lines of magnetic flux within a magnetic field is a measure of the flux density. Strong magnetic fields have a high flux density. The symbol used for flux density is *B*, and the unit of flux density is the tesla (symbol T), to commemorate the work of the Croatian-born American physicist Nikola Tesla (1857–1943).
9 The places on a magnetic material where the lines of flux are concentrated are called the magnetic poles.
10 Like poles repel; unlike poles attract. These two statements are sometimes called the 'first laws of magnetism' and are shown in Fig. 8.16.

EXAMPLE

The magnetizing coil of a radio speaker induces a magnetic flux of 360 μWb in an iron core of cross-sectional area 300 mm². Calculate the flux density in the core.

$$\text{Flux density } B = \frac{\Phi}{\text{area}} \text{ (tesla)}$$

$$B = \frac{360 \times 10^{-6} \text{ (Wb)}}{300 \times 10^{-6} \text{ (m}^2\text{)}}$$

$$B = 1.2 \text{ T}$$

MAGNETIC FIELDS

If a permanent magnet is placed on a surface and covered by a piece of paper, iron filings can be shaken on to the paper from a dispenser. Gently tapping the paper then causes the filings to take up the shape of the magnetic field surrounding the permanent magnet. The magnetic fields around a permanent magnet are shown in Figs 8.15 and 8.16.

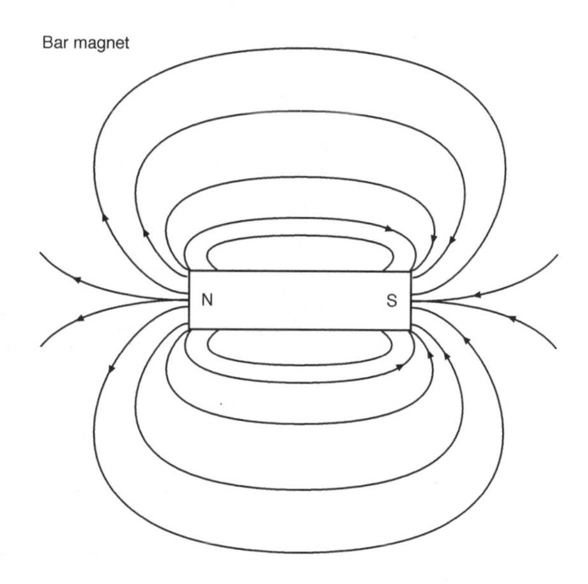

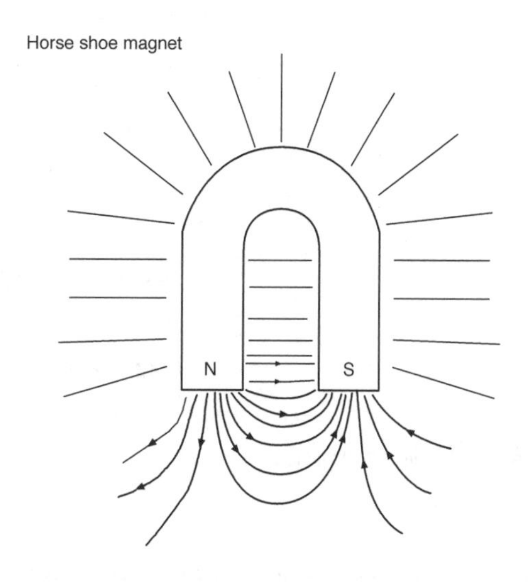

Fig. 8.15 Magnetic fields around a permanent magnet.

Electromagnetism

Electricity and magnetism have been inseparably connected since the experiments by Oersted and Faraday in the early nineteenth century. An electric current flowing in a conductor produces a magnetic field

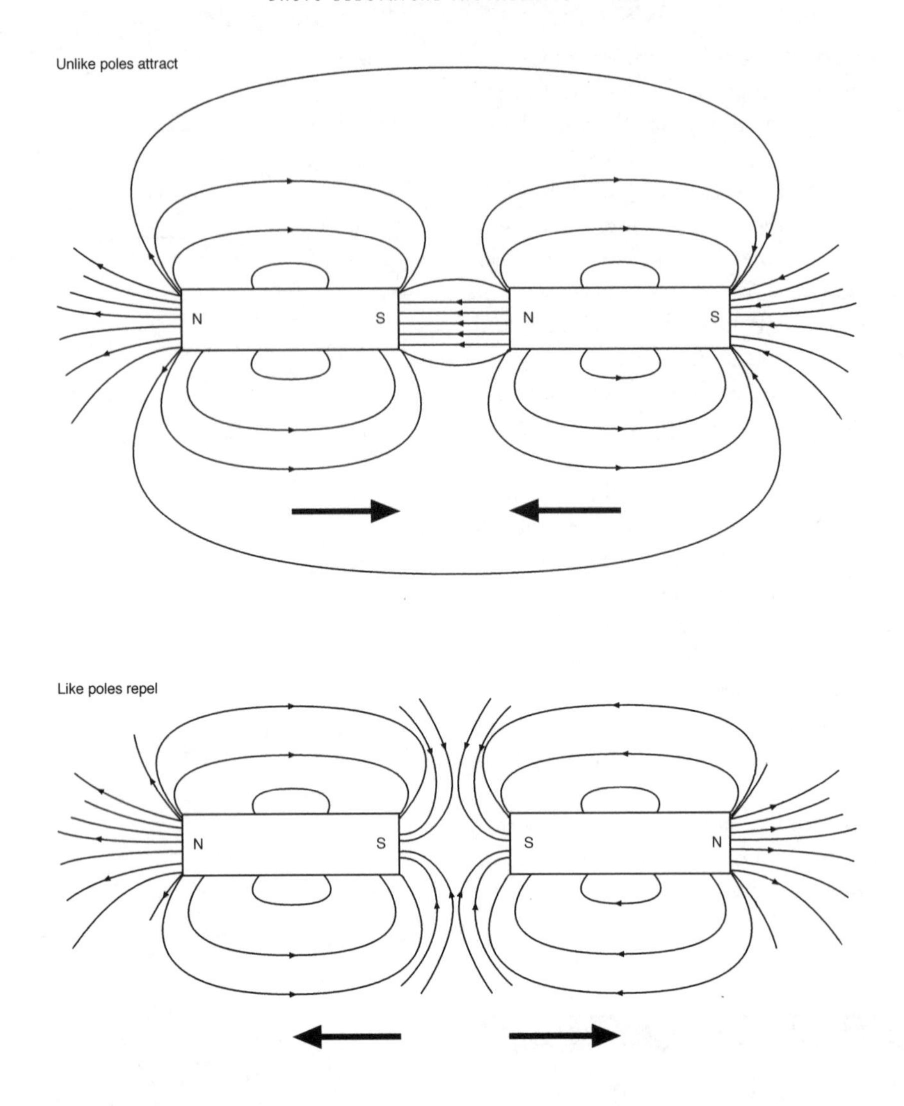

Fig. 8.16 The first laws of magnetism.

'around' the conductor which is proportional to the current. Thus a small current produces a weak magnetic field, while a large current will produce a strong magnetic field. The magnetic field 'spirals' around the conductor, as shown in Fig. 8.17 and its direction can be determined by the 'dot' or 'cross' notation and the 'screw rule'. To do this, we think of the current as being represented by a dart or arrow inside the conductor. The dot represents current coming towards us when we would see the point of the arrow or dart inside the conductor. The cross represents current going away from us when we would see the flights of the dart or arrow. Imagine a corkscrew or screw being turned so that it will move in the direction of the current. Therefore, if the current was coming out of the paper, as shown in Fig. 8.17(a), the magnetic field would be spiralling anticlockwise around the conductor. If the current was going into the paper, as shown by Figure 8.17(b), the magnetic field would spiral clockwise around the conductor.

A current flowing in a *coil* of wire or solenoid establishes a magnetic field which is very similar to that of a bar magnet. Winding the coil around a soft iron core increases the flux density because the lines of magnetic flux concentrate on the magnetic material. The advantage of the electromagnet when compared with the permanent magnet is that the magnetism of the electromagnet can be switched on and off by a

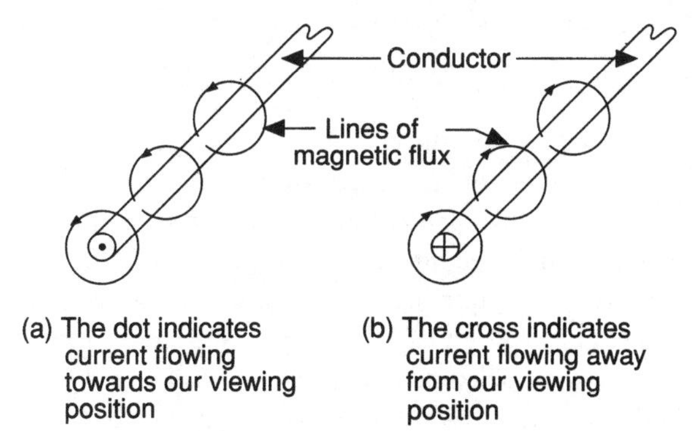

Fig. 8.17 Magnetic fields around a current carrying conductor.

functional switch controlling the coil current. This effect is put to practical use in the electrical relay as used in a motor starter or alarm circuit. Figure 8.18 shows the structure and one application of the solenoid.

Inductance

If a coil of wire is wound on to an iron core as shown in Fig. 8.19, a magnetic field will become established in the core when a current flows in the coil due to the switch being closed.

When the switch is opened the current stops flowing and, therefore, the magnetic flux collapses. The collapsing magnetic flux cuts the electrical conductors of the coil and induces an emf into it. This voltage appears across the switch contacts. The effect is known as *inductance* and is one property of any coil. The unit of inductance is the henry (symbol H), to commemorate the work of the American physicist Joseph Henry (1797–1878) who, quite independently, discovered electromagnetic induction just one year after Michael Faraday in 1831.

Faraday's law states that when a conductor cuts or is cut by a magnetic field, an emf is induced in that conductor. The amount of induced emf is proportional to the rate or speed at which the magnetic field cuts the conductor. This is the principle of operation of the simple generator shown in Fig. 9.2 in Chapter 9. Rotating a coil of wire in a magnetic field induces an emf or voltage in the coil. Increasing the speed of rotation will increase the generated voltage.

Modern power station generators work on this principle and, indeed, the foundations of all our modern knowledge of electricity were laid down by Michael Faraday. He is one of the most famous English scientists, and when Robert Peel, the Prime Minister of the day, asked Faraday 'What use will electricity be?', Faraday replied 'I know not, sir, but I'll wager that one day you will tax it!'

Any circuit in which a change of magnetic flux induces an emf is said to be 'inductive' or to possess 'inductance'.

Fluorescent light fittings contain a choke or inductive coil in series with the tube and starter lamp. The starter lamp switches on and off very quickly, causing

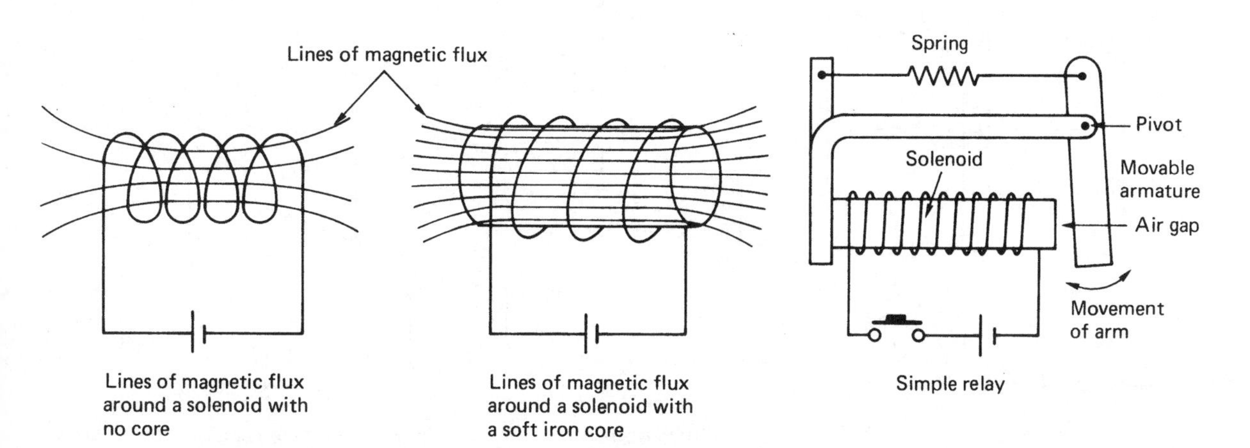

Fig. 8.18 The solenoid and one practical application: the relay.

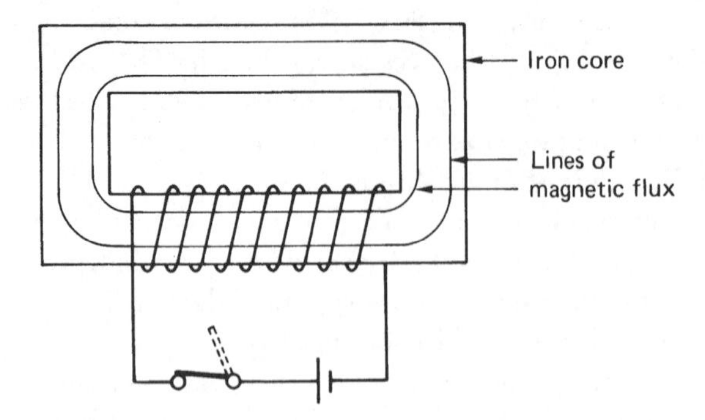

Fig. 8.19 An inductive coil or choke.

rapid current changes which induce a large voltage across the tube electrodes sufficient to strike an arc in the tube.

Further information on fluorescent lighting circuits, and the regulations associated with inductive circuits, are given in Chapter 10 of *Advanced Electrical Installation Work* (3rd edition).

Electrostatics

If a battery is connected between two insulated plates, the emf of the battery forces electrons from one plate to another until the p.d. between the plates is equal to the battery emf.

The electrons flowing through the battery constitute a current, I (in amperes), which flows for a time, t (in seconds). The plates are then said to be charged. The amount of charge transferred is given by

$$Q = It \text{ (coulomb [Symbol C])}$$

Figure 8.20 shows the charges on a capacitor's plates.

When the voltage is removed the charge Q is trapped on the plates, but if the plates are joined together, the same quantity of electricity, $Q + It$, will flow back from one plate to the other, so discharging them. The property of a pair of plates to store an electric charge is called its *capacitance*.

By definition, a capacitor has a capacitance (C) of one farad (symbol F) when a p.d. of one volt maintains a charge of one coulomb on that capacitor, or

$$C = \frac{Q}{V} \text{ (F)}$$

Collecting these important formulae together, we have

$$Q = It = CV$$

CAPACITORS

A capacitor consists of two metal plates, separated by an insulating layer called the dielectric. It has the ability of storing a quantity of electricity as an excess of electrons on one plate and a deficiency on the other.

EXAMPLE

A 100 μF capacitor is charged by a steady current of 2 mA flowing for 5 seconds. Calculate the total charge stored by the capacitor and the p.d. between the plates.

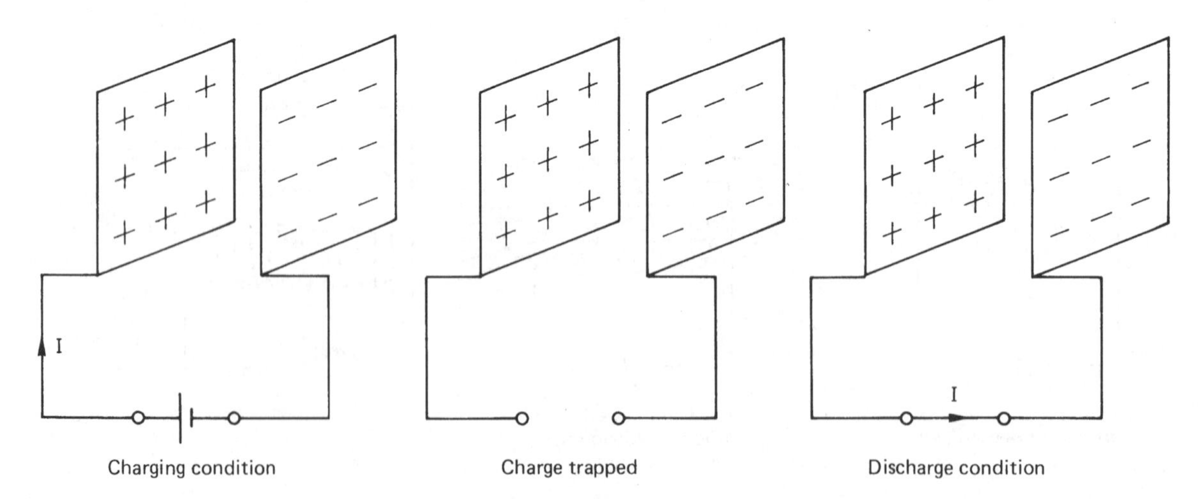

Fig. 8.20 The charge on a capacitor's plates.

$$Q = It \text{ (C)}$$
$$\therefore Q = 2 \times 10^{-3}\,\text{A} \times 5\,\text{s} = 10\,\text{mC}$$
$$Q = CV$$
$$\therefore V = \frac{Q}{C} \text{ (V)}$$
$$V = \frac{10 \times 10^{-3}\,\text{C}}{100 \times 10^{-6}\,\text{F}} = 100\,\text{V}$$

The p.d. which may be maintained across the plates of a capacitor is determined by the type and thickness of the dielectric medium. Capacitor manufacturers usually indicate the maximum safe working voltage for their products.

Capacitors are classified by the type of dielectric material used in their construction. Figure 8.21 shows the general construction and appearance of some capacitor types to be found in installation work.

Air-dielectric capacitors

Air-dielectric capacitors are usually constructed of multiple aluminium vanes of which one section moves to make the capacitance variable. They are often used for radio tuning circuits.

Mica-dielectric capacitors

Mica-dielectric capacitors are constructed of thin aluminium foils separated by a layer of mica. They are expensive, but this dielectric is very stable and has low dielectric loss. They are often used in high-frequency electronic circuits.

Paper-dielectric capacitors

Paper-dielectric capacitors usually consist of thin aluminium foils separated by a layer of waxed paper. This paper–foil sandwich is rolled into a cylinder and usually contained in a metal cylinder. These capacitors are used in fluorescent lighting fittings and motor circuits.

Electrolytic capacitors

The construction of these is similar to that of the paper-dielectric capacitors, but the dielectric material in this case is an oxide skin formed electrolytically by the manufacturers. Since the oxide skin is very thin, a large capacitance is achieved for a small physical size,

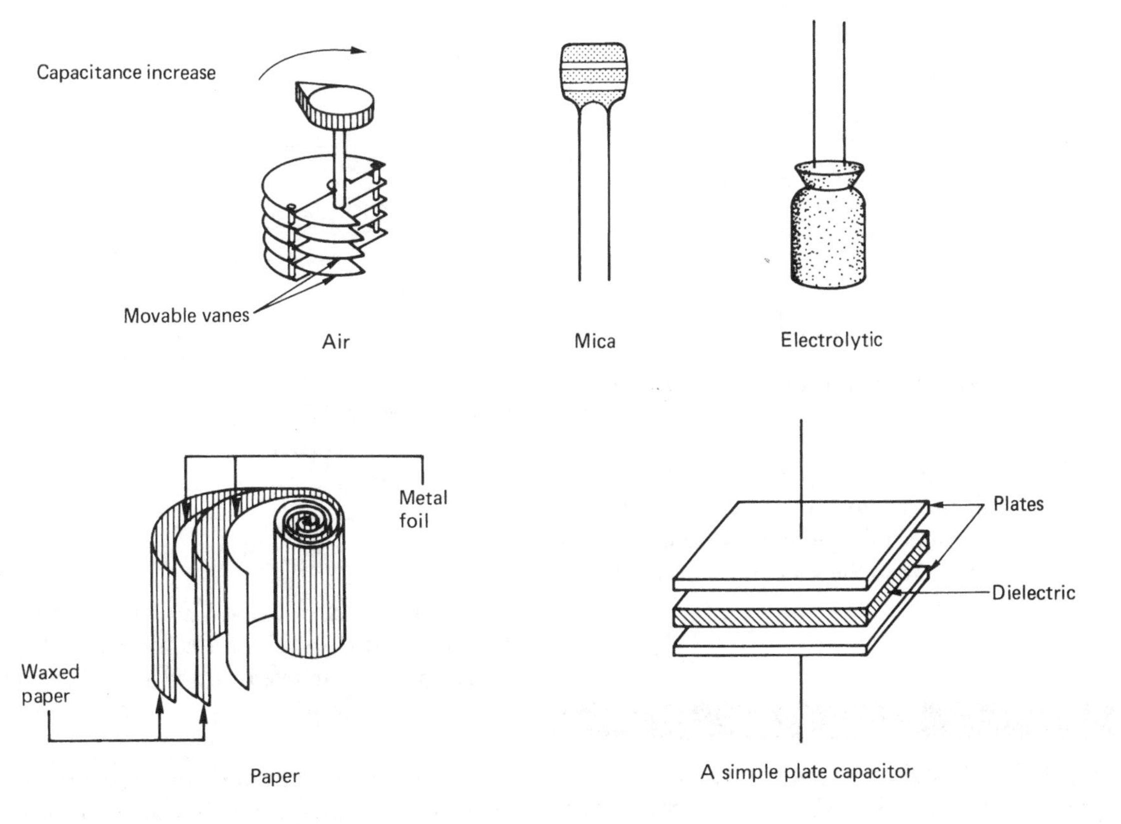

Fig. 8.21 Construction and appearance of capacitors.

but if a voltage of the wrong polarity is applied, the oxide skin is damaged and the gas inside the sealed container explodes. For this reason electrolytic capacitors must be connected to the correct voltage polarity. They are used where a large capacitance is required from a small physical size and where the terminal voltage never reverses polarity.

CAPACITORS IN COMBINATION

Capacitors, like resistors, may be joined together in various combinations of series or parallel connections (see Figure 8.22). The equivalent capacitance, C_T, of a number of capacitors is found by the application of similar formulae to those used for resistors and discussed earlier in this chapter. *Note* that the form of the formulae is the opposite way round to that used for series and parallel resistors.

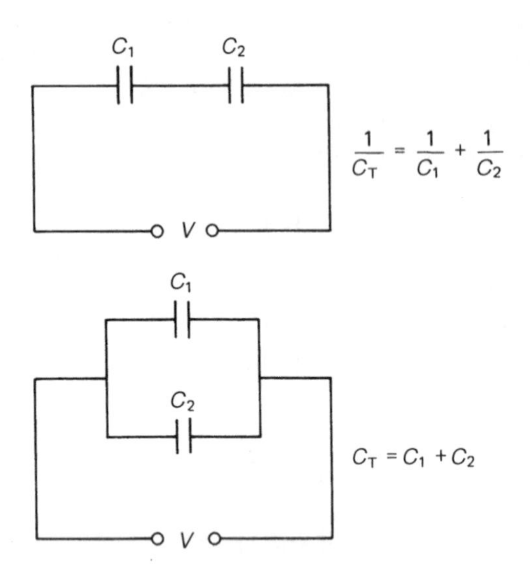

Fig. 8.22 Connection of and formulae for series and parallel capacitors.

The most complex arrangement of capacitors may be simplified into a single equivalent capacitor by applying the separate rules for series or parallel capacitors in a similar way to the simplification of resistive circuits.

EXAMPLE

Capacitors of 10 μF and 20 μF are connected first in series, and then in parallel, as shown in Figs 8.23 and 8.24. Calculate the effective capacitance for each connection. For connection in series,

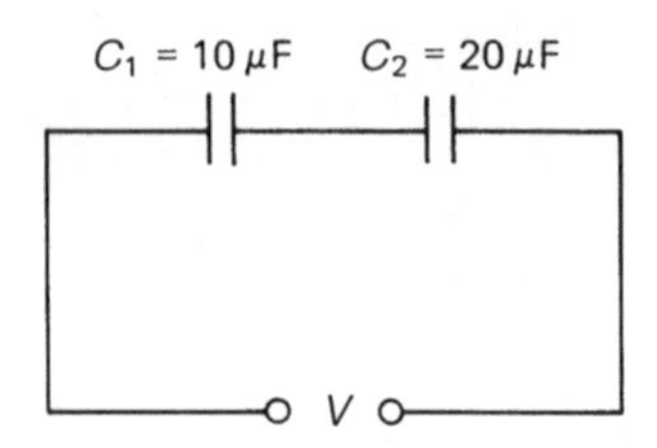

Fig. 8.23 Series capacitors.

$$\frac{1}{C_T} = \frac{1}{C_1} + \frac{1}{C_2}$$

$$\frac{1}{C_T} = \frac{1}{10\ \mu\text{F}} + \frac{1}{20\ \mu\text{F}}$$

$$\frac{1}{C_T} = \frac{2+1}{20\ \mu\text{F}} = \frac{3}{20\ \mu\text{F}}$$

$$\therefore C_T = \frac{20\ \mu\text{F}}{3} = 6.66\ \mu\text{F}$$

For connection in parallel,

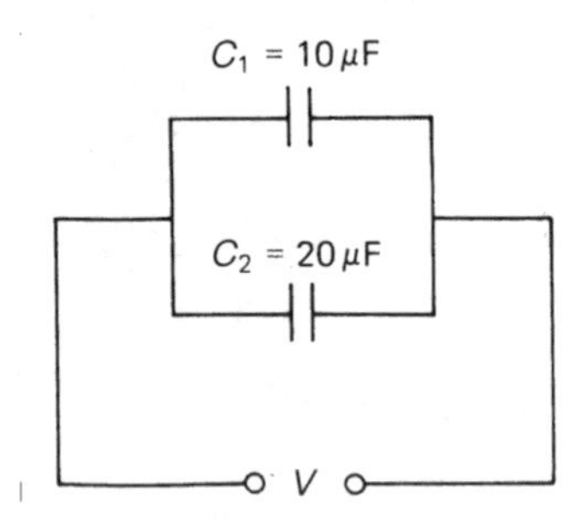

Fig. 8.24 Parallel capacitors.

$$C_T = C_1 + C_2$$

$$C_T = 10\ \mu\text{F} + 20\ \mu\text{F} = 30\ \mu\text{F}$$

Therefore, when capacitors of 10 μF and 20 μF are connected in series their combined effect is equivalent to a capacitor of 6.66 μF. But, when the same capacitors are connected in parallel their combined effect is equal to a capacitor of 30 μF.

The practical considerations of capacitors and the use of colour codes to determine capacitor values are dealt with in Chapter 10.

Mechanics

Mechanics is the scientific study of 'machines', where a machine is defined as a device which transmits motion or force from one place to another. An engine is one particular type of machine, an energy-transforming machine, converting fuel energy into a more directly useful form of work.

Most modern machines can be traced back to the five basic machines described by the Greek inventor Hero of Alexandria who lived about the time of Christ. The machines described by him were the wedge, the screw, the wheel and axle, the pulley, and the lever. Originally they were used for simple purposes, to raise water and move objects which man alone could not lift, but today their principles are of fundamental importance to our scientific understanding of mechanics. Let us now consider some fundamental mechanical principles and calculations.

MASS

This is a measure of the amount of material in a substance, such as metal, plastic, wood, brick or tissue, which is collectively known as a body. The mass of a body remains constant and can easily be found by comparing it on a set of balance scales with a set of standard masses. The SI unit of mass is the kilogram (kg).

WEIGHT

This is a measure of the force which a body exerts on anything which supports it. Normally it exerts this force because it is being attracted toward the earth by the force of gravity.

For scientific purposes the weight of a body is *not* constant, because gravitational force varies from the equator to the poles; in space a body would be 'weightless' but here on earth under the influence of gravity a 1 kg mass would have a weight of approximately 9.81 newtons (see also the definition of 'force').

SPEED

The feeling of speed is something with which we are all familiar. If we travel in a motor vehicle we know that an increase in speed would, excluding accidents, allow us to arrive at our destination more quickly. Therefore, speed is concerned with distance travelled and time taken. Suppose we were to travel a distance of 30 miles in one hour; our speed would be an average of 30 miles per hour:

$$\text{Speed} = \frac{\text{Distance (m)}}{\text{Time (s)}}$$

VELOCITY

In everyday conversation we often use the word velocity to mean the same as speed, and indeed the units are the same. However, for scientific purposes this is not acceptable since velocity is also concerned with direction. Velocity is speed in a given direction. For example, the speed of an aircraft might be 200 miles per hour, but its velocity would be 200 miles per hour in, say, a westerly direction. Speed is a scalar quantity, while velocity is a vector quantity.

$$\text{Velocity} = \frac{\text{Distance (m)}}{\text{Time (s)}}$$

ACCELERATION

When an aircraft takes off, it starts from rest and increases its velocity until it can fly. This change in velocity is called its acceleration. By definition, acceleration is the rate of change in velocity with time.

$$\text{Acceleration} = \frac{\text{Velocity}}{\text{Time}} = (\text{m/s}^2)$$

EXAMPLE

If an aircraft accelerates from a velocity of 15 m/s to 35 m/s in 4 s, calculate its average acceleration.

$$\text{Average velocity} = 35\ \text{m/s} - 15\ \text{m/s} = 20\ \text{m/s}$$

$$\text{Average acceleration} = \frac{\text{Velocity}}{\text{Time}} = \frac{20}{4} = 5\ \text{m/s}^2$$

Thus, the average acceleration is 5 metres per second per second.

FORCE

The presence of a force can only be detected by its effect on a body. A force may cause a stationary object to move or bring a moving body to rest. For example, a number of people pushing a broken-down motor car exert a force which propels it forward, but applying the motor car brakes applies a force on the brake drums which slows down or stops the vehicle. Gravitational force causes objects to fall to the ground. The apple fell from the tree on to Isaac Newton's head as a result of gravitational force. The standard rate of acceleration due to gravity is accepted as 9.81 m/s^2. Therefore, an apple weighing 1 kg will exert a force of 9.81 N since

$$\text{Force} = \text{Mass} \times \text{Acceleration (N)}$$

The SI unit of force is the newton, symbol N, to commemorate the great English scientist Sir Isaac Newton (1642–1727).

EXAMPLE

A 50 kg bag of cement falls from a forklift truck while being lifted to a storage shelf. Determine the force with which the bag will strike the ground:

$$\text{Force} = \text{Mass} \times \text{Acceleration (N)}$$
$$\text{Force} = 50\ \text{kg} \times 9.81\ \text{m/s}^2 = 490.5\ \text{N}$$

A force can manifest itself in many different ways. Let us consider a few examples:

- 'Inertial force' is the force required to get things moving, to change direction or stop, like the motor car discussed above.
- 'Cohesive or adhesive force' is the force required to hold things together.
- 'Tensile force' is the force pulling things apart.
- Compressive force' is the force pushing things together.
- 'Friction force' is the force which resists or prevents the movement of two surfaces in contact.
- 'Shearing force' is the force which moves one face of a material over another.
- 'Centripetal force' is the force acting towards the centre when a mass attached to a string is rotated in a circular path.
- 'Centrifugal force' is the force acting away from the centre, the opposite to centripetal force.
- 'Gravitational force' is the force acting towards the centre of the earth due to the effect of gravity.
- 'Magnetic force' is the force created by a magnetic field.
- 'Electrical force' is the force created by an electrical field.

PRESSURE OR STRESS

To move a broken-down motor car I might exert a force on the back of the car to propel it forward. My hands would apply a pressure on the body panel at the point of contact with the car. Pressure or stress is a measure of the force per unit area.

$$\text{Pressure or stress} = \frac{\text{Force}}{\text{Area}}\ (\text{N/m}^2)$$

EXAMPLE 1

A young woman of mass 60 kg puts all her weight on to the heel of one shoe which has an area of 1 cm^2. Calculate the pressure exerted by the shoe on the floor (assuming the acceleration due to gravity to be 9.81 m/s^2).

$$\text{Pressure} = \frac{\text{Force}}{\text{Area}}\ (\text{N/m}^2)$$

$$\text{Pressure} = \frac{60\ \text{kg} \times 9.81\ \text{m/s}^2}{1 \times 10^{-4}\ \text{m}^2} = 5886\ \text{kN/m}^2$$

EXAMPLE 2

A small circus elephant of mass 1 tonne (1000 kg) puts all its weight on to one foot which has a surface area of 400 cm^2. Calculate the pressure exerted by the elephant's foot on the floor, assuming the acceleration due to gravity to be 9.81 m/s^2.

$$\text{Pressure} = \frac{\text{Force}}{\text{Area}}\ (\text{N/m}^2)$$

$$\text{Pressure} = \frac{1000\ \text{kg} \times 9.81\ \text{m/s}^2}{400 \times 10^{-4}\ \text{m}^2} = 245.3\ \text{kN/m}^2$$

These two examples show that the young woman exerts 24 times more pressure on the ground than the elephant. This is because her mass exerts a force over a much smaller area than the elephant's foot, and is the reason why many wooden dance floors are damaged by high-heeled shoes.

WORK DONE

Suppose a broken-down motor car was to be pushed along a road; work would be done on the car by applying the force necessary to move it along the road. Heavy breathing and perspiration would be evidence of the work done:

Work done = Force × Distance moved in the direction of the force (J)

The SI unit of work done is the newton metre or joule (symbol J). The joule is the preferred unit and it commemorates an English physicist, James Prescot Joule (1818–89).

EXAMPLE

A building hoist lifts ten 50 kg bags of cement through a vertical distance of 30 m to the top of a high rise building. Calculate the work done by the hoist, assuming the acceleration due to gravity to be $9.81\ m/s^2$.

$$\text{Work done} = \text{Force} \times \text{Distance moved (J)}$$

but

$$\text{Force} = \text{Mass} \times \text{Acceleration (N)}$$

$\therefore$

$$\text{Work done} = \text{Mass} \times \text{Acceleration} \times \text{Distance moved (J)}$$

$$\text{Work done} = 10 \times 50\ kg \times 9.81\ m/s^2 \times 30\ m$$

$$\text{Work done} = 147.15\ kJ.$$

POWER

If one motor car can cover the distance between two points more quickly than another car, we say that the faster car is more powerful. It can do a given amount of work more quickly. By definition, power is the rate of doing work.

$$\text{Power} = \frac{\text{Work done}}{\text{Time taken}}\ (W)$$

The SI unit of power, both electrical and mechanical, is the watt (symbol W). This commemorates the name of James Watt (1736–1819), the inventor of the steam engine.

EXAMPLE 1

A building hoist lifts ten 50 kg bags of cement to the top of a 30 m high building. Calculate the rating (power) of the motor to perform this task in 60 seconds if the acceleration due to gravity is taken as $9.81\ m/s^2$.

$$\text{Power} = \frac{\text{Work done}}{\text{Time taken}}\ (W),$$

but Work done = Force × Distance moved (J)
and Force = Mass × Acceleration (N)

By substitution,

$$\text{Power} = \frac{\text{Mass} \times \text{Acceleration} \times \text{Distance moved}}{\text{Time taken}}\ (W)$$

$$\text{Power} = \frac{10 \times 50\ kg \times 9.81\ m/s^2 \times 30\ m}{60\ s}$$

$$\text{Power} = 2452.5\ W$$

The rating of the building hoist motor will be 2.45 kW.

EXAMPLE 2

A hydroelectric power station pump motor working continuously during a 7 hour period raises 856 tonnes of water through a vertical distance of 60 m. Determine the rating (power) of the motor, assuming the acceleration due to gravity is $9.81\ m/s^2$.

From Example 1,

$$\text{Power} = \frac{\text{Mass} \times \text{Acceleration} \times \text{Distance moved}}{\text{Time taken}}\ (W)$$

$$\text{Power} = \frac{856 \times 1000\ kg \times 9.81\ m/s^2 \times 60\ m}{7 \times 60 \times 60\ s}$$

$$\text{Power} = 20\,000\ W$$

The rating of the pump motor is 20 kW.

EXAMPLE 3

An electric hoist motor raises a load of 500 kg at a velocity of 2 m/s. Calculate the rating (power) of the motor if the acceleration due to gravity is $9.81\ m/s^2$.

$$\text{Power} = \frac{\text{Mass} \times \text{Acceleration} \times \text{Distance moved}}{\text{Time taken}}\ (W)$$

$$\text{but Velocity} = \frac{\text{Distance}}{\text{Time}}\ (m/s)$$

$$\therefore \text{Power} = \text{Mass} \times \text{Acceleration} \times \text{Velocity}$$

$$\text{Power} = 500\ kg \times 9.81\ m/s^2 \times 2\ m/s$$

$$\text{Power} = 9810\ W.$$

The rating of the hoist motor is 9.81 kW.

EFFICIENCY

In any machine the power available at the output is less than that which is put in because losses occur in the machine. The losses may result from friction in the bearings, wind resistance to moving parts, heat, noise or vibration.

The ratio of the output power to the input power is known as the *efficiency* of the machine. The symbol for efficiency is the Greek letter 'eta' (η). In general,

$$\eta = \frac{\text{Power output}}{\text{Power input}}$$

Since efficiency is usually expressed as a percentage we modify the general formula as follows.

$$\eta = \frac{\text{Power output}}{\text{Power input}} \times 100$$

EXAMPLE

A transformer feeds the 9.81 kW motor driving the mechanical hoist of the previous example. The input power to the transformer was found to be 10.9 kW. Find the efficiency of the transformer.

$$\eta = \frac{\text{Power output}}{\text{Power input}} \times 100$$

$$\eta = \frac{9.81\ \text{kW}}{10.9\ \text{kW}} \times 100 = 90\%$$

Thus the transformer is 90% efficient. Note that efficiency has no units, but is simply expressed as a percentage.

LEVERS

Every time we open a door, turn on a tap or tighten a nut with a spanner, we exert a lever-action turning force. A lever is any rigid body which pivots or rotates about a fixed axis or fulcrum. The simplest form of lever is the crowbar, which is useful because it enables a person to lift a load at one end which is greater than the effort applied through his or her arm muscles at the other end. In this way the crowbar is said to provide a 'mechanical advantage'. A washbasin tap and a spanner both provide a mechanical advantage through the simple lever action. The mechanical advantage of a simple lever is dependent upon the length of lever on either side of the fulcrum. Applying the principle of turning forces to a lever, we obtain the formula:

Load force × Distance from fulcrum =
Effort force × Distance from fulcrum

This formula can perhaps better be understood by referring to Fig. 8.25. A small effort at a long distance from the fulcrum can balance a large load at a short distance from the fulcrum. Thus a 'turning force' or 'turning moment' depends upon the distance from the fulcrum and the magnitude of the force.

EXAMPLE

Calculate the effort required to raise a load of 500 kg when the effort is applied at a distance of five times the load distance from the fulcrum (assume the acceleration due to gravity to be 10 m/s^2).

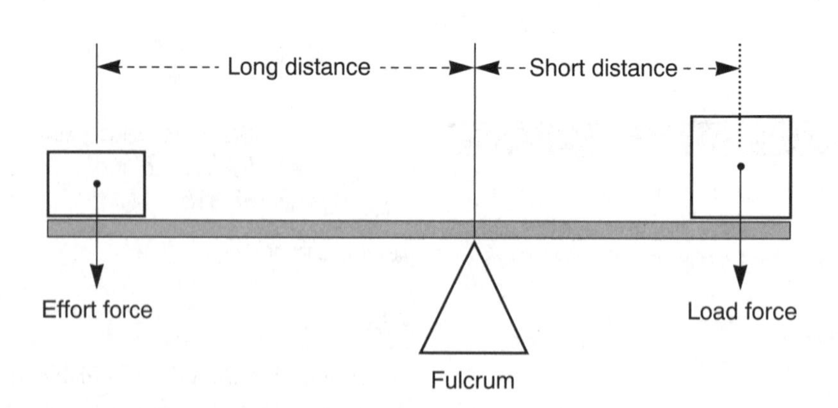

Fig. 8.25 Turning forces of a simple lever.

$$\text{Load force} = \text{Mass} \times \text{Acceleration (N)}$$
$$\text{Load force} = 500\ \text{kg} \times 10\ \text{m/s}^2 = 5000\ \text{N}$$

$$\text{Load force} \times \text{Distance from fulcrum} = \text{Effort force} \times \text{Distance from fulcrum}$$

$$5000\ \text{N} \times 1\ \text{m} = \text{Effort force} \times 5\ \text{m}$$

$$\therefore \text{Effort force} = \frac{5000\ \text{N} \times 1\ \text{m}}{5\ \text{m}} = 1000\ \text{N}$$

Thus an effort force of 1000 N can overcome a load force of 5000 N using the mechanical advantage of this simple lever.

Every lever has one pivot point, the fulcrum, and is acted upon by two forces, the load and the effort. There are three classes or types of lever, according to the position of the load, effort and fulcrum (Fig. 8.26).
In a first-class lever the load is applied to one side of the fulcrum and the effort to the other, as shown in Fig. 8.26(a). A typical example of a first-class lever is a crowbar or a sack truck. A crocodile clip, a pair of pliers or side cutters, are examples of two first-class levers acting together as shown in Fig. 8.27.

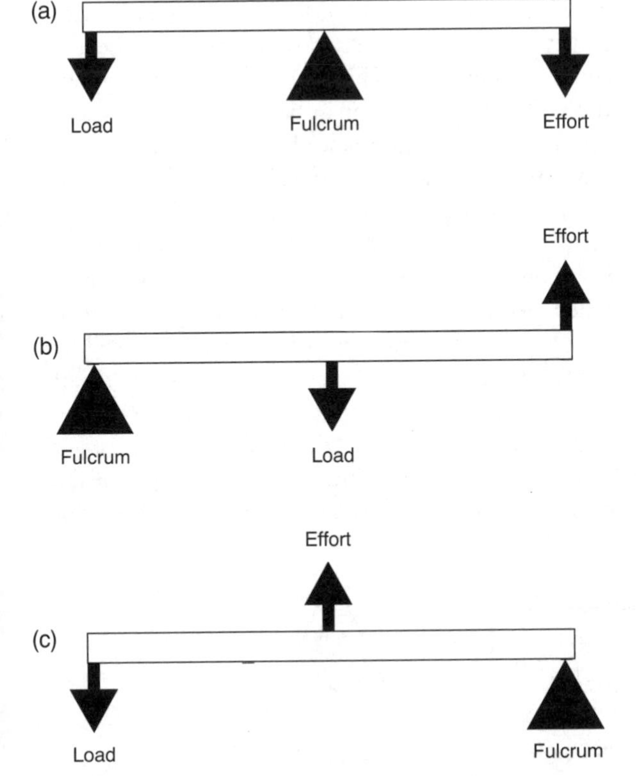

Fig. 8.26 The three classes of lever: (a) first class; (b) second class; (c) third class.

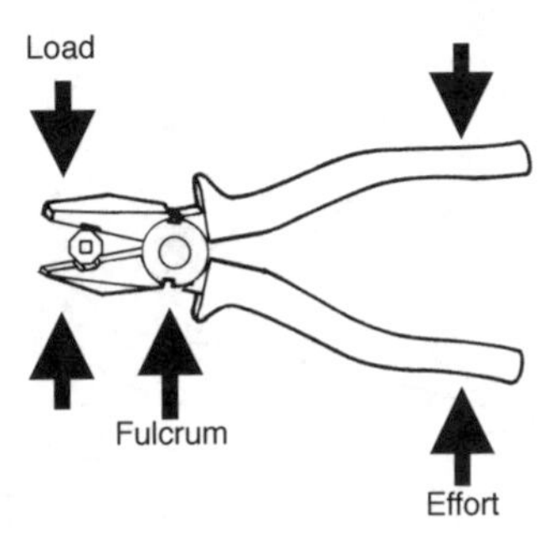

Fig. 8.27 A pair of pliers, an example of two first-class levers acting together.

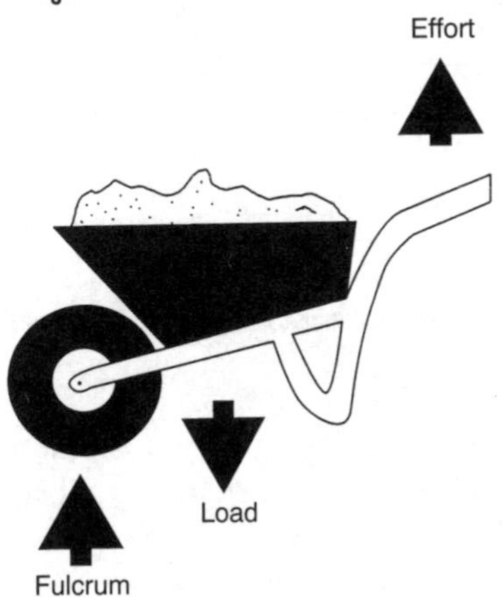

Fig. 8.28 A wheelbarrow, an example of a second-class lever.

A second-class lever has the load and effort applied to one side of the fulcrum, as shown in Fig. 8.26(b). A typical example of a second class lever is a beer bottle opener or nutcracker, a conduit bending machine or wheelbarrow, as shown in Fig. 8.28. The general principle is just the same, that a small effort applied at the end of a long arm can be used to overcome a much larger load at the end of a short arm.

The third class of lever gives us the least mechanical advantage because the effort is closer to the fulcrum than the load, as shown in Fig. 8.26(c). However, it does have many useful applications. The forearm operates as a third-class lever with the fulcrum at the elbow. A load in the hand is raised by the effort exerted by the biceps acting on the forearm close to the elbow. Other applications are sugar or coal tongs, or the heat shunt used when soldering electronic components, as shown in Fig. 8.29.

CENTRE OF GRAVITY

We have discussed earlier in this chapter that a body is attracted to the centre of the earth by the force of gravity. This statement, however, says nothing about

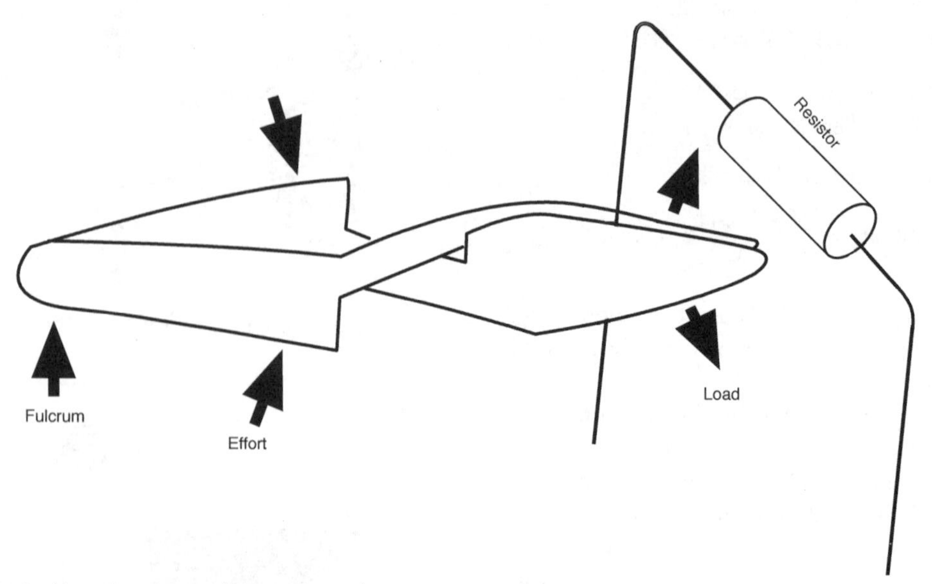

Fig. 8.29 A soldering heat shunt used when soldering electronic components, an example of two third-class levers acting together.

where the force is acting, so let us assume that any body is made up of a very large number of tiny particles. The force of gravity will act equally on every particle, thus creating a large number of parallel forces. These forces will have a resultant force equal to the total force of gravity and will act through a point called the *centre of gravity*. By definition, the centre of gravity of a body is the point through which the total weight of the body appears to act. The centre of gravity of a 30 cm rule will act through the 15 cm point. If you place the rule on your finger at this point, it will balance because equal forces are acting on both sides of your finger.

The centre of gravity of a disc is at the centre, as shown in Fig. 8.30. The centre of gravity of a ring is also at the centre, even though there is no material at the centre. This is the point through which the resultant forces act. The centre of gravity of a square, a rectangle or a triangle is at the point where the intersecting diagonal lines meet, as shown in Fig. 8.30.

The point at which the centre of gravity acts on an object is important to the 'stability' of that object. If a cone is placed on its point it will have a high centre of gravity and the slightest movement will cause the cone to topple over. This is called 'unstable equilibrium'. Placing the cone on its base gives it a lower centre of gravity and a broader base. The cone is now more difficult to knock over and is, therefore, said to be in 'stable equilibrium'. Laying the cone on its side gives it a lower centre of gravity, a very broad base and it becomes impossible to topple. This position is called 'neutral equilibrium'. In general, if the centre of gravity acts within the base width, the object will be stable. If a small displacement brings the centre of gravity outside the base width it becomes unstable. Figure 8.31 shows these three effects. The risk of unstable equilibrium is increased as the height of the centre of gravity is increased. The nearer the centre of

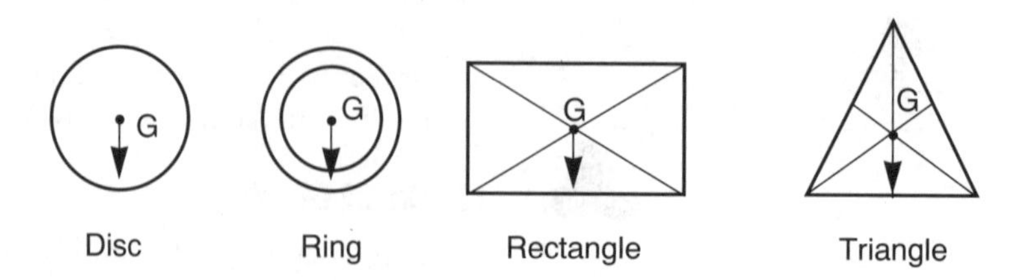

Fig. 8.30 The centre of gravity (point G) of some regular shapes.

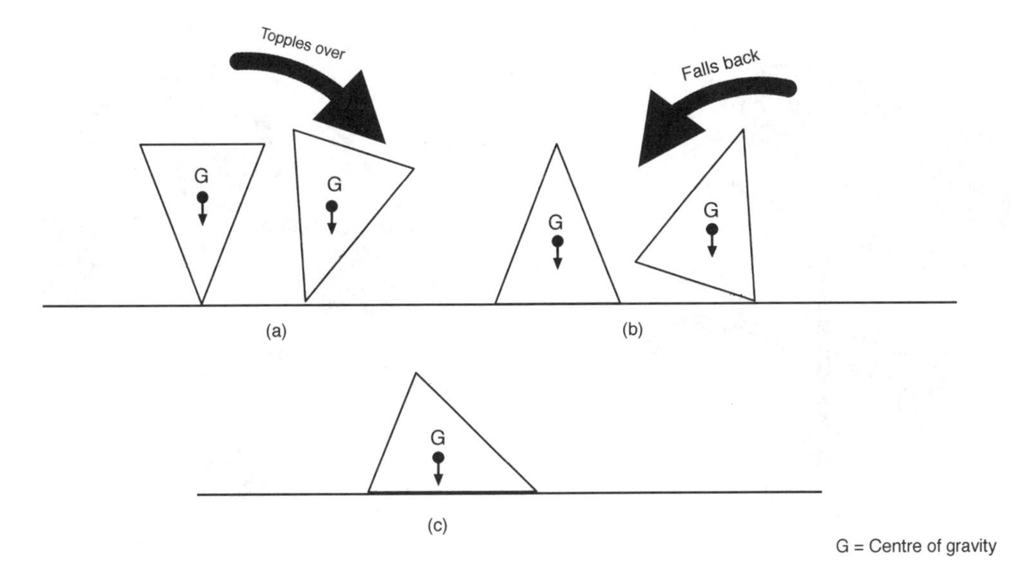

Fig. 8.31 Equilibrium of a cone: (a) unstable equilibrium; (b) stable equilibrium; (c) neutral equilibrium.

gravity is to the ground, the more stable the equilibrium is likely to be.

In designing a car or a ship the engineer must take into account its stability under normal operating conditions. A car turning a sharp corner at high speed may become unstable if its centre of gravity is high. Racing cars have a very low centre of gravity, as shown in Fig. 8.32. A ship is liable to considerable rolling and lurching in heavy seas and must be designed to be in stable equilibrium even in rough weather.

Temperature and heat

Heat has the capacity to do work and is a form of energy. Temperature is not an energy unit but describes the hotness or coldness of a substance or material.

HEAT TRANSFER

Heat energy is transferred by three separate processes which can occur individually or in combination. The processes are convection, radiation and conduction.

Convection

Air which passes over a heated surface expands, becomes lighter and warmer and rises, being replaced by descending cooler air. These circulating currents of air are called *convection currents*. In circulating, the warm air gives up some of its heat to the surfaces over which it passes and so warms a room and its contents.

Radiation

Molecules on a metal surface vibrating with thermal energies generate electromagnetic waves. The waves travel away from the surface at the speed of light, taking energy with them and leaving the surface cooler. If the waves meet another material they produce a disturbance of the surface molecules which raises the temperature of that material. Radiated heat requires no intervening medium between the transmitter and receiver, obeys the same laws as light energy and is the method by which the energy from the sun reaches the earth.

Conduction

Heat transfer through a material by conduction occurs because there is direct contact between the vibrating molecules of the material. The application of a heat source to the end of a metal bar causes the atoms to vibrate rapidly within the lattice framework of the material. This violent shaking causes adjacent atoms to vibrate and liberates any loosely bound electrons, which also pass on the heat energy. Thus the heat energy travels *through* the material by conduction.

Fig. 8.32 Unstable and stable vehicles.

TEMPERATURE SCALES

When planning any scale of measurement a set of reference points must first be established. Two obvious reference points on any temperature scale are the temperatures at which ice melts and water boils. Between these points the scale is divided into a convenient number of divisions. In the case of the Celsius or Centigrade scale, the lower fixed point is zero degrees and the upper fixed point 100 degrees Celsius. The Kelvin scale takes as its lower fixed point the lowest possible temperature which has a value of –273°C, called absolute zero. A temperature change of one kelvin is exactly the same as one degree Celsius and so we can say that

$$0°C = 273\ K \quad \text{or } 0\ K = -273°C$$

EXAMPLE 1

Convert the following Celsius temperatures into kelvin: –20°C, 20°C and 200°C.

$$-20°C = -20 + 273 = 253\ K$$
$$20°C = 20 + 273 = 293\ K$$
$$200°C = 200 + 273 = 473\ K$$

EXAMPLE 2

Convert into degrees Celsius: 250 K, 300 K and 500 K.

$$250\ K = 250 - 273 = -23°C$$
$$300\ K = 300 - 273 = 27°C$$
$$500\ K = 500 - 273 = 227°C$$

Temperature measurement

One instrument which measures temperature is a thermometer. This uses the properties of an expanding liquid in a glass tube to indicate a temperature level. Most materials change their dimensions when heated and this property is often used to give a measure of temperature. Many materials expand with an increase in temperature and the *rate* of expansion varies with different materials. A bimetal strip is formed from two dissimilar metals joined together. As the temperature increases the metals expand at different rates and the bimetal strip bends or bows.

THERMOSTATS

A thermostat is a device for maintaining a constant temperature at some predetermined value. The operation of a thermostat is often based upon the principle of differential expansion between dissimilar metals which causes a contact to make or break at a chosen temperature. Figure 8.33 shows the principle of a rod-type thermostat which is often used with water heaters. An Invar rod, which has minimal expansion when heated, is housed within a copper tube and the two metals are brazed together at one end. The other end of the copper tube is secured to one end of the switch mechanism. As the copper expands and contracts under the influence of a varying temperature, the switch contacts are activated. In this case the contact breaks the electrical circuit when the temperature setting is reached.

SIMMERSTATS

A simmerstat is a device used to control the temperature of an electrical element, typically the boiling ring of a cooker. A snap-action switch is opened and closed at time intervals by passing current through a heater wrapped around a bimetal strip, as shown in Fig. 8.34.

With the switch contact made, current flows through the load and the heating coil. The heater warms the bimetal strip which expands and opens out, pushing against the spring steel strip and the control knob, so opening the switch contacts. The load and the heating coil are then switched off and the bimetal strip cools to its original shape, which allows the contacts to close, and the process repeats. The load and heating coil are switched on and off frequently if the control knob is arranged to allow little movement of the bimetal strip. Alternatively, the heater will remain switched on for longer periods if the control knob is adjusted to allow a larger movement. In this way the temperature of the load is controlled.

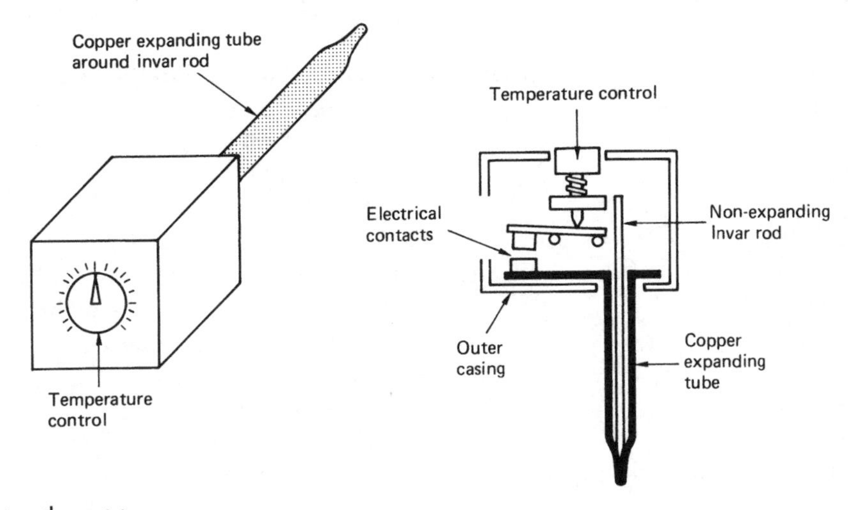

Fig. 8.33 A rod-type thermostat.

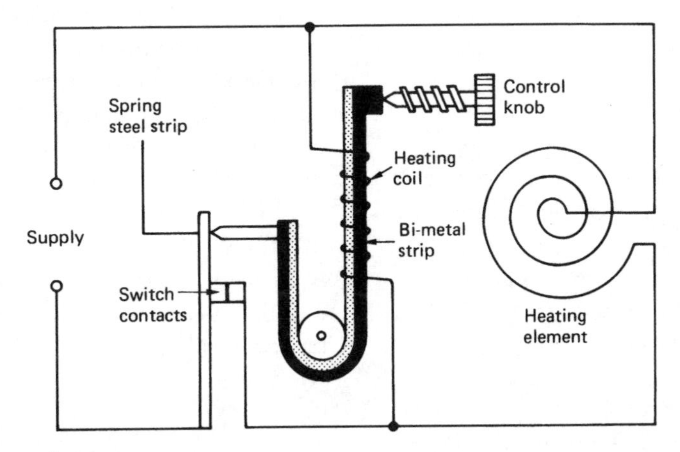

Fig. 8.34 Simmerstat control arrangement.

Exercises

1 Solar energy heats the earth by:
(a) conduction only
(b) convection only
(c) radiation only
(d) conduction and convection.

2 The heat which is transferred to a room from an oil-filled radiator is by:
(a) conduction only
(b) convection only
(c) radiation only
(d) radiation and convection.

3 A temperature of 25° on the Celsius scale is equal to a temperature on the Kelvin scale of:
(a) –248 K
(b) 187 K
(c) 248 K
(d) 298 K.

4 The absolute zero, 0 K, is equal to a temperature reading on the Celsius scale of:
(a) –273°C
(b) –100°C
(c) 32°C
(d) 212°C.

5 The pre-set temperature of an immersion heater is maintained by a:
(a) thermometer
(b) rheostat
(c) thermostat
(d) simmerstat.

6 The temperature of a boiling ring is controlled by a:
(a) thermometer
(b) rheostat
(c) thermostat
(d) simmerstat.

7 Describe briefly how heat is transferred by conduction, convection and radiation.

8 Describe the operation of a thermostat and a simmerstat, and give one practical application for each.

9 The SI units of length, resistance, and power are:
(a) millimetre, ohm, kilowatt
(b) centimetre, ohm, watt
(c) metre, ohm, watt
(d) kilometre, ohm, kilowatt.

10 The current taken by a 10 Ω resistor when connected to a 230 V supply is:
(a) 41 mA
(b) 2.3 A
(c) 23 A
(d) 230 A.

11 The resistance of an element which takes 12 A from a 230 V supply is:
(a) 2.88 Ω
(b) 5 Ω
(c) 12.24 Ω
(d) 19.16 Ω.

12 A 12 Ω lamp was found to be taking a current of 2 A at full brilliance. The voltage across the lamp under these conditions was:
(a) 6 V
(b) 12 V
(c) 24 V
(d) 240 V.

13 The resistance of 100 m of 1 mm^2 cross-section copper cable of resistivity 17.5×10^{-9} Ω m will be:
(a) 1.75 mΩ
(b) 1.75 Ω
(c) 17.5 Ω
(d) 17.5 kΩ.

14 The resistance of a motor field winding at 0°C was found to be 120 Ω. Find its new resistance at 20°C if the temperature coefficient of the winding is 0.004 Ω/Ω°C.
(a) 116.08 Ω
(b) 120.004 Ω
(c) 121.08 Ω
(d) 140.004 Ω.

15 The resistance of a motor field winding was found to be 120 Ω at an ambient temperature of 20°C. If the temperature coefficient of resistance is 0.004 Ω/Ω°C the resistance of the winding at 60°C will be approximately:
(a) 102 Ω
(b) 120 Ω
(c) 130 Ω
(d) 138 Ω.

16 A capacitor is charged by a steady current of 5 mA for 10 s. The total charge stored on the capacitor will be:
(a) 5 mC
(b) 50 mC
(c) 5 C
(d) 50 C.

17 When 100 V was connected to a 20 μF capacitor the charge stored was:
(a) 2 mC
(b) 5 mC
(c) 20 mC
(d) 100 mC.

18 An air dielectric capacitor is often used:
(a) for power-factor correction of fluorescents
(b) for tuning circuits
(c) when correct polarity connections are essential
(d) when only a very small physical size can be accommodated by the circuit enclosure.

19 An electrolytic capacitor:
(a) is used for power-factor correction in fluorescents
(b) is used for tuning circuits
(c) must only be connected to the correct polarity
(d) has a small capacitance for a large physical size.

20 A paper dielectric capacitor is often used:
(a) for power-factor correction in fluorescents
(b) for tuning circuits
(c) when correct polarity connections are essential
(d) when only a small physical size can be accommodated in the circuit enclosure.

21 A current flowing through a solenoid sets up a magnetic flux. If an iron core is added to the solenoid while the current is maintained at a constant value the magnetic flux will:
(a) remain constant
(b) totally collapse
(c) decrease in strength
(d) increase in strength.

22 Describe and give one practical example of the three effects of an electric current.

23 Resistors of 6 Ω and 3 Ω are connected in series. The combined resistance value will be:
(a) 2 Ω
(b) 3.6 Ω
(c) 6.3 Ω
(d) 9 Ω.

24 Resistors of 3 Ω and 6 Ω are connected in parallel. The equivalent resistance will be:
(a) 2 Ω
(b) 3.6 Ω
(c) 6.3 Ω
(d) 9 Ω.

25 Three resistors of 24, 40 and 60 Ω are connected in series. The total resistance will be:
(a) 12 Ω
(b) 26.4 Ω
(c) 44 Ω
(d) 124 Ω.

26 Resistors of 24, 40 and 60 Ω are connected together in parallel. The effective resistance of this combination will be:
(a) 12 Ω
(b) 26.4 Ω
(c) 44 Ω
(d) 124 Ω.

27 Two identical resistors are connected in series across a 12 V battery. The voltage drop across each resistor will be:
(a) 2 V
(b) 3 V
(c) 6 V
(d) 12 V.

28 Two identical resistors are connected in parallel across a 24 V battery. The voltage drop across each resistor will be:
(a) 6 V
(b) 12 V
(c) 24 V
(d) 48 V.

29 A 6 Ω resistor is connected in series with a 12 Ω resistor across a 36 V supply. The current flowing through the 6 Ω resistor will be:
(a) 2 A
(b) 3 A
(c) 6 A
(d) 9 A.

30 A 6 Ω resistor is connected in parallel with a 12 Ω resistor across a 36 V supply. The current flowing through the 12 Ω resistor will be:
(a) 2 A
(b) 3 A
(c) 6 A
(d) 9 A.

31 The total power dissipated by a 6 Ω and 12 Ω resistor connected in parallel across a 36 V supply will be:
(a) 72 W
(b) 324 W
(c) 576 W
(d) 648 W.

32 Three resistors are connected in series and a

current of 10 A flows when they are connected to a 100 V supply. If another resistor of 10 Ω is connected in series with the three series resistors the current carried by this resistor will be:
(a) 4 A
(b) 5 A
(c) 10 A
(d) 100 A.

33 The rms value of a sinusoidal waveform whose maximum value is 100 V will be:
(a) 63.7 V
(b) 70.71 V
(c) 100 V
(d) 100.67 V.

34 The average value of a sinusoidal alternating current whose maximum value is 10 A will be:
(a) 6.37 A
(b) 7.071 A
(c) 10 A
(d) 10.67 A.

35 Capacitors of 24, 40 and 60 μF are connected in series. The equivalent capacitance will be:
(a) 12 μF
(b) 44 μF
(c) 76 μF
(d) 124 μF.

36 Capacitors of 24, 40 and 60 μF are connected in parallel. The total capacitance will be:
(a) 12 μF
(b) 44 μF
(c) 76 μF
(d) 124 μF.

37 Describe with sketches the meaning of the terms *frequency* and *period* as applied to an a.c. waveform.

38 If we assume the acceleration due to gravity to be 10 m/s^2 a 50 kg bag of cement falling to the ground will exert a force of:
(a) 5 N
(b) 50 N
(c) 100 N
(d) 500 N.

39 The work done by a man carrying a 50 kg bag of cement up a 10 m ladder, assuming the acceleration due to gravity to be 10 m/s^2, will be:
(a) 50 J
(b) 500 J
(c) 5000 J
(d) 10 000 J.

40 A building hoist is to be used to raise sixty 50 kg bags of cement to the top of a 100 m high building in 1 minute. Assuming the acceleration due to gravity to be 10 m/s^2, the size of the hoist motor would be:
(a) 10 kW
(b) 50 kW
(c) 60 kW
(d) 100 kW.

41 A passenger lift has the capacity to raise 500 kg at the rate of 2 m/s. Assuming the acceleration due to gravity to be 10 m/s^2, the rating of the lift motor will be:
(a) 5 kW
(b) 10 kW
(c) 50 kW
(d) 100 kW.

42 A conduit bending machine operates on the principle of a lever of the
(a) first class
(b) second class
(c) third class
(d) fourth class.

43 A sack truck is a machine which operates on the principle of a lever of the
(a) first class
(b) second class
(c) third class
(d) fourth class.

44 A vehicle with a high centre of gravity will exhibit
(a) high stability
(b) low stability
(c) neutral equilibrium
(d) stable equilibrium.

45 A lift motor is to be used to raise a constant load of 2000 kg at a speed of 0.3 m/s. The motor is supplied at 400 V and works at a power-factor of 0.8 lagging. Find the current taken by the motor, assuming the acceleration due to gravity to be 9.81 m/s^2.

9

ELECTRICAL MACHINES

Introduction

Electrical machines are energy converters. If the machine input is mechanical energy and the output electrical energy then that machine is a generator, as shown in Fig. 9.1(a). Alternatively, if the machine input is electrical energy and the output mechanical energy then the machine is a motor, as shown in Fig. 9.1(b).

An electrical machine may be used as a motor or a generator, although in practice the machine will operate more efficiently when operated in the mode for which it was designed.

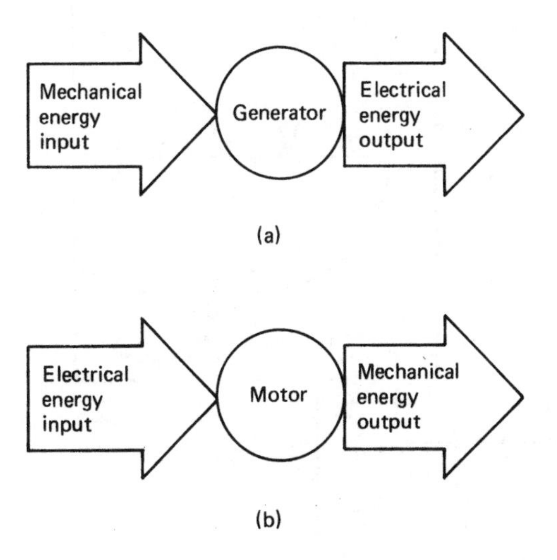

Fig. 9.1 Electrical machines as energy converters.

Generators

SIMPLE ALTERNATING CURRENT GENERATOR

If a simple loop of wire is rotated between the poles of a permanent magnet, as shown in Fig. 9.2, the loop of wire will cut the lines of magnetic flux between the north and south poles. This flux cutting will induce an emf in the wire by Faraday's law which states that *when a conductor cuts or is cut by a magnetic field, an emf is induced in that conductor.* If the generated emf is collected by carbon brushes at the slip rings and displayed on the screen of a cathode ray oscilloscope, the waveform will be seen to be approximately sinusoidal.

SIMPLE DIRECT CURRENT GENERATOR

If the slip rings of Fig. 9.2 are replaced by a single split ring, called a commutator, the generated emf will be seen to be in one direction, as shown in Fig. 9.3. The action of the commutator is to reverse the generated emf every half cycle, rather like an automatic change-over switch. However, this arrangement produces a very bumpy d.c. output similar to that produced by semiconductor rectifiers, as shown in Fig. 10.44 in Chapter 10. In a practical machine the commutator would contain many segments and many windings to produce a smoother d.c. output.

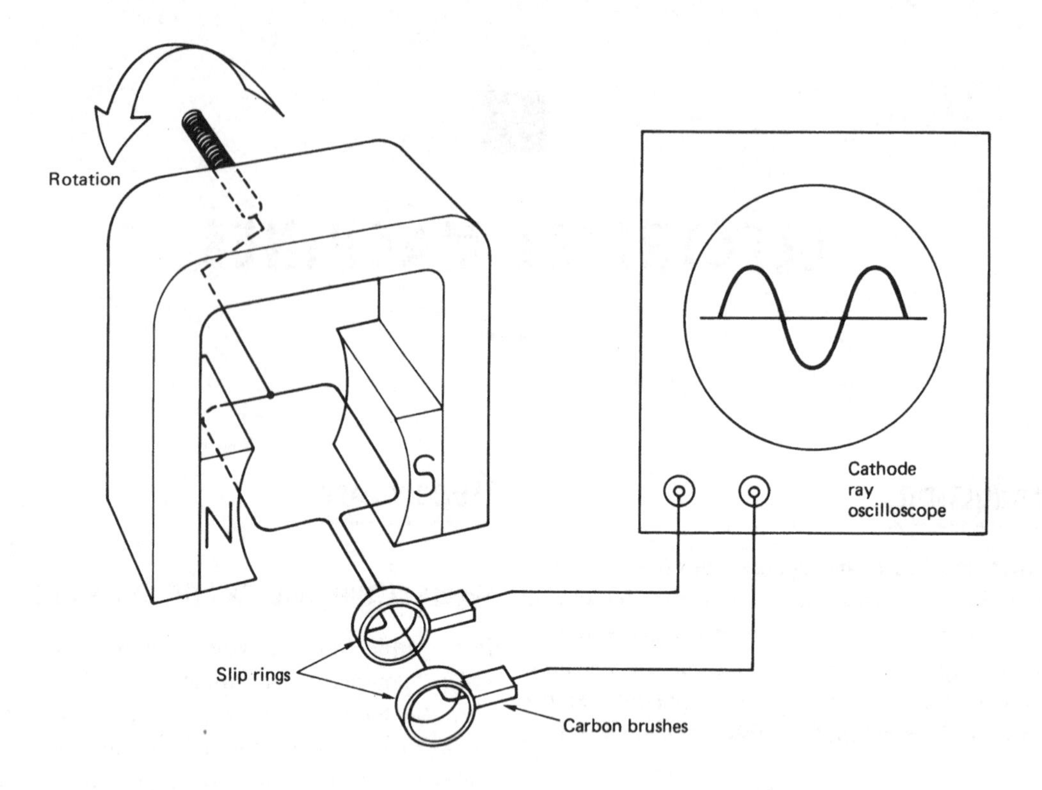

Fig. 9.2 Simple a.c. generator.

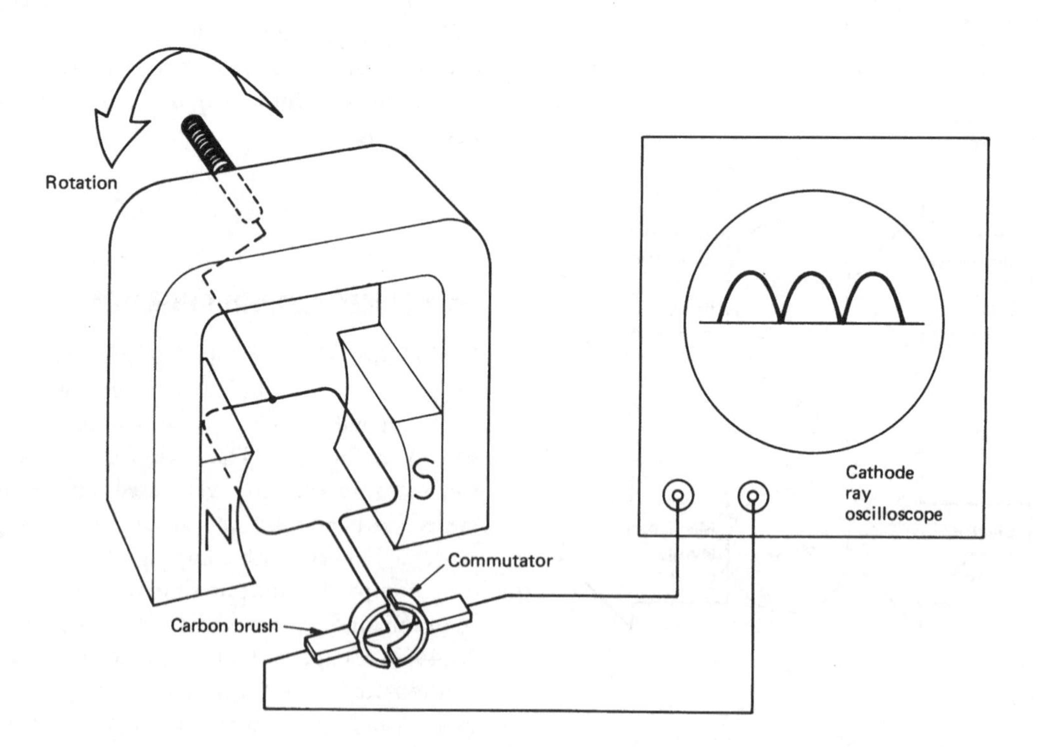

Fig. 9.3 Simple d.c. generator.

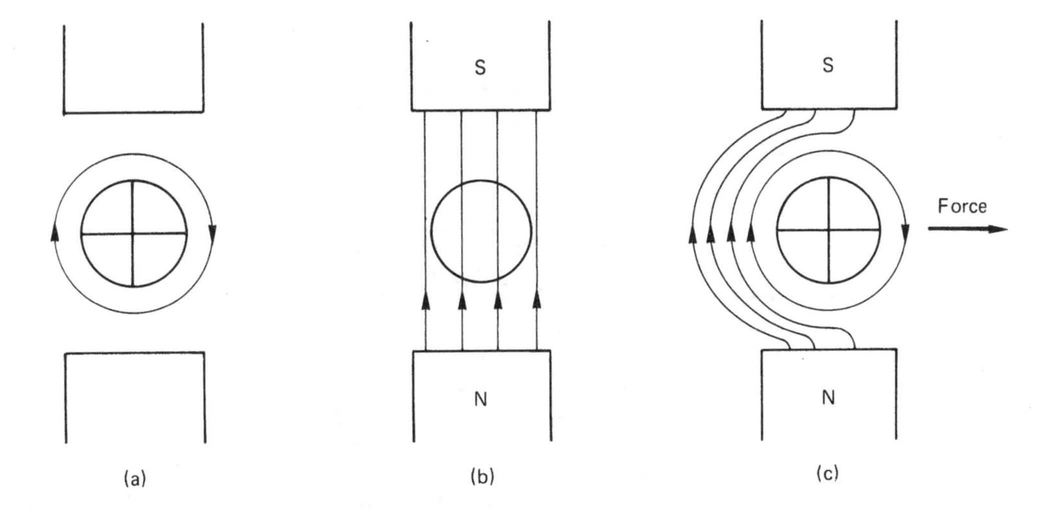

Fig. 9.4 Force on a conductor in a magnetic field.

Motors

THE D.C. MOTOR

If a current carrying conductor is placed into the field of a permanent magnet, as shown in Fig. 9.4(c), a force will be exerted on the conductor to push it out of the magnetic field. To understand the force, let us consider each magnetic field acting alone. Figure 9.4(a) shows the magnetic field due to the current carrying conductor only, shown as a cross-section. Figure 9.4(b) shows the magnetic field due to the permanent magnet, in which is placed the conductor carrying no current. Figure 9.4(c) shows the effect of the two magnetic fields and the force exerted on the conductor.

The magnetic field of the permanent magnet is distorted by the magnetic field from the current carrying conductor. Since lines of magnetic flux behave like stretched elastic bands, always trying to find the shorter distance between the north and south poles, a force is exerted on the conductor, pushing it out of the permanent magnetic field.

This is the basic principle which produces the rotation in a d.c. machine and a moving coil instrument. Current fed into the single coil winding of Fig. 9.3, through the commutator, will set up magnetic fluxes as shown in Fig. 9.5. The resultant forces will cause the coil to rotate – in the case of Fig. 9.5 in an anticlockwise direction. Reversing the current in the coil, or the polarity of the permanent field, would cause the coil to rotate in the opposite direction.

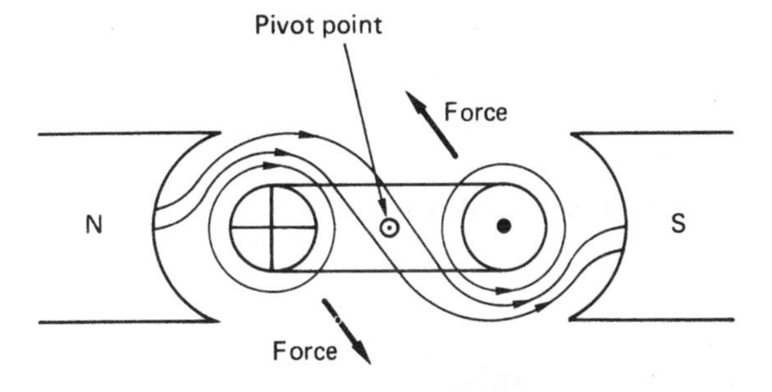

Fig. 9.5 Forces exerted on a current carrying coil in a magnetic field.

Direct current machines work on this basic principle but the permanent field magnets of Fig. 9.5 are usually replaced by electromagnets. This arrangement gives greater control of the magnetic field strength and the motor performance.

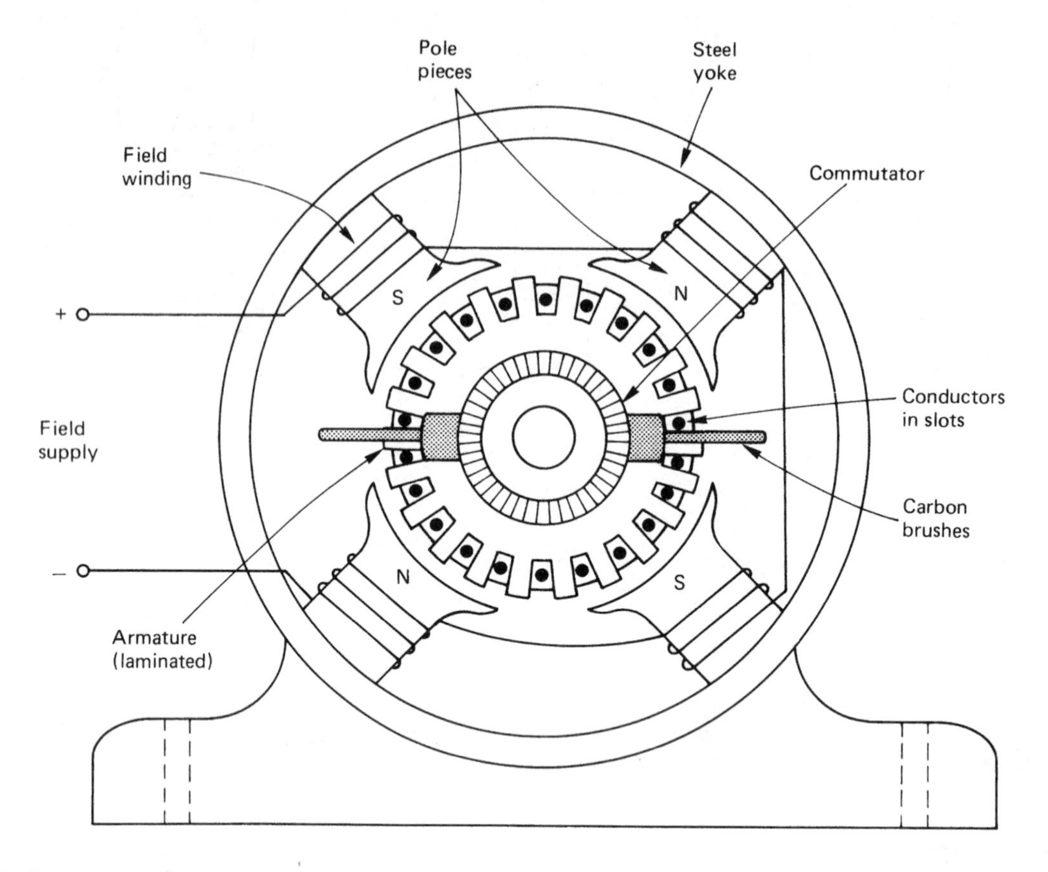

Fig. 9.6 Direct current machine construction.

PRACTICAL D.C. MOTORS

Practical motors are constructed as shown in Fig. 9.6. All d.c. motors contain a field winding wound on pole pieces attached to a steel yoke. The armature winding rotates between the poles and is connected to the commutator. Contact with the external circuit is made through carbon brushes rubbing on the commutator segments. Direct current motors are classified by the way in which the field and armature windings are connected, which may be in series or in parallel.

Series motor

The field and armature windings are connected in series and consequently share the same current. The series motor has the characteristics of a high starting torque but a speed which varies with load. Theoretically the motor would speed up to self-

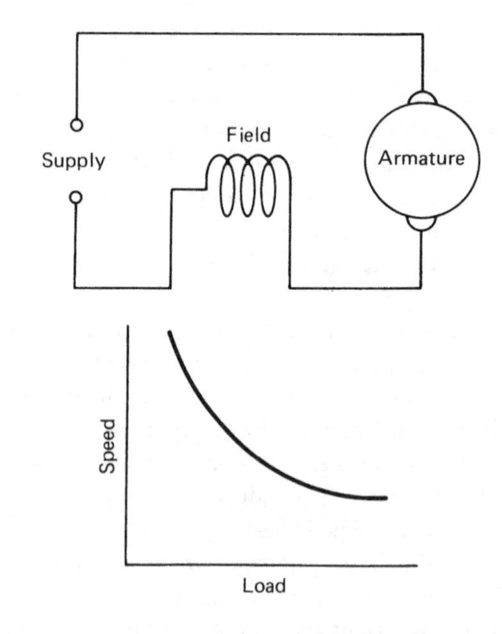

Fig. 9.7 Series motor connections and characteristics.

destruction, limited only by the windage of the rotating armature and friction, if the load were completely removed. Figure 9.7 shows series motor connections and characteristics. For this reason the motor is only suitable for direct coupling to a load, except in very small motors, such as vacuum cleaners and hand drills, and is ideally suited for applications where the machine must start on load, such as electric trains, cranes and hoists.

Reversal of rotation may be achieved by reversing the connections of either the field or armature windings but not both. This characteristic means that the machine will run on both a.c. or d.c. and is, therefore, sometimes referred to as a 'universal' motor.

Shunt motor

The field and armature windings are connected in parallel (see Fig. 9.8). Since the field winding is across the supply, the flux and motor speed are considered constant under normal conditions. In practice, however, as the load increases the field flux distorts and there is a small drop in speed of about 5% at full load, as shown in Fig. 9.8. The machine has a low starting torque and it is advisable to start with the load disconnected. The shunt motor is a very desirable d.c. motor because of its constant speed characteristics. It is used for driving power tools, such as lathes and drills. Reversal of rotation may be achieved by reversing the connections to either the field or armature winding but not both.

Compound motor

The compound motor has two field windings – one in series with the armature and the other in parallel. If the field windings are connected so that the field flux acts in opposition, the machine is known as a *short shunt* and has the characteristics of a series motor. If the fields are connected so that the field flux is strengthened, the machine is known as a *long shunt* and has constant speed characteristics similar to a shunt motor. The arrangement of compound motor connections is given in Fig. 9.9. The compound motor may be designed to possess the best characteristics of both series and shunt motors, that is, good

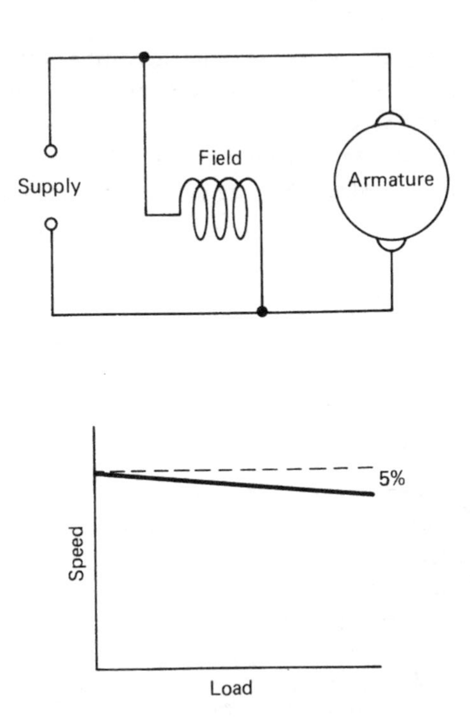

Fig. 9.8 Shunt motor connections and characteristics.

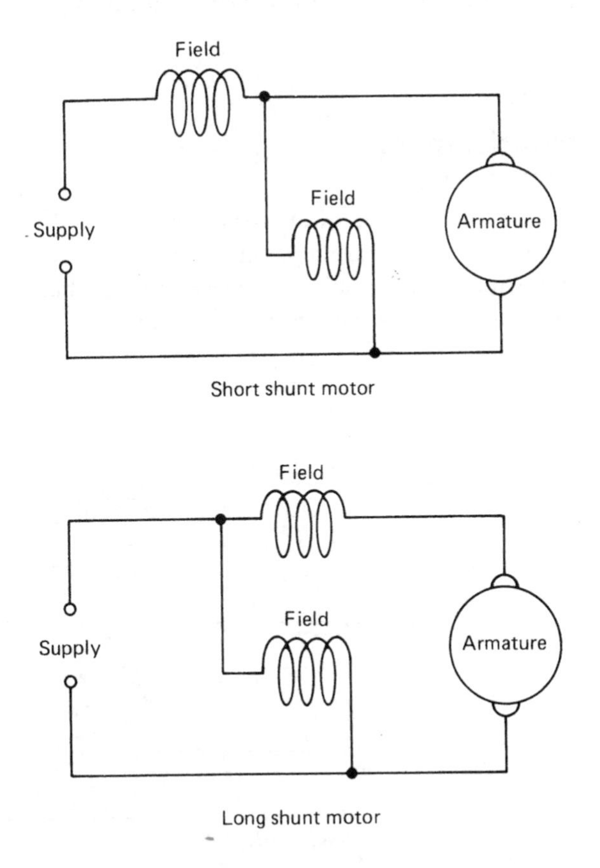

Fig. 9.9 Compound motor connections.

starting torque together with almost constant speed. Typical applications are for electric motors in steel rolling mills, where a constant speed is required under varying load conditions.

SPEED CONTROL OF D.C. MACHINES

One of the advantages of a d.c. machine is the ease with which the speed may be controlled. The speed of a d.c. motor is inversely proportional to the strength of the magnetic flux in the field winding. The magnetic flux in the field winding can be controlled by the field current and, as a result, controlling the field current will control the motor speed.

A variable resistor connected into the field circuit, as shown in Fig. 9.10, provides one method of controlling the field current and the motor speed. This method has the disadvantage that much of the input energy is dissipated in the variable resistor, and an alternative, when an a.c. supply is available, is to use thyristor control, which is described in Chapter 10. The control effect of the thyristor is shown in Fig. 10.35.

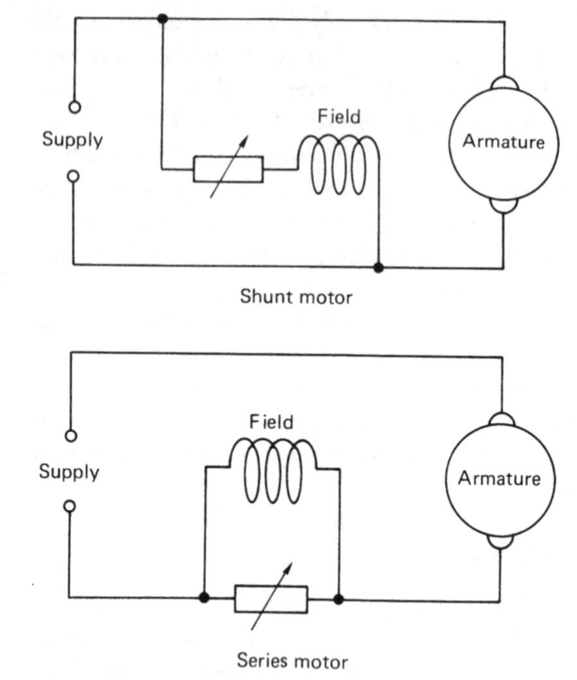

Fig. 9.10 Speed control of a d.c. motor.

Transformers

A transformer is an electrical machine which is used to change the value of an alternating voltage. They vary in size from miniature units used in electronics to huge power transformers used in power stations. A transformer will only work when an alternating voltage is connected. It will not normally work from a d.c. supply such as a battery.

A transformer, as shown in Fig. 9.11, consists of two coils, called the primary and secondary coils, or

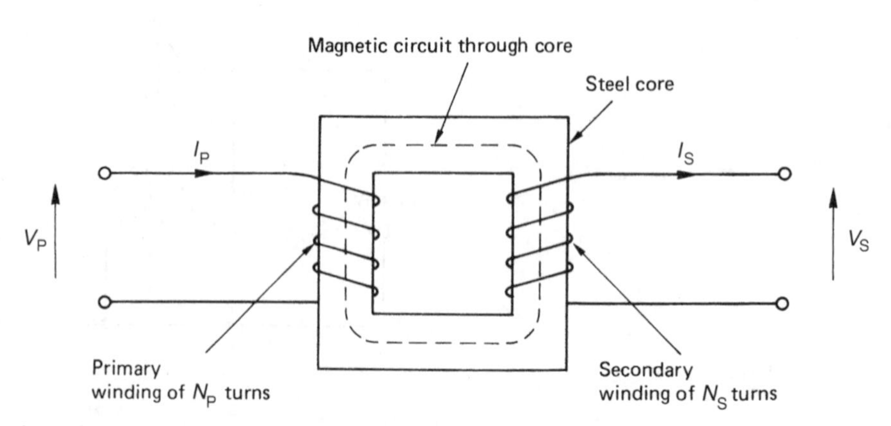

Fig. 9.11 A simple transformer.

windings, which are insulated from each other and wound on to the same steel or iron core.

An alternating voltage applied to the primary winding produces an alternating current, which sets up an alternating magnetic flux throughout the core. This magnetic flux induces an emf in the secondary winding, as described by Faraday's law, which says that when a conductor is cut by a magnetic field, an emf is induced in that conductor. Since both windings are linked by the same magnetic flux, the induced emf per turn will be the same for both windings. Therefore, the emf in both windings is proportional to the number of turns. In symbols,

$$\frac{V_P}{N_P} = \frac{V_S}{N_S} \qquad (1)$$

Most practical power transformers have a very high efficiency, and for an ideal transformer having 100% efficiency the primary power is equal to the secondary power:

Primary power = Secondary power

and, since

Power = Voltage × Current

then

$$V_P \times I_P = V_S \times I_S \qquad (2)$$

Combining equations (1) and (2), we have

$$\frac{V_P}{V_S} = \frac{N_P}{N_S} = \frac{I_S}{I_P}$$

EXAMPLE

A 230 V to 12 V bell transformer is constructed with 800 turns on the primary winding. Calculate the number of secondary turns and the primary and secondary currents when the transformer supplies a 12 V 12 W alarm bell.

Collecting the information given in the question into a usable form, we have

$$V_P = 230\ \text{V}$$
$$V_S = 12\ \text{V}$$
$$N_P = 800$$
$$\text{Power} = 12\ \text{W}$$

Information required: N_S, I_S and I_P

Secondary turns

$$N_S = \frac{N_P\, V_S}{V_P}$$

$$\therefore\ N_S = \frac{800 \times 12\ \text{V}}{230\ \text{V}} = 42\ \text{turns}$$

Secondary current

$$I_S = \frac{\text{Power}}{V_S}$$

$$\therefore\ I_S = \frac{12\ \text{W}}{12\ \text{V}} = 1\ \text{A}$$

Primary current

$$I_P = \frac{I_S \times V_S}{V_P}$$

$$\therefore\ I_P = \frac{1\ \text{A} \times 12\ \text{V}}{230\ \text{V}} = 0.052\ \text{A}$$

TRANSFORMER LOSSES

As they have no moving parts causing frictional losses, most transformers have a very high efficiency, usually better than 90%. However, the losses which do occur in a transformer can be grouped under two general headings: copper losses and iron losses.

Copper losses occur because of the small internal resistance of the windings. They are proportional to the load, increasing as the load increases because copper loss is an 'I^2R' loss.

Iron losses are made up of *hysteresis loss* and *eddy current loss*. The hysteresis loss depends upon the type of iron used to construct the core and consequently core materials are carefully chosen. Transformers will only operate on an alternating supply. Thus, the current which establishes the core

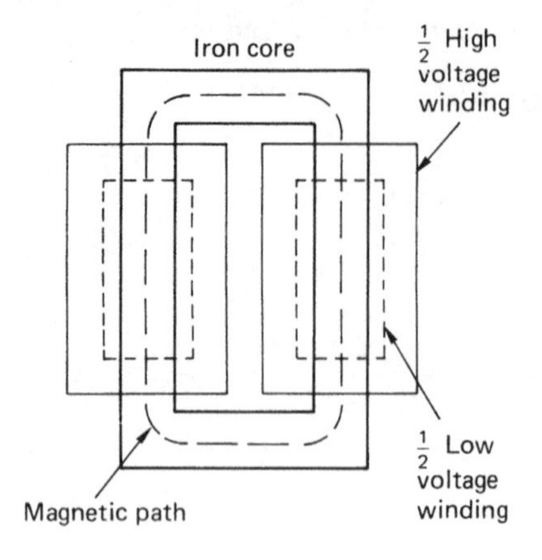

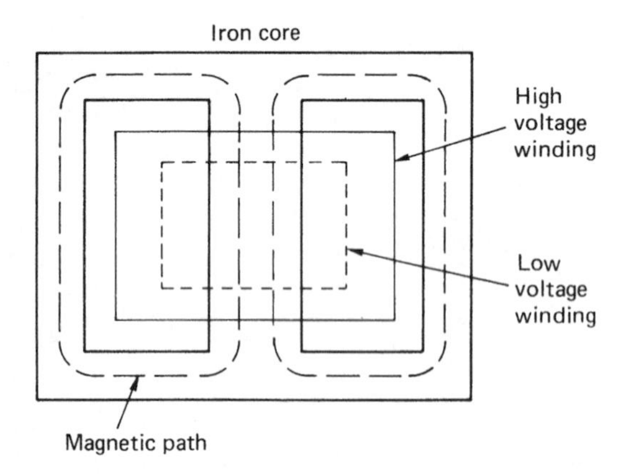

Fig. 9.12 Transformer construction.

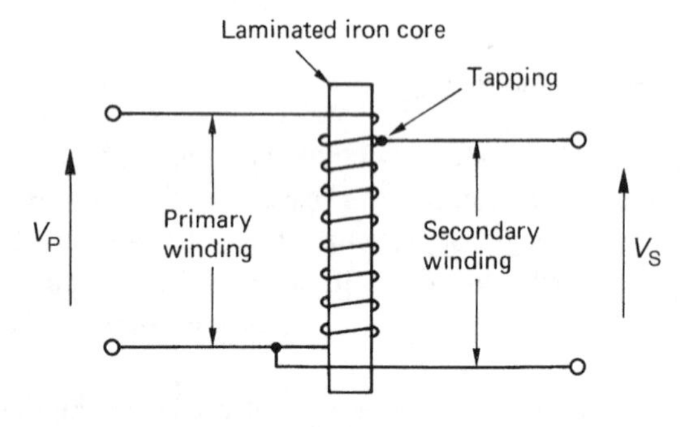

Fig. 9.13 An auto-transformer.

flux is constantly changing from positive to negative. Each time there is a current reversal, the magnetic flux reverses and it is this build-up and collapse of magnetic flux in the core material which accounts for the hysteresis loss.

Eddy currents are circulating currents created in the core material by the changing magnetic flux. These are reduced by building up the core of thin slices or laminations of iron and insulating the separate laminations from each other. The iron loss is a constant loss consuming the same power from no load to full load.

TRANSFORMER CONSTRUCTION

Transformers are constructed in a way which reduces the losses to a minimum. The core is usually made of silicon-iron laminations, because at fixed low frequencies silicon-iron has a small hysteresis loss and the laminations reduce the eddy current loss. The primary and secondary windings are wound close to each other on the same limb. If the windings are spread over two limbs there will usually be half of each winding on each limb, as shown in Fig. 9.12.

AUTO-TRANSFORMERS

Transformers having a separate primary and secondary winding, as shown in Fig. 9.12, are called double-wound transformers, but it is possible to construct a transformer which has only one winding which is common to the primary and secondary circuits. The secondary voltage is supplied by means of a 'tapping' on the primary winding. An arrangement such as this is called an auto-transformer.

The auto-transformer is much cheaper and lighter than a double-wound transformer because less copper and iron are used in its construction. However, the primary and secondary windings are not electrically separate and a short circuit on the upper part of the winding shown in Fig. 9.13 would result in the primary voltage appearing across the secondary terminals. For this reason auto-transformers are mostly used where only a small difference is required between the primary and secondary voltages. When installing transformers, the regulations of Section 555 must be complied with, in addition to any other regulations relevant to the particular installation.

Exercises

1 Any motor converts:
(a) electrical energy to power
(b) electrical energy to mechanical energy
(c) mechanical energy to power
(d) mechanical energy to electrical energy.

2 Any generator converts:
(a) electrical energy to power
(b) electrical energy to mechanical energy
(c) mechanical energy to power
(d) mechanical energy to electrical energy.

3 An electric motor operates on the principle of:
(a) Ohm's law
(b) the three effects of an electric current
(c) the forces acting upon a conductor in a magnetic field
(d) a cage rotor induction motor.

4 A centre zero voltmeter is connected to the commutator of a simple single-loop generator. If the loop remains stationary the meter will:
(a) read zero
(b) give a positive value
(c) give a negative value
(d) oscillate about the zero point.

5 A centre zero voltmeter is connected to the slip rings of a simple single-loop generator. If the loop is rotated through one complete revolution the meter will:
(a) continue to read zero
(b) move from zero to positive and back to zero twice
(c) move from zero to positive, then to negative and finally return to zero
(d) move from zero to positive and maintain the positive value.

6 An oscilloscope is connected to the slip rings of a simple generator. If the coil is continuously rotated the oscilloscope will show:
(a) a zero voltage
(b) a positive d.c. voltage
(c) a unidirectional waveform
(d) a sinusoidal waveform.

7 A series d.c. motor has the characteristic of:
(a) constant speed about 5% below synchronous speed
(b) start winding 90° out of phase with the run winding
(c) low starting torque but almost constant speed
(d) high starting torque and a speed which varies with load.

8 A shunt motor has the characteristic of:
(a) constant speed about 5% below synchronous speed
(b) start winding 90° out of phase with the run winding
(c) low starting torque but almost constant speed
(d) high starting torque and a speed which varies with load.

9 One advantage of all d.c. machines is:
(a) that they are almost indestructable
(b) that starters are never required
(c) that they may be operated on a.c. or d.c. supplies
(d) the ease with which speed may be controlled.

10 One advantage of a series d.c. motor is that:
(a) it is almost indestructable
(b) starters are never required
(c) it may be operated on a.c. or d.c. supplies
(d) speed is constant at all loads.

11 A d.c. shunt motor would normally be used for a:
(a) domestic oven fan motor
(b) portable electric drill motor
(c) constant speed lathe motor
(d) record turntable drive motor.

12 A d.c. series motor would normally be used for a:
(a) domestic oven fan motor
(b) portable electric drill motor
(c) constant speed lathe motor
(d) record turntable drive motor.

13 Describe with the aid of a circuit diagram how speed control may be achieved with:
(a) a d.c. series motor
(b) a d.c. shunt motor.

14 The core of a transformer is laminated to:
(a) reduce cost
(b) reduce copper losses
(c) reduce hysteresis loss
(d) reduce eddy current loss.

15 The transformation ratio of a step-down

transformer is 20 : 1. If the primary voltage is 230 V the secondary voltage will be:
(a) 2.3 V
(b) 11.5 V
(c) 20 V
(d) 23 V.

16 With the aid of sketches, describe the construction of:
(a) a double-wound transformer
(b) an auto-transformer.
State the losses which occur in a transformer.

10

ELECTRONIC COMPONENTS

There are numerous types of electronic component – diodes, transistors, thyristors and integrated circuits – each with its own limitations, characteristics and designed application. When repairing electronic circuits it is important to replace a damaged component with an identical or equivalent component. Manufacturers issue comprehensive catalogues with details of working voltage, current, power dissipation etc., and the reference numbers of equivalent components, and some of this information is included in the Appendices. These catalogues of information, together with a high-impedance multimeter as described in Chapter 12, should form a part of the extended tool-kit for an installation electrician proposing to repair electronic circuits.

Electronic circuit symbols

The British Standard BS EN 60617 recommends that particular graphical symbols be used to represent a range of electronic components on circuit diagrams. The same British Standard recommends a range of symbols suitable for electrical installation circuits with which electricians will already be familiar. Figure 10.1 shows a selection of electronic symbols.

Resistors

All materials have some resistance to the flow of an electric current but, in general, the term *resistor* describes a conductor specially chosen for its resistive properties.

Resistors are the most commonly used electronic component and they are made in a variety of ways to suit the particular type of application. They are usually manufactured as either carbon composition or carbon film. In both cases the base resistive material is carbon and the general appearance is of a small cylinder with leads protruding from each end, as shown in Fig. 10.2(a).

If subjected to overload, carbon resistors usually decrease in resistance since carbon has a negative temperature coefficient. This causes more current to flow through the resistor, so that the temperature rises and failure occurs, usually by fracturing. Carbon resistors have a power rating of between 0.1 W and 2 W which should not be exceeded.

When a resistor of a larger power rating is required a wire-wound resistor should be chosen. This consists of a resistance wire of known value wound on a small ceramic cylinder which is encapsulated in a vitreous enamel coating, as shown in Fig. 10.2(b). Wire-wound resistors are designed to run hot and have a power rating up to 20 W. Care should be taken when mounting wire-wound resistors to prevent the high operating temperature affecting any surrounding components.

A variable resistor is one which can be varied continuously from a very low value to the full rated resistance. This characteristic is required in tuning circuits to adjust the signal or voltage level for brightness, volume or tone. The most common type used in electronic work has a circular carbon track contacted by a metal wiper arm. The wiper arm can be adjusted by means of an adjusting shaft (rotary type) or by placing a screwdriver in a slot (preset type), as shown in Fig. 10.3. Variable resistors are also known as potentiometers because they can be used to adjust the poten-

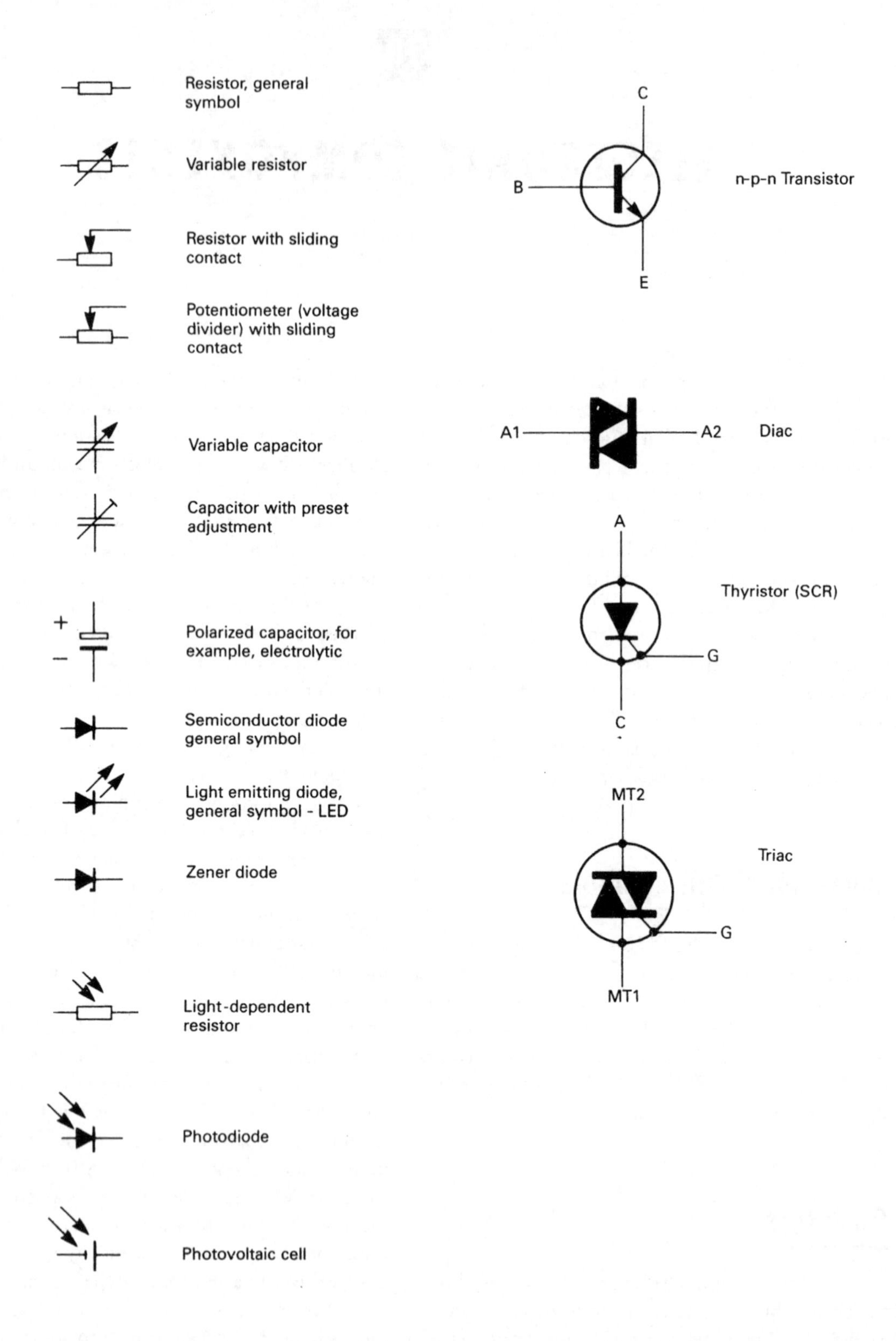

Fig. 10.1 Some BS EN 60617 graphical symbols used in electronics.

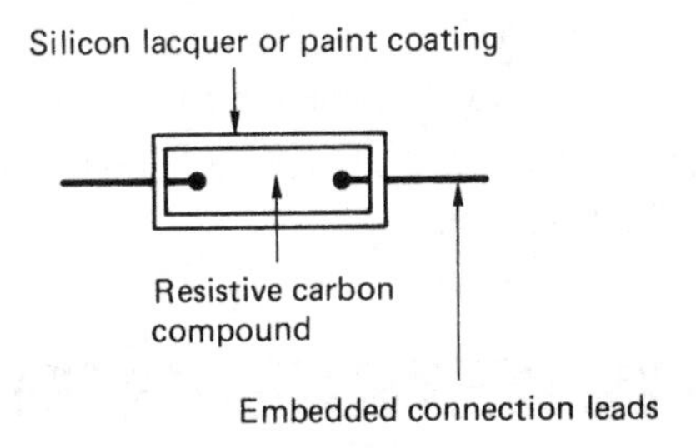

(a) Carbon composition resistor

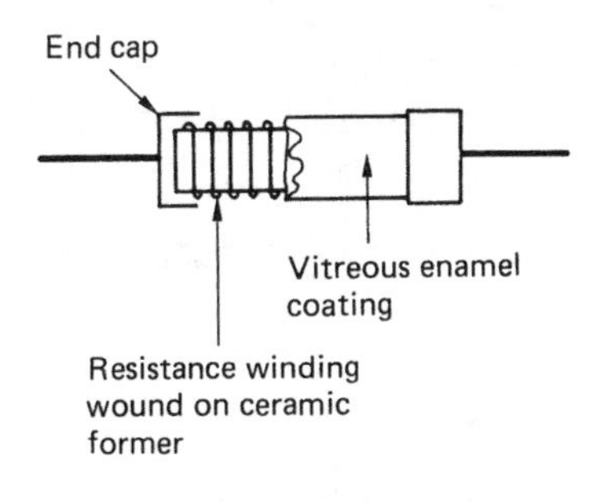

(b) Wire-wound resistor

Fig. 10.2 Construction of resistors.

tial difference (voltage) in a circuit. The variation in resistance can be to either a logarithmic or a linear scale.

The value of the resistor and the tolerance may be marked on the body of the component either by direct numerical indication or by using a standard colour code. The method used will depend upon the type, physical size and manufacturer's preference, but in general the larger components have values marked directly on the body and the smaller components use the standard resistor colour code.

ABBREVIATIONS USED IN ELECTRONICS

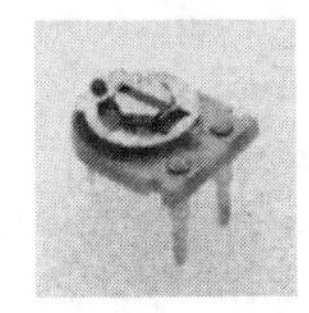

Fig. 10.3 Types of variable resistor.

Where the numerical value of a component includes a decimal point, it is standard practice to include the prefix for the multiplication factor in place of the decimal point, to avoid accidental marks being mistaken for decimal points. Multiplication factors and prefixes are dealt with in Chapter 8.

The abbreviation R means × 1
k means × 1000
M means × 1 000 000

Therefore, a 4.7 kΩ resistor would be abbreviated to 4k7, a 5.6 Ω resistor to 5R6 and a 6.8 MΩ resistor to 6M8.

Tolerances may be indicated by adding a letter at the end of the printed code.

The abbreviation F means ± 1%, G means ± 2%, J means ± 5%, K means ± 10% and M means ± 20%. Therefore a 4.7 kΩ resistor with a tolerance of 2% would be abbreviated to 4k7G. A 5.6 Ω resistor with a tolerance of 5% would be abbreviated to 5R6J. A 6.8 MΩ resistor with a 10% tolerance would be abbreviated to 6M8K.

This is the British Standard BS 1852 code which is recommended for indicating the values of resistors on circuit diagrams and components when their physical size permits.

THE STANDARD COLOUR CODE

Small resistors are marked with a series of coloured bands, as shown in Table 10.1. These are read according to the standard colour code to determine the resistance. The bands are located on the component towards one end. If the resistor is turned so that this end is towards the left, the bands are then read from left to right. Band (a) gives the first number of the component value, band (b) the second number, band (c) the number of zeros to be added after the first two numbers and band (d) the resistor tolerance. If the bands are not clearly oriented towards one end, first identify the tolerance band and turn the resistor so that this is towards the right before commencing to read the colour code as described.

The tolerance band indicates the maximum tolerance variation in the declared value of resistance. Thus a 100 Ω resistor with a 5% tolerance will have a value somewhere between 95 Ω and 105 Ω, since 5% of 100 Ω is 5 Ω.

Table 10.1 The resistor colour code

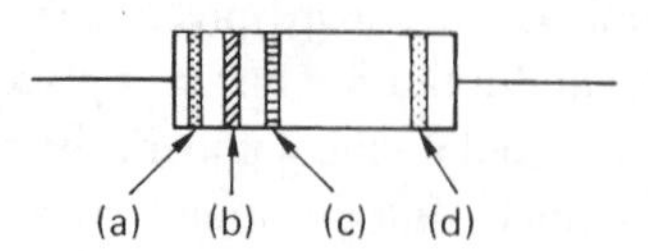

Colour	Band (a) first number	Band (b) second number	Band (c) number of zeros	Band (d) tolerance band
Black	0	0	None	–
Brown	1	1	1	1%
Red	2	2	2	2%
Orange	3	3	3	–
Yellow	4	4	4	–
Green	5	5	5	–
Blue	6	6	6	–
Violet	7	7	7	–
Grey	8	8	–	–
White	9	9	–	–
Gold	–	–	÷ 10	5%
Silver	–	–	÷ 100	10%
None	–	–	–	20%

EXAMPLE 1

A resistor is colour coded yellow, violet, red, gold. Determine the value of the resistor.

Band (a) – yellow has a value of 4.
Band (b) – violet has a value of 7.
Band (c) – red has a value of 2.
Band (d) – gold indicates a tolerance of 5%.

The value is therefore 4700 ± 5%.
This could be written as 4.7 kΩ ± 5% or 4k7J.

EXAMPLE 2

A resistor is colour coded green, blue, brown, silver. Determine the value of the resistor.

Band (a) – green has a value of 5.
Band (b) – blue has a value of 6.
Band (c) – brown has a value of 1.
Band (d) – silver indicates a tolerance of 10%.

The value is therefore 560 ± 10% and could be written as 560 Ω ± 10% or 560RK.

EXAMPLE 3

A resistor is colour coded blue, grey, green, gold. Determine the value of the resistor.

Band (a) – blue has a value of 6.
Band (b) – grey has a value of 8.
Band (c) – green has a value of 5.
Band (d) – gold indicates a tolerance of 5%.

The value is therefore 6 800 000 ± 5% and could be written as 6.8 MΩ ± 5% or 6M8J.

EXAMPLE 4

A resistor is colour coded orange, white, silver, silver. Determine the value of the resistor.

Band (a) – orange has a value of 3.
Band (b) – white has a value of 9.
Band (c) – silver indicates divide by 100 in this band.
Band (d) – silver indicates a tolerance of 10%.

The value is therefore 0.39 ± 10% and could be written as 0.39 Ω ± 10% or R39K.

PREFERRED VALUES

It is difficult to manufacture small electronic resistors to exact values by mass production methods. This is not a disadvantage as in most electronic circuits the value of the resistors is not critical. Manufacturers produce a limited range of *preferred* resistance values rather than an overwhelming number of individual resistance values. Therefore, in electronics, we use the preferred value closest to the actual value required.

A resistor with a preferred value of 100 Ω and a 10% tolerance could have any value between 90 Ω and 110 Ω. The next larger preferred value which would give the maximum possible range of resistance values without too much overlap would be 120 Ω. This could have any value between 108 Ω and 132 Ω. Therefore, these two preferred value resistors cover all possible resistance values between 90 Ω and 132 Ω. The next preferred value would be 150 Ω, then 180 Ω, 220 Ω and so on.

There is a series of preferred values for each tolerance level, as shown in Table 10.2, so that every possible numerical value is covered. Table 10.2 indicates values between 10 and 100 but larger values can be obtained by multiplying these preferred values by some multiplication factor. Resistance values of 47 Ω, 470 Ω, 4.7 kΩ, 470 kΩ, 4.7 MΩ, etc., are available in this way.

TESTING RESISTORS

The resistor being tested should have a value close to the preferred value and within the tolerance stated by

Table 10.2 Preferred values

E6 series 20% tolerance	E12 series 10% tolerance	E24 series 5% tolerance
10	10	10
		11
	12	12
		13
15	15	15
		16
	18	18
		20
22	22	22
		24
	27	27
		30
33	33	33
		36
	39	39
		43
47	47	47
		51
	56	56
		62
68	68	68
		75
	82	82
		91

the manufacturer. To measure the resistance of a resistor which is not connected into a circuit, the leads of a suitable ohmmeter should be connected to each resistor connection lead and a reading obtained. The ohmmeter and its use are discussed in Chapter 12.

If the resistor to be tested is connected into an electronic circuit it is *always necessary* to disconnect one lead from the circuit before the test leads are connected, otherwise the components in the circuit will provide parallel paths, and an incorrect reading will result.

Capacitors

The fundamental principles of capacitors are discussed in Chapter 8 under the subheading '*Electrostatics*'. In this chapter we shall consider the practical aspects associated with capacitors in electronic circuits.

A capacitor stores a small amount of electric charge; it can be thought of as a small rechargeable battery which can be quickly recharged. In electronics we are not only concerned with the amount of charge stored by the capacitor but in the way the value of the capacitor determines the performance of timers and oscillators by varying the time constant of a simple capacitor–resistor circuit.

CAPACITORS IN ACTION

If a test circuit is assembled as shown in Fig. 10.4 and the changeover switch connected to d.c. the signal lamp will only illuminate for a very short pulse as the capacitor charges. The charged capacitor then blocks any further d.c. current flow. If the changeover switch is then connected to a.c. the lamp will illuminate at full brilliance because the capacitor will charge and discharge continuously at the supply frequency. Current is *apparently* flowing through the capacitor because electrons are moving to and fro in the wires joining the capacitor plates to the a.c. supply.

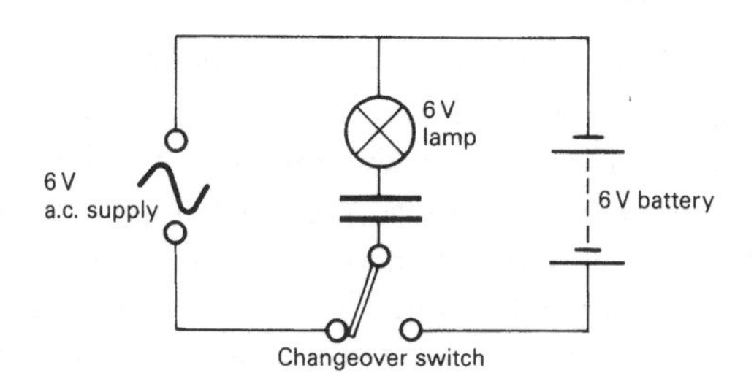

Fig. 10.4 Test circuit showing capacitors in action.

COUPLING AND DECOUPLING CAPACITORS

Capacitors can be used to separate a.c. and d.c. in an electronic circuit. If the output from circuit A, shown in Fig. 10.5(a), contains both a.c. and d.c. but only an a.c. input is required for circuit B then a *coupling* capacitor is connected between them. This blocks the d.c. while offering a low reactance to the a.c. component. Alternatively, if it is required that only d.c. be connected to circuit B, shown in Fig. 10.5(b), a *decoupling* capacitor can be connected in parallel with circuit B.

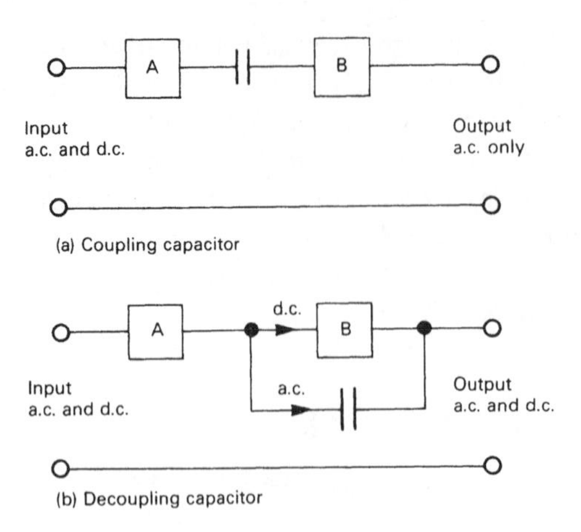

Fig. 10.5 (a) Coupling and (b) decoupling capacitors.

This will provide a low reactance path for the a.c. component of the supply and only d.c. will be presented to the input of B. This technique is used to *filter out* unwanted a.c. in, for example, d.c. power supplies.

TYPES OF CAPACITOR

There are two broad categories of capacitor, the non-polarized and polarized type. The non-polarized type can be connected either way round, but polarized capacitors *must* be connected to the polarity indicated otherwise a short circuit and consequent destruction of the capacitor will result. There are many different types of capacitor, each one being distinguished by the type of dielectric used in its construction. Figure 10.6 shows some of the capacitors used in electronics.

Polyester capacitors

Polyester capacitors are an example of the plastic film capacitor. Polypropylene, polycarbonate and polystyrene capacitors are other types of plastic film capacitor. The capacitor value may be marked on the plastic film, or the capacitor colour code given in Table 10.3 may be used. This dielectric material gives a compact capacitor with good electrical and temperature characteristics. They are used in many electronic circuits, but are not suitable for high-frequency use.

Mica capacitors

Mica capacitors have excellent stability and are accurate to ± 1% of the marked value. Since costs usually increase with increased accuracy, they tend to be more expensive than plastic film capacitors. They are used where high stability is required, for example in tuned circuits and filters.

Ceramic capacitors

Ceramic capacitors are mainly used in high-frequency circuits subjected to wide temperature variations. They have high stability and low loss.

Electrolytic capacitors

Electrolytic capacitors are used where a large value of capacitance coupled with a small physical size is required. They are constructed on the 'Swiss roll' principle as are the paper dielectric capacitors used for power-factor correction in electrical installation circuits. The electrolytic capacitors' high capacitance for very small volume is derived from the extreme thinness of the dielectric coupled with a high dielectric

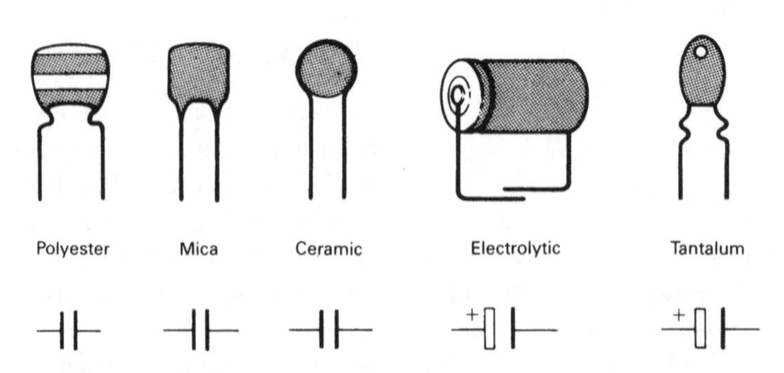

Fig. 10.6 Capacitors and their symbols used in electronic circuits.

Table 10.3 Colour code for plastic film capacitors (values in picofarads)

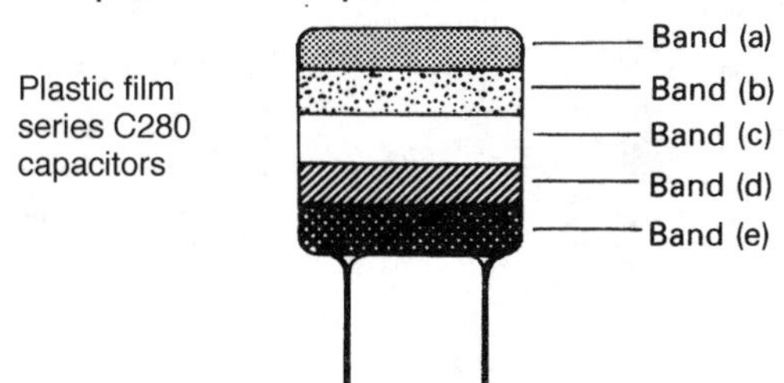

Colour	Band (a) first number	Band (b) second number	Band (c) number of zeros to be added	Band (d) tolerance	Band (e) maximum voltage
Black	–	0	None	20%	–
Brown	1	1	1	–	100 V
Red	2	2	2	–	250 V
Orange	3	3	3	–	–
Yellow	4	4	4	–	400 V
Green	5	5	5	5%	–
Blue	6	6	6	–	–
Violet	7	7	7	–	–
Grey	8	8	8	–	–
White	9	9	9	10%	–

strength. Electrolytic capacitors have a size gain of approximately 100 times over the equivalent non-electrolytic type. Their main disadvantage is that they are polarized and must be connected to the correct polarity in a circuit. Their large capacity makes them ideal as smoothing capacitors in power supplies.

Tantalum capacitors

Tantalum capacitors are a new type of electrolytic capacitor using tantalum and tantalum oxide to give a further capacitance/size advantage. They look like a 'raindrop' or 'blob' with two leads protruding from the bottom. The polarity and values may be marked on the capacitor, or the colour code shown in Table 10.4 may be used. The voltage ratings available tend to be low, as with all electrolytic capacitors. They are also extremely vulnerable to reverse voltages in excess of 0.3 V. This means that even when testing with an ohmmeter, extreme care must be taken to ensure correct polarity.

Variable capacitors

Variable capacitors are constructed so that one set of metal plates moves relative to another set of fixed metal plates as shown in Fig. 10.7. The plates are separated by air or sheet mica, which acts as a dielectric. Air dielectric variable capacitors are used to tune radio receivers to a chosen station, and small variable capacitors called *trimmers* or *presets* are used to make fine, infrequent adjustments to the capacitance of a circuit.

SELECTING A CAPACITOR

When choosing a capacitor for a particular application, three factors must be considered: value, working voltage and leakage current.

The unit of capacitance is the *farad* (symbol F), to commemorate the name of the English scientist Michael Faraday. However, for practical purposes the farad is much too large and in electrical installation work and electronics we use fractions of a farad as follows:

$$1 \text{ microfarad} = 1\ \mu\text{F} = 1 \times 10^{-6}\ \text{F}$$
$$1 \text{ nanofarad} = 1\ \text{nF} = 1 \times 10^{-9}\ \text{F}$$
$$1 \text{ picofarad} = 1\ \text{pF} = 1 \times 10^{-12}\ \text{F}$$

The power-factor correction capacitor used in a domestic fluorescent luminaire would typically have a value of 8 μF at a working voltage of 400 V. In an electronic filter circuit a typical capacitor value might be 100 pF at 63 V.

Table 10.4 Colour code for tantalum polarized capacitors (values in microfarads)

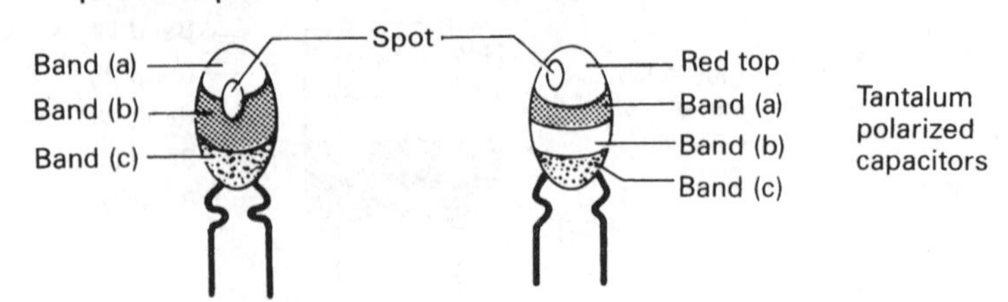

Colour	Band (a) first number	Band (b) second number	Spot number of zeros to be added	Band (c) maximum voltage
Black	–	0	None	10 V
Brown	1	1	1	–
Red	2	2	2	–
Orange	3	3	–	–
Yellow	4	4	–	6.3 V
Green	5	5	–	16 V
Blue	6	6	–	20 V
Violet	7	7	–	–
Grey	8	8	÷ 100	25 V
White	9	9	÷ 1000	30 V
Pink				35 V

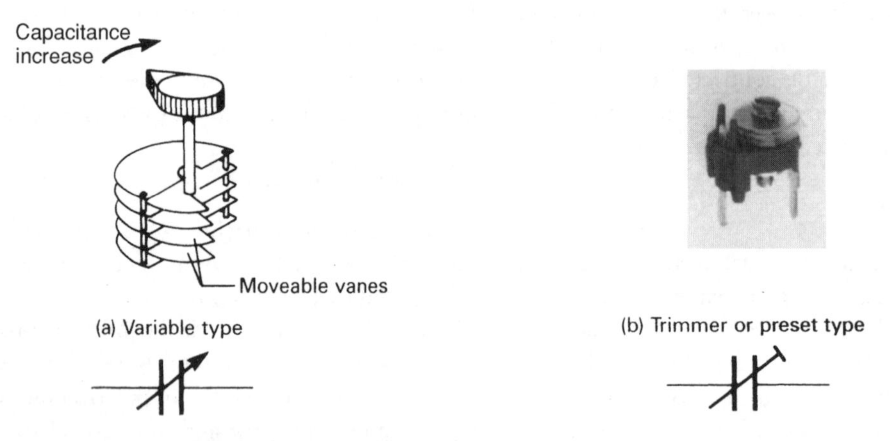

Fig. 10.7 Variable capacitors and their symbols: (a) variable type; (b) trimmer or preset type.

One microfarad is one million times greater than one picofarad. It may be useful to remember that

$$1000 \text{ pF} = 1 \text{ nF}$$
$$1000 \text{ nF} = 1 \text{ µF}$$

The working voltage of a capacitor is the *maximum* voltage that can be applied between the plates of the capacitor without breaking down the dielectric insulating material. This is a d.c. rating and, therefore, a capacitor with a 200 V rating must only be connected across a maximum of 200 V d.c. Since a.c. voltages are usually given as rms values, a 200 V a.c. supply would have a maximum value of about 283 V, which would damage the 200 V capacitor. When connecting a capacitor to the 230 V mains supply we must choose a working voltage of about 400 V because 230 V rms.

is approximately 325 V maximum. The 'factor of safety' is small and, therefore, the working voltage of the capacitor must not be exceeded.

An ideal capacitor which is isolated will remain charged forever, but in practice no dielectric insulating material is perfect, and the charge will slowly *leak* between the plates, gradually discharging the capacitor. The loss of charge by leakage through it should be very small for a practical capacitor.

Capacitor colour code

The actual value of a capacitor can be identified by using the colour codes given in Tables 10.3 and 10.4 in the same way that the resistor colour code was applied to resistors.

EXAMPLE 1

A plastic film capacitor is colour coded, from top to bottom, brown, black, yellow, black, red. Determine the value of the capacitor, its tolerance and working voltage.

From Table 10.3

Band (a) – brown has a value 1.
Band (b) – black has a value 0.
Band (c) – yellow indicates multiply by 10 000.
Band (d) – black indicates 20%.
Band (e) – red indicates 250 V.

The capacitor has a value of 100 000 pF or 0.1 μF with a tolerance of 20% and a maximum working voltage of 250 V.

EXAMPLE 2

Determine the value, tolerance and working voltage of a polyester capacitor colour-coded, from top to bottom, yellow, violet, yellow, white, yellow.

From Table 10.3

Band (a) – yellow has a value 4.
Band (b) – violet has a value 7.
Band (c) – yellow indicates multiply by 10 000.
Band (d) – white indicates 10%.
Band (e) – red indicates 400 V.

The capacitor has a value of 470 000 pF or 0.47 μF with a tolerance of 10% and a maximum working voltage of 400 V.

EXAMPLE 3

A plastic film capacitor has the following coloured bands from its top down to the connecting leads: blue, grey, orange, black, brown. Determine the value, tolerance and voltage of this capacitor.

From Table 10.3 we obtain the following:

Band (a) – blue has a value 6.
Band (b) – grey has a value 8.
Band (c) – orange indicates multiply by 1000.
Band (d) – black indicates 20%.
Band (e) – brown indicates 100 V.

The capacitor has a value of 68 000 pF or 68 nF with a tolerance of 20% and a maximum working voltage of 100 V.

CAPACITANCE VALUE CODES

Where the numerical value of the capacitor includes a decimal point, it is standard practice to use the prefix for the multiplication factor in place of the decimal point. This is the same practice as we used earlier for resistors.

The abbreviation μ means microfarad, n means nanofarad and p means picofarad. Therefore, a 1.8 pF capacitor would be abbreviated to 1p8, a 10 pF capacitor to 10p, a 150 pF capacitor to 150p or n15, a 2200 pF capacitor to 2n2 and a 10 000 pF capacitor to 10n.

1000 pF = 1nF = 0.001μF

TESTING CAPACITORS

The discussion earlier in this chapter about *ideal* and *leaky* capacitors provides us with a basic principle to test for a faulty capacitor.

Non-polarized capacitors

Using an ohmmeter as described in Chapter 12, connect the leads of the capacitor to the ohmmeter and observe the reading. If the resistance is less than about 1 MΩ, it is allowing current to pass from the ohmmeter and, therefore, the capacitor is leaking and is faulty. With large-value capacitors (in the microfarad range) there may be a short initial burst of current as the capacitor charges up.

Polarized capacitors

It is essential to connect the *true positive* of the ohmmeter to the positive lead of the capacitor, as shown in Fig. 12.4 in Chapter 12. When first connected, the resistance is low but rises to a steady value as the dielectric forms between the capacitor plates.

Inductors and transformers

An inductor is a coil of wire wound on a former (to give it a specific shape) having a core of air or iron. When a current flows through the coil a magnetic field is established. A transformer consists of two coils wound on a common magnetic core and, therefore, in this sense, is also an inductor. Simple transformer theory is discussed in Chapter 9. A small electronic transformer and the aerial of a radio receiver comprising a coil wound on a ferrite core are shown in Fig. 10.8.

Inductors such as the radio receiver aerial can be connected in parallel with a variable capacitor and *tuned* for maximum response so that a particular radio station can be listened to while excluding all others.

Most electronic circuits require a voltage between 5 and 12 V and the transformer provides an ideal way of initially reducing the mains voltage to a value which is suitable for the particular electronic circuit.

Compared with other individual electronic components, inductors are large. The magnetic fields produced by industrial electronic equipment such as electromagnets, relays and transformers can cut across other electronic components and cause undesirable emfs to be induced. This causes electrical interference – called *electrical noise* – and may prevent the normal operation of the electronic circuit. This interference can be avoided by magnetically *screening* the inductive components from the remaining circuits.

Electromagnetic relays

An electromagnetic relay is simply an electromagnet operating a number of switch contacts, as shown in Fig. 10.9. When a current is passed through the coil, the soft iron core becomes magnetized, attracts the iron armature and closes the switch contacts. The relay coil is electrically insulated from the switch contacts and, therefore, a relay is able to switch circuits operating at a different voltage than the coil operating voltage. The small current which energizes the coil is also able to switch larger currents at the switch contacts. The switch part of the relay may have many poles controlling several circuits at once.

Miniature plug-in relays are popular in electronic circuits and intruder alarm circuits. However, all mechanical-electrical switches are limited in their speed of operation by the time taken physically for a movable contact to make or break a switch contact. Where extremely high-speed operations are required, the switching action must take place without physical movement. This is only possible using the properties of semiconductor materials in devices such as transistors and thyristors. They permit extremely high-speed switching without arcing and are considered later in this chapter.

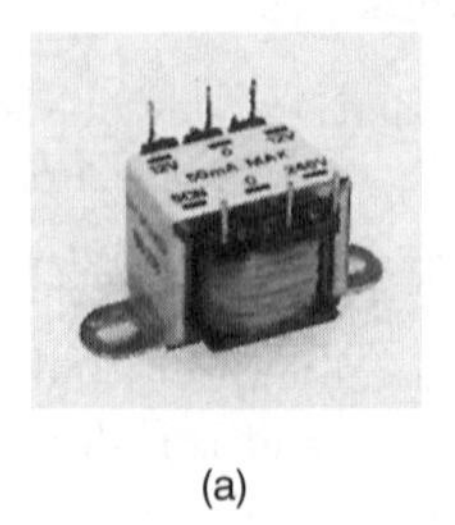

(a)

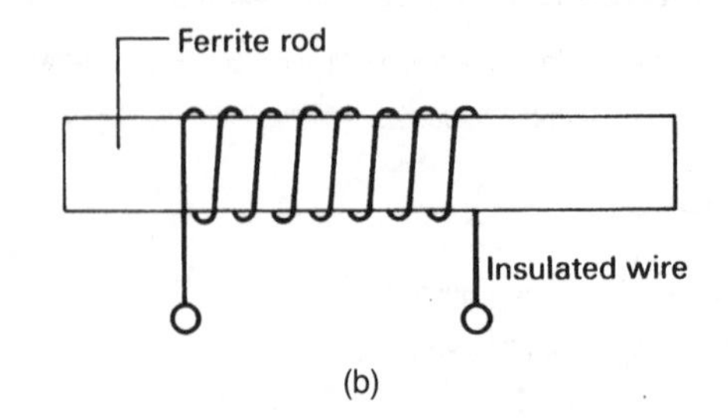

(b)

Fig. 10.8 Examples of an inductor: (a) transformer; (b) radio receiver aerial.

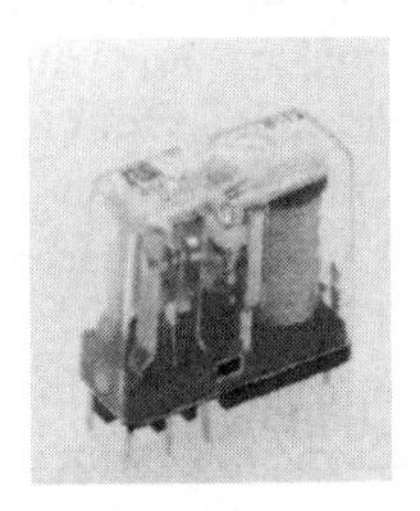

Fig. 10.9 An electromagnetic relay.

Fig. 10.10 A fuse holder for cartridge fuses up to 15 A.

Overcurrent protection

Every piece of electronic equipment must incorporate some means of overcurrent protection. The term 'overcurrent' can be subdivided into *overload* current and *short-circuit* current. An overload can be defined as a current which exceeds the rated value in an otherwise healthy circuit and a short circuit as an overcurrent resulting from a fault of negligible impedance between conductors. An overload may result in currents of two or three times the rated current flowing in the circuit, while short-circuit currents may be hundreds of times greater than the rated current. In both cases the basic requirement for safety is that the fault current should be interrupted quickly and the circuit isolated from the supply. Fuses provide overcurrent protection when connected in the live conductor; they must not be connected in the neutral conductor. Circuit breakers may be used in place of fuses and the best protection of all is obtained when the equipment is connected to a residual current device. Figure 10.10 shows a cartridge fuse holder. Protection from excess current is covered in some detail in Chapter 6.

Packaging electronic components

When we talk about packaging electronic components we are not referring to the parcel or box which contains the components for storage and delivery, but to the type of encapsulation in which the tiny semiconductor material is contained. Figure 10.11 shows three different package outlines for just one type of discrete component, the transistor. Identification of the pin connections for different packages is given within the text as each separate or discrete component is considered, particularly later in this chapter when we discuss semiconductor devices. However, the Appendices aim to draw together all the information on pin connections and packages for easy reference.

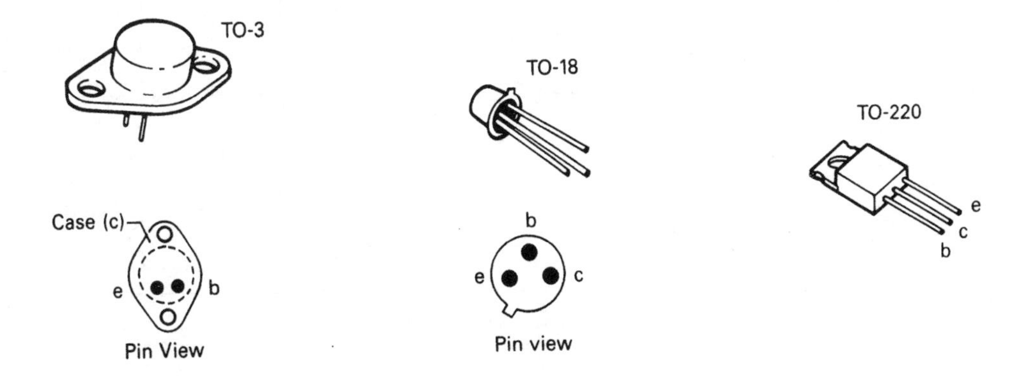

Fig. 10.11 Three different package outlines for transistors.

Obtaining information and components

Electricians use electrical wholesalers and suppliers to purchase electrical cable, equipment and accessories. Similar facilities are available in most towns and cities for the purchase of electronic components and equipment. There are also a number of national suppliers who employ representatives who will call at your workshop to offer technical advice and take your order. Some of these national companies also offer a 24-hour telephone order and mail order service. Their full-colour, fully illustrated catalogues also contain an enormous amount of technical information. The names and addresses of these national companies are given in Appendix A. For local suppliers you must consult your local phone book and *Yellow Pages*. The Appendices of this book also contain some technical reference information.

Semiconductor devices

SEMICONDUCTOR MATERIALS

Modern electronic devices use the semiconductor properties of materials such as silicon or germanium. The atoms of pure silicon or germanium are arranged in a lattice structure, as shown in Fig. 10.12. The outer electron orbits contain four electrons known as *valence* electrons. These electrons are all linked to other valence electrons from adjacent atoms, forming a covalent bond. There are no free electrons in pure silicon or germanium and, therefore, no conduction can take place unless the bonds are broken and the lattice framework is destroyed.

To make conduction possible without destroying the crystal it is necessary to replace a four-valent atom with a three- or five-valent atom. This process is known as *doping*.

If a three-valent atom is added to silicon or germanium a hole is left in the lattice framework. Since the material has lost a negative charge, the material becomes positive and is known as a p-type material (p for positive).

If a five-valent atom is added to silicon or germanium, only four of the valence electrons can form a bond and one electron becomes mobile or free to carry charge. Since the material has gained a negative charge it is known as an n-type material (n for negative).

Bringing together a p-type and n-type material allows current to flow in one direction only through the p-n junction. Such a junction is called a diode, since it is the semiconductor equivalent of the vacuum diode valve used by Fleming to rectify radio signals in 1904.

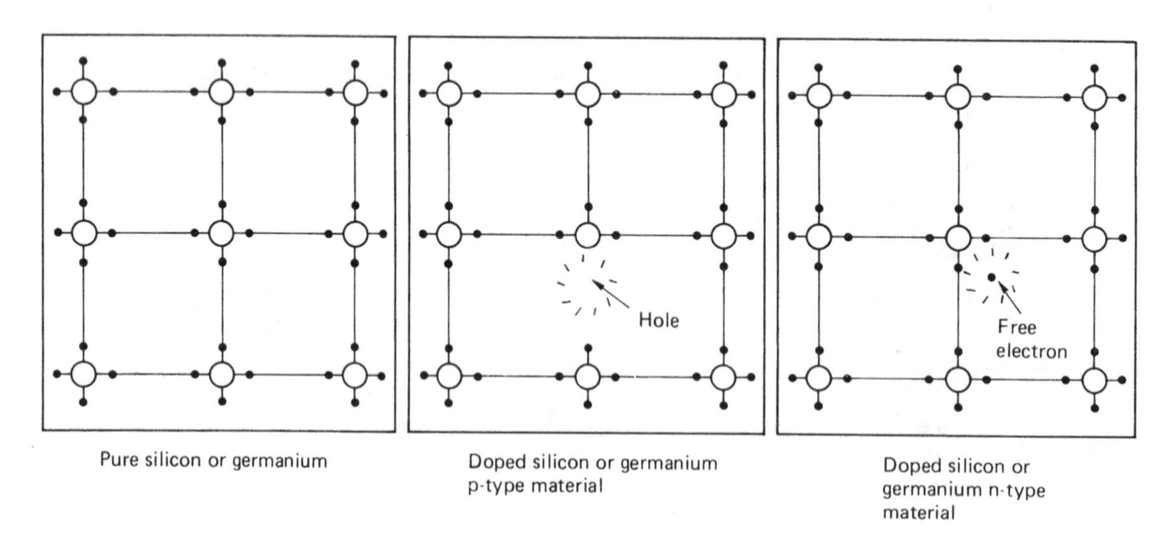

Fig. 10.12 Semiconductor material.

SEMICONDUCTOR DIODE

A semiconductor or junction diode consists of a p-type and n-type material formed in the same piece of silicon or germanium. The p-type material forms the anode and the n-type the cathode, as shown in Fig. 10.13. If the anode is made positive with respect to the cathode, the junction will have very little resistance and current will flow. This is referred to as forward bias. However, if reverse bias is applied, that is, the anode is made negative with respect to the cathode, the junction resistance is high and no current can flow, as shown in Fig. 10.14. The characteristics for a forward and reverse bias p-n junction are given in Fig. 10.15.

It can be seen that a small voltage is required to forward bias the junction before a current can flow. This is approximately 0.6 V for silicon and 0.2 V for germanium. The reverse bias potential of silicon is about 1200 V and for germanium about 300 V. If the reverse bias voltage is exceeded the diode will break down and current will flow in both directions. Similarly, the diode will break down if the current rating is exceeded, because excessive heat will be generated. Manufacturer's information therefore gives maximum voltage and current ratings for individual diodes which must not be exceeded. However, it is possible to connect a number of standard diodes in series or parallel, thereby sharing current or voltage, as shown in Fig. 10.16, so that the manufacturers' maximum values are not exceeded by the circuit.

DIODE TESTING

The p-n junction of the diode has a low resistance in one direction and a very high resistance in the reverse direction.

Connecting an ohmmeter, as described in Chapter 12, with the red positive lead to the anode of the junction diode and the black negative lead to the cathode, would give a very low reading. Reversing the lead connections would give a high resistance reading in a 'good' component.

ZENER DIODE

A Zener diode is a silicon junction diode but with a different characteristic than the semiconductor diode considered previously. It is a special diode with a predetermined reverse breakdown voltage, the

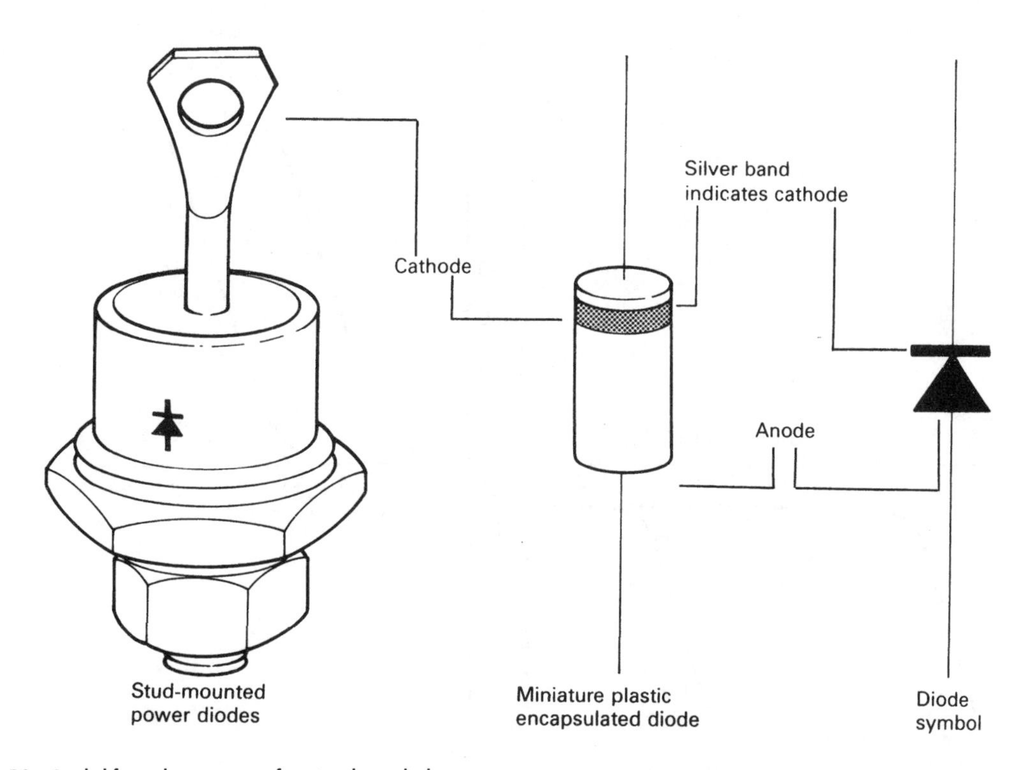

Fig. 10.13 Symbol for and appearance of semiconductor diodes.

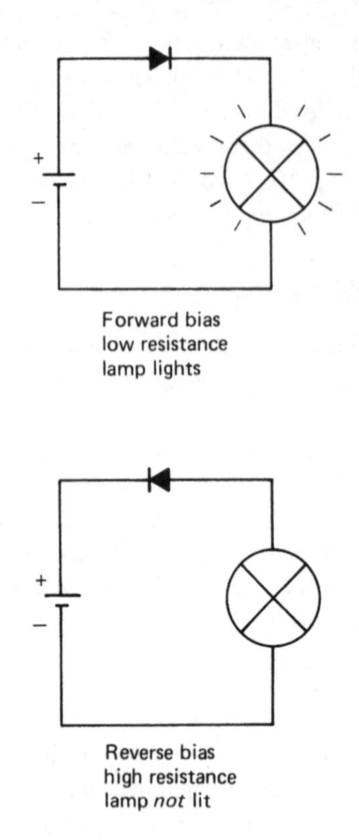

Fig. 10.14 Forward and reverse bias of a diode.

mechanism for which was discovered by Carl Zener in 1934. Its symbol and general appearance are shown in Fig. 10.17. In its forward bias mode, that is, when the anode is positive and the cathode negative, the Zener diode will conduct at about 0.6 V, just like an ordinary diode, but it is in the reverse mode that the Zener diode is normally used. When connected with the anode made negative and the cathode positive, the reverse current is zero until the reverse voltage reaches a predetermined value, when the diode switches on, as shown by the characteristics given in Fig. 10.18. This is called the Zener voltage or reference voltage. Zener diodes are manufactured in a range of preferred values, for example, 2.7 V, 4.7 V, 5.1 V, 6.2 V, 6.8 V, 9.1 V, 10 V, 11 V, 12 V etc., up to 200 V at various ratings. The diode may be damaged by overheating if the current is not limited by a series resistor, but when this is connected, the voltage across the diode remains constant. It is this property of the Zener diode which makes it useful for stabilizing power supplies and these circuits are considered at the end of this chapter.

If a test circuit is constructed as shown in Fig. 10.19, the Zener action can be observed. When the supply is less than the Zener voltage (5.1 V in this case) no current will flow and the output voltage will

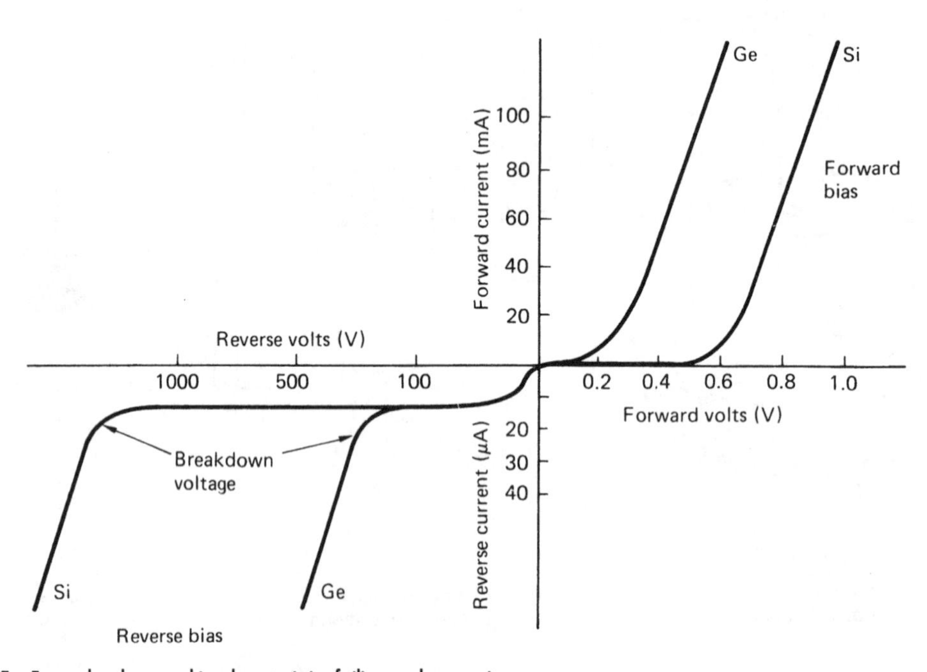

Fig. 10.15 Forward and reverse bias characteristic of silicon and germanium.

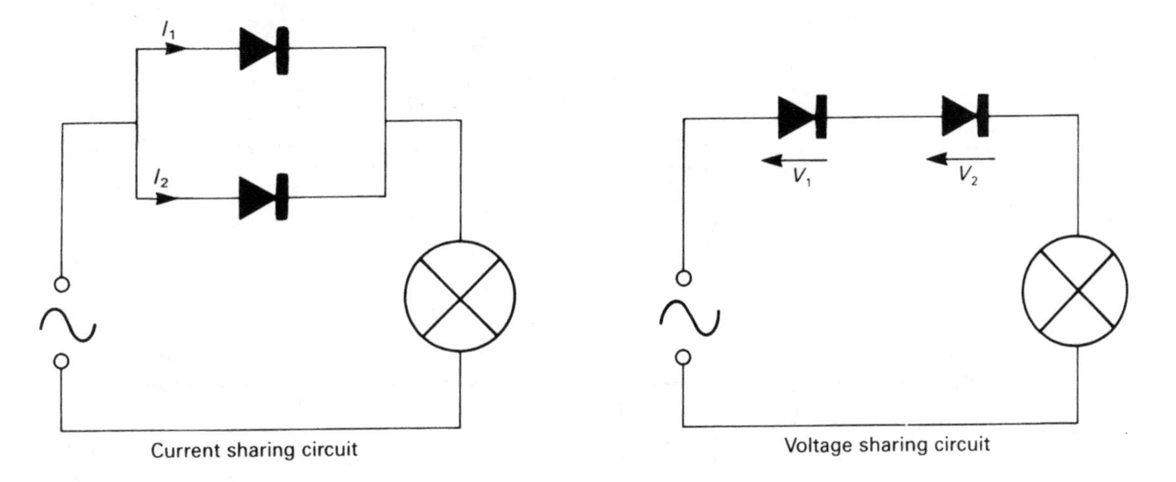

Fig. 10.16 Using two diodes to reduce the current or voltage applied to a diode.

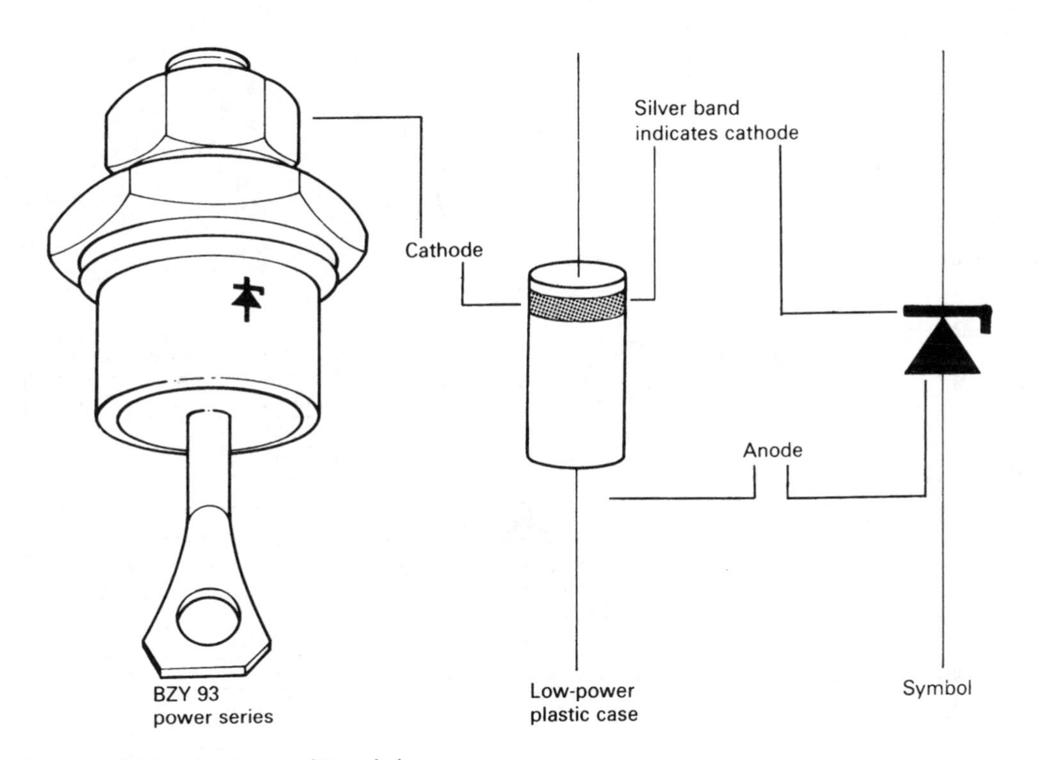

Fig. 10.17 Symbol for and appearance of Zener diodes.

be equal to the input voltage. When the supply is equal to or greater than the Zener voltage, the diode will conduct and any excess voltage will appear across the 680 Ω resistor, resulting in a very stable voltage at the output. When connecting this and other electronic circuits you must take care to connect the polarity of the Zener diode as shown in the diagram. Note that current must flow through the diode to enable it to stabilize.

LIGHT-EMITTING DIODE (LED)

The light-emitting diode is a p-n junction especially manufactured from a semiconducting material which

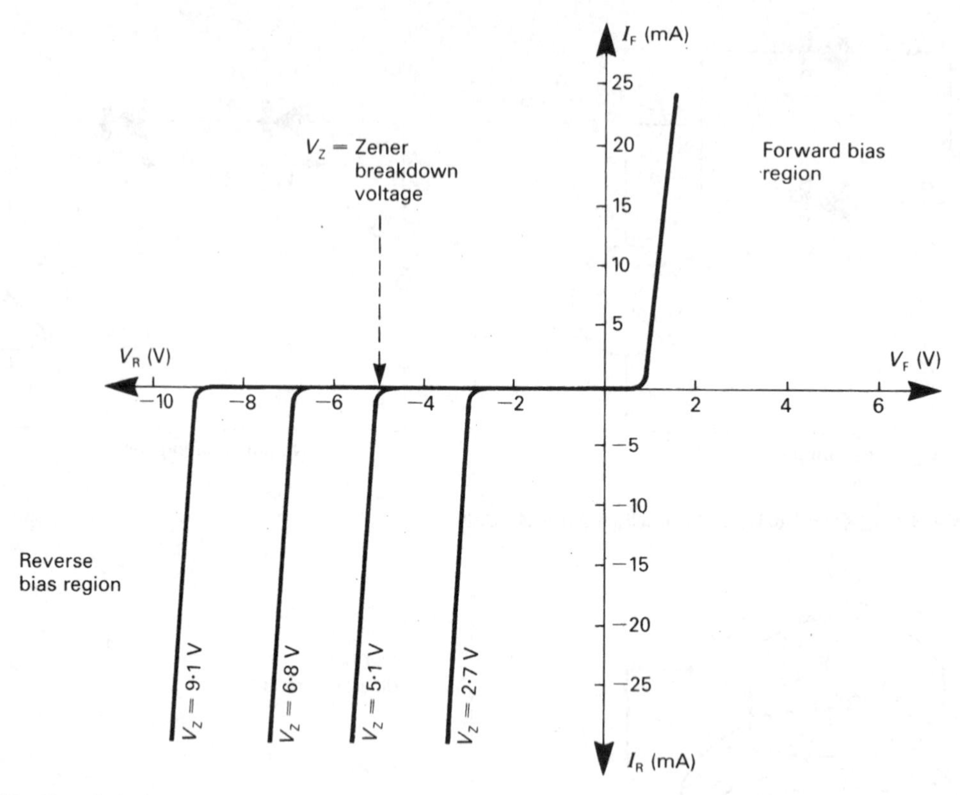

Fig. 10.18 Zener diode characteristics.

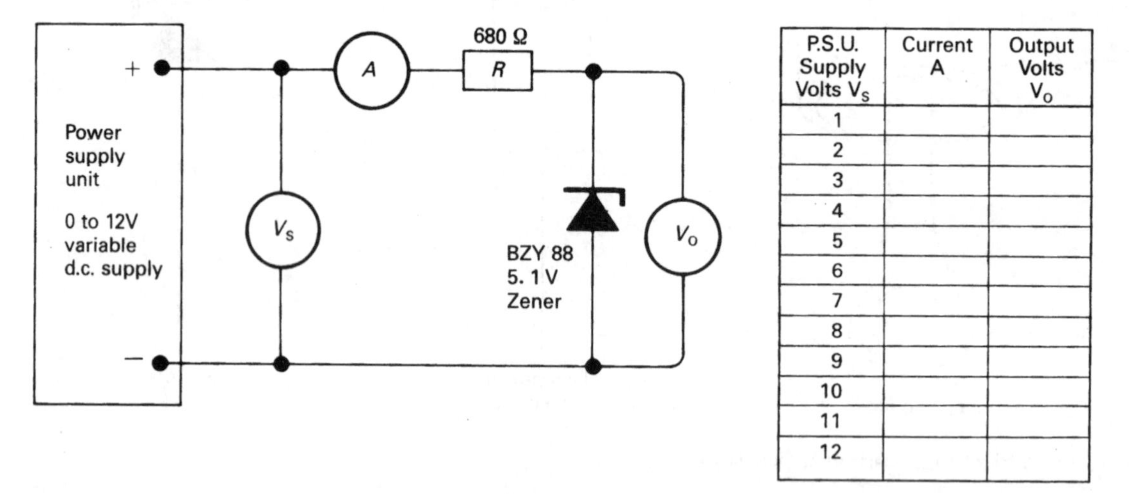

P.S.U. Supply Volts V_S	Current A	Output Volts V_O
1		
2		
3		
4		
5		
6		
7		
8		
9		
10		
11		
12		

Fig. 10.19 Experiment to demonstrate the operation of a Zener diode.

emits light when a current of about 10 mA flows through the junction.

No light is emitted when the junction is reverse biased and if this exceeds about 5 V the LED may be damaged.

The general appearance and circuit symbol are shown in Fig. 10.20.

The LED will emit light if the voltage across it is about 2V. If a voltage greater than 2 V is to be used then a resistor must be connected in series with the LED.

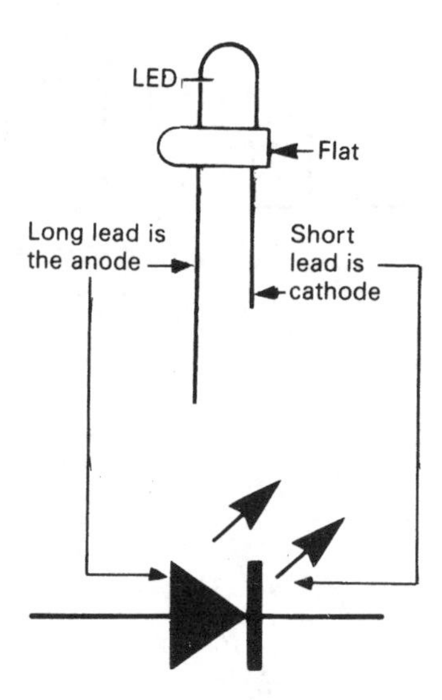

Fig. 10.20 Symbol for and general appearance of an LED.

To calculate the value of the series resistor we must ask ourselves what we know about LEDs. We know that the diode requires a forward voltage of about 2 V and a current of about 10 mA must flow through the junction to give sufficient light. The value of the series resistor R will, therefore, be given by

$$R = \frac{\text{Supply voltage} - 2\text{ V}}{10\text{ mA}}\ \Omega$$

EXAMPLE

Calculate the value of the series resistor required when an LED is to be used to show the presence of a 12 V supply.

$$R = \frac{12\text{ V} - 2\text{ V}}{10\text{ mA}}\ \Omega$$

$$R = \frac{10\text{ V}}{10\text{ mA}} = 1\text{ k}\Omega$$

The circuit is, therefore, as shown in Fig. 10.21.

LEDs are available in red, yellow and green and, when used with a series resistor, may replace a filament lamp. They use less current than a filament lamp, are smaller, do not become hot and last indefinitely. A filament lamp, however, is brighter and emits white light. LEDs are often used as indicator

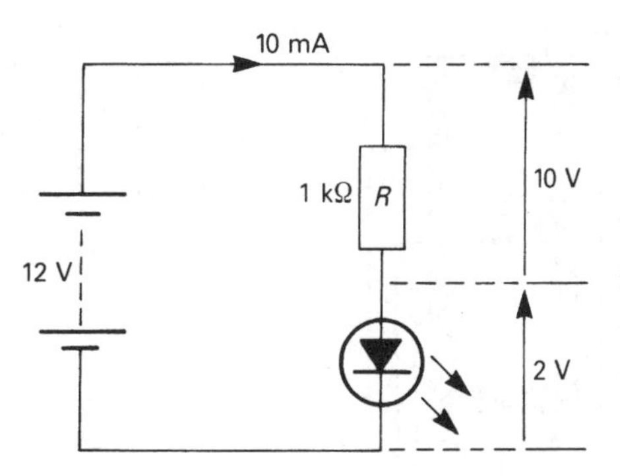

Fig. 10.21 Circuit diagram for LED example.

lamps, to indicate the presence of a voltage. They do not, however, indicate the *precise* amount of voltage present at that point.

Another application of the LED is the seven-segment display used as a numerical indicator in calculators, digital watches and measuring instruments. Seven LEDs are arranged as a figure 8 so that when various segments are illuminated, the numbers 0 to 9 are displayed as shown in Fig. 10.22.

LIGHT-DEPENDENT RESISTOR (LDR)

Almost all materials change their resistance with a change in temperature. Light energy falling on a suitable semiconductor material also causes a change in resistance. The semiconductor material of an LDR is encapsulated as shown in Fig. 10.23 together with the circuit symbol. The resistance of an LDR in total darkness is about 10 MΩ, in normal room lighting about 5 kΩ and in bright sunlight about 100 Ω. They can carry tens of milliamperes, an amount which is sufficient to operate a relay. The LDR uses this characteristic to switch on automatically street lighting and security alarms.

PHOTODIODE

The photodiode is a normal junction diode with a transparent window through which light can enter. The circuit symbol and general appearance are shown in Fig. 10.24. It is operated in reverse bias mode and the leakage current increases in proportion to the amount of light falling on the junction. This is due to the light energy breaking bonds in the crystal lattice of the semiconductor material to produce holes and electrons.

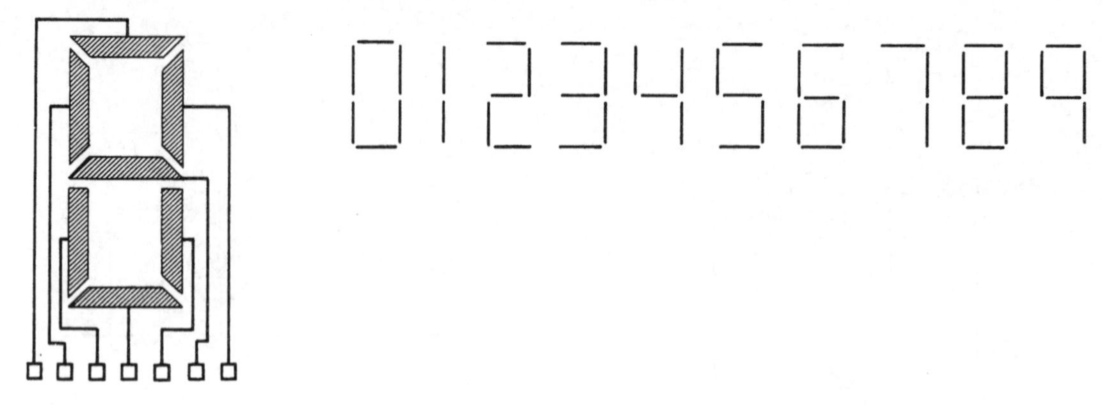

Fig. 10.22 LED used in seven-segment display.

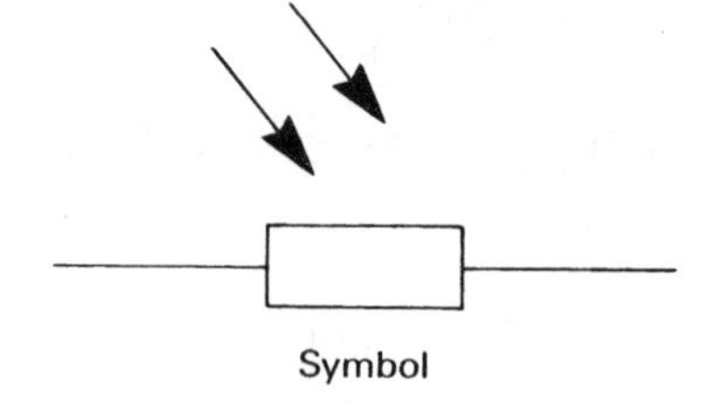

Fig. 10.23 Symbol and appearance of a light-dependent resistor.

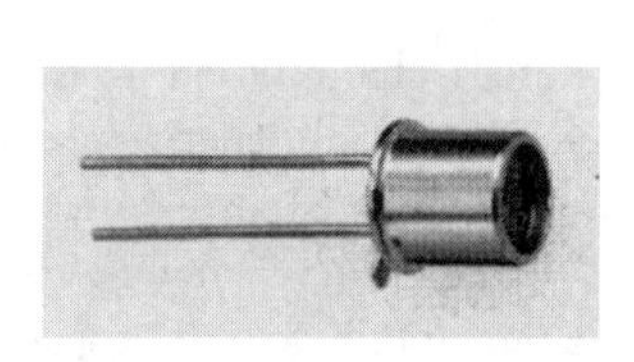

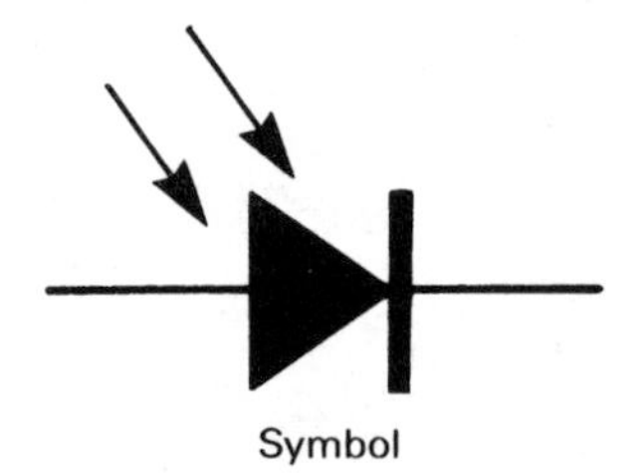

Fig. 10.24 Symbol for, pin connections of and appearance of a photodiode.

Photodiodes will only carry microamperes of current but can operate much more quickly than LDRs and are used as 'fast' counters when the light intensity is changing rapidly.

THERMISTOR

The thermistor is a thermal resistor, a semiconductor device whose resistance varies with temperature. Its circuit symbol and general appearance are shown in Fig. 10.25. They can be supplied in many shapes and are used for the measurement and control of temperature up to their maximum useful temperature limit of about 300°C. They are very sensitive and because the bead of semiconductor material can be made very small, they can measure temperature in the most inaccessible places with very fast response times. Thermistors are embedded in high-voltage underground transmission cables in order to monitor the temperature of the cable. Information about the temperature of a cable allows engineers to load the cables more efficiently. A particular cable can carry a larger load in winter for example, when heat from the cable is being dissipated more efficiently. A thermistor is also used to monitor the water temperature of a motor car.

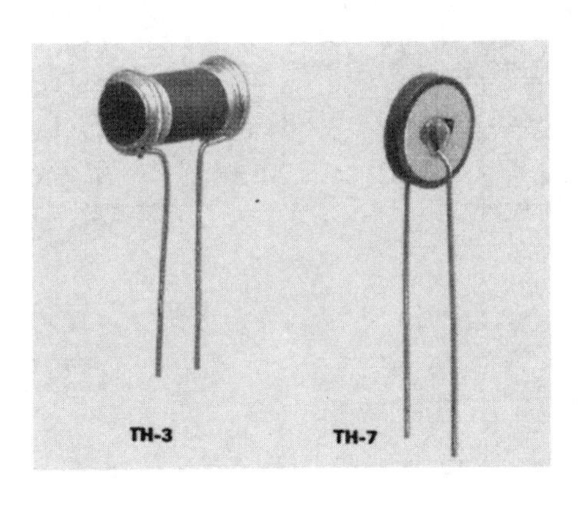

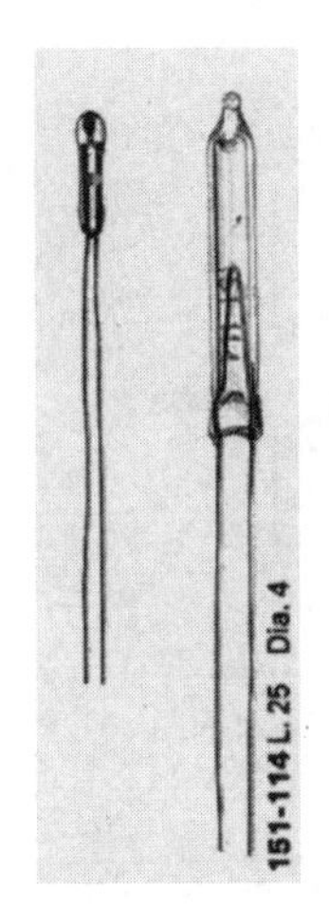

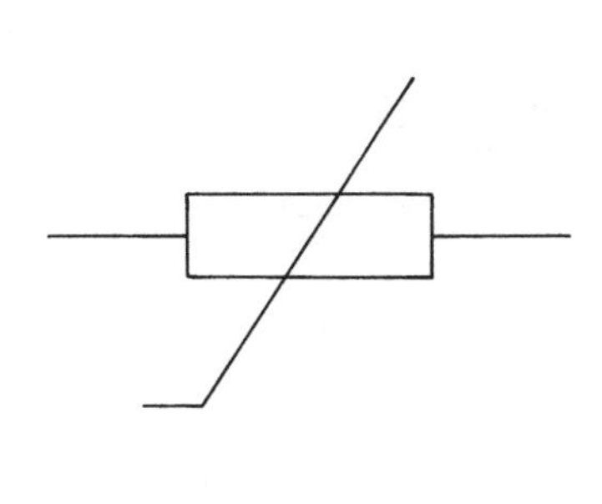

Fig. 10.25 Symbol for and appearance of a thermistor.

TRANSISTORS

The transistor has become the most important building block in electronics. It is the modern, miniature, semiconductor equivalent of the thermionic valve and was invented in 1947 by Bardeen, Shockley and Brattain at the Bell Telephone Laboratories in the USA. Transistors are packaged as separate or *discrete* components, as shown in Fig. 10.26.

There are two basic types of transistor, the *bipolar* or junction transistor and the *field-effect transistor* (FET).

The FET has some characteristics which make it a better choice in electronic switches and amplifiers. It uses less power and has a higher resistance and frequency response. It takes up less space than a bipolar transistor and, therefore, more of them can be packed together on a given area of silicon chip. It is, therefore, the FET which is used when many transistors are integrated on to a small area of silicon chip as in the *integrated circuit* (IC) discussed later.

When packaged as a discrete component the FET looks much the same as the bipolar transistor. Its circuit symbol and connections are given in Appendix F. However, it is the bipolar transistor which is much more widely used in electronic circuits as a discrete component.

The bipolar transistor

The bipolar transistor consists of three pieces of semiconductor material sandwiched together as shown in Fig. 10.27. The structure of this transistor makes it a three-terminal device having a base, collector and

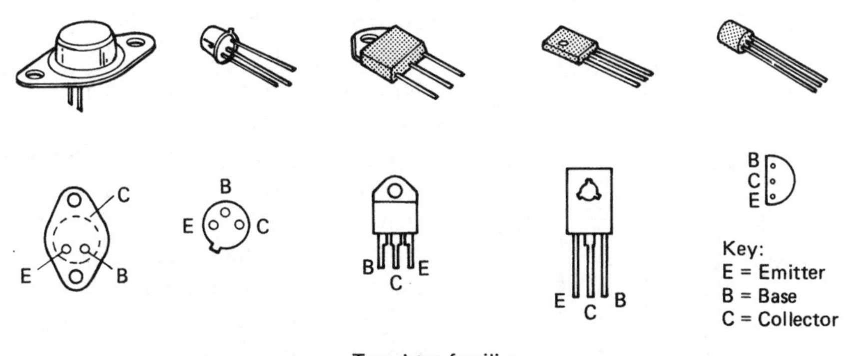

Fig. 10.26 The appearance and pin connections of the transistor family.

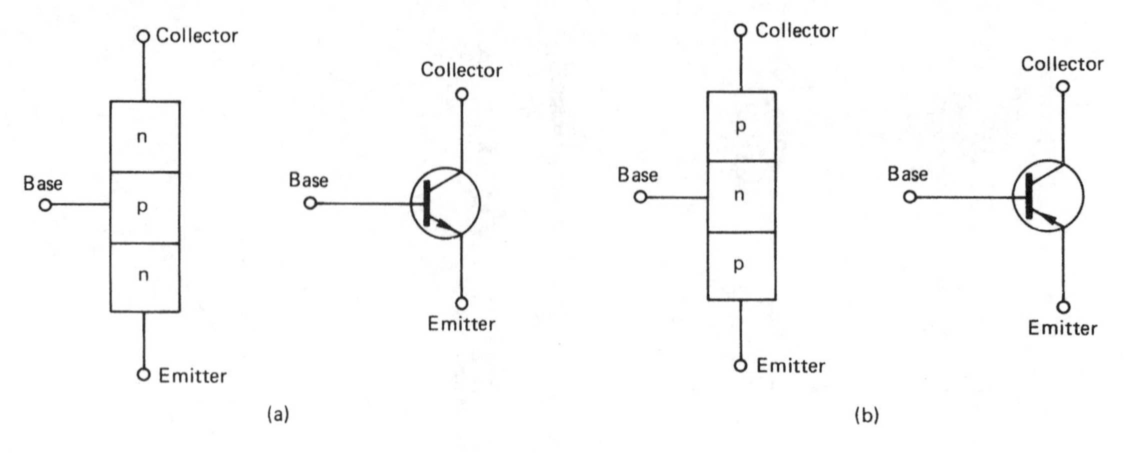

Fig. 10.27 Structure of and symbol for (a) n-p-n and (b) p-n-p transistors.

emitter terminal. By varying the current flowing into the base connection a much larger current flowing between collector and emitter can be controlled. Apart from the supply connections, the n-p-n and p-n-p types are essentially the same but the n-p-n type is more common.

A transistor is generally considered a current-operated device. There are two possible current paths through the transistor circuit, shown in Fig. 10.28: the base–emitter path when the switch is closed; and the collector–emitter path. Initially, the positive battery supply is connected to the n-type material of the collector, the junction is reverse biased and, therefore, no current will flow. Closing the switch will forward bias the base–emitter junction and current flowing through this junction causes current to flow across the collector–emitter junction and the signal lamp will light.

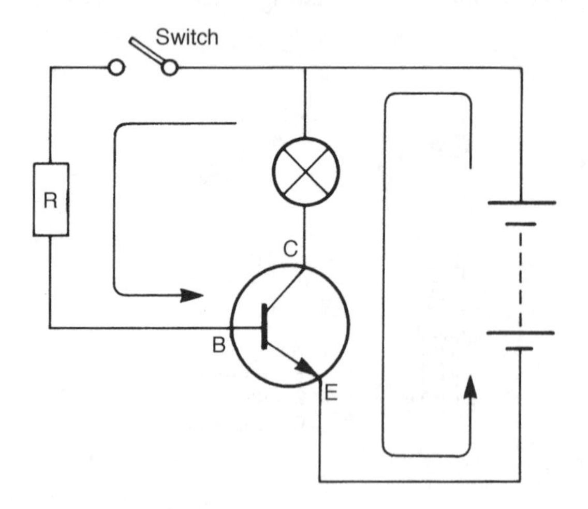

Fig. 10.28 Operation of the transistor.

A small base current can cause a much larger collector current to flow. This is called the *current gain* of the transistor, and is typically about 100. When I say a much larger collector current, I mean a large current in electronic terms, up to about half an ampere.

We can, therefore, regard the transistor as operating in two ways: as a switch because the base current turns on and controls the collector current; and as a current amplifier because the collector current is greater than the base current.

We could also consider the transistor to be operating in a similar way to a relay. However, transistors have many advantages over electrically operated switches such as relays. They are very small, reliable, have no moving parts and, in particular, they can switch millions of times a second without arcing occurring at the contacts.

Transistor testing

A transistor can be thought of as two diodes connected together and, therefore, a transistor can be tested using an ohmmeter in the same way as was described for the diode.

Assuming that the red lead of the ohmmeter is positive, as described in Chapter 12, the transistor can be tested in accordance with Table 10.5.

When many transistors are to be tested, a simple test circuit can be assembled as shown in Fig. 10.29.

With the circuit connected, as shown in Fig. 10.29, a 'good' transistor will give readings on the voltmeter of 6 V with the switch open and about 0.5 V when the switch is made. The voltmeter used for the test should

Table 10.5 Transistor testing using an ohmmeter

A 'good' n-p-n transistor will give the following readings:
Red to base and black to collector = low resistance Red to base and black to emitter = low resistance
Reversed connections on the above terminals will result in a high resistance reading, as will connections of either polarity between the collector and emitter terminals.

A 'good' p-n-p transistor will give the following readings:
Black to base and red to collector = low resistance Black to base and red to emitter = low resistance
Reversed connections on the above terminals will result in a high resistance reading, as will connections of either polarity between the collector and emitter terminals.

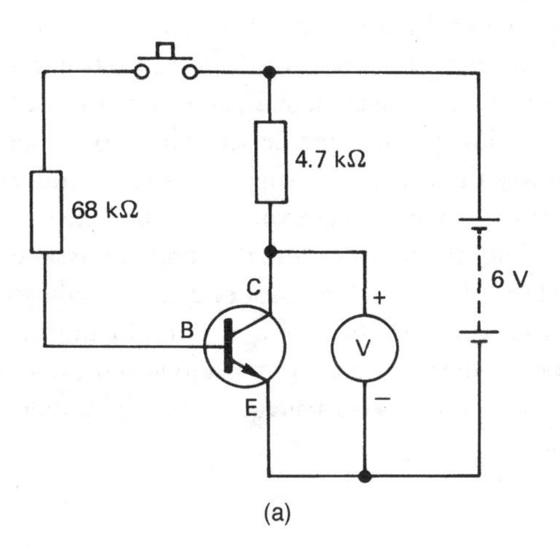

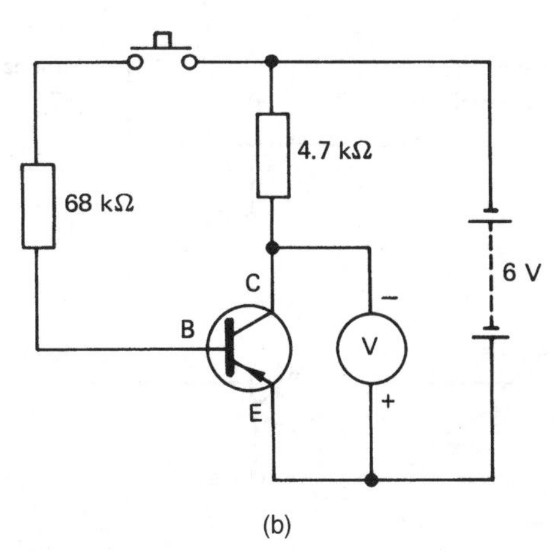

Fig. 10.29 Transistor test circuits (a) n-p-n transistor test; (b) p-n-p transistor test.

have a high internal resistance, about ten times greater than the value of the resistor being tested – in this case 4.7 kΩ – and this is usually indicated on the back of a multirange meter or in the manufacturers' information supplied with a new meter.

INTEGRATED CIRCUITS

Integrated circuits (ICs) were first developed in the 1960s. They are densely populated miniature electronic circuits made up of hundreds and sometimes thousands of microscopically small transistors, resistors, diodes and capacitors, all connected together on a single chip of silicon no bigger than a baby's fingernail. When assembled in a single package, as shown in Fig. 10.30, we call the device an integrated circuit.

There are two broad groups of integrated circuit: digital ICs and linear ICs. Digital ICs contain simple switching-type circuits used for logic control and calculators, discussed in Chapter 13. Linear ICs incorporate amplifier-type circuits which can respond to audio and radio frequency signals. The most versatile linear IC is the operational amplifier which has applications in electronics, instrumentation and control.

The integrated circuit is an electronic revolution. ICs are more reliable, cheaper and smaller than the same circuit made from discrete or separate transistors, and electronically superior. One IC behaves differently than another because of the arrangement of the transistors within the IC.

Manufacturers' data sheets describe the characteristics of the different ICs, which have a reference number stamped on the top.

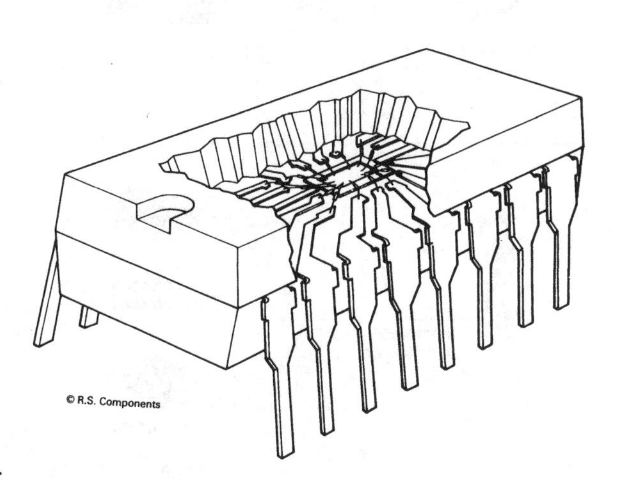

Fig. 10.30 Exploded view of an integrated circuit.

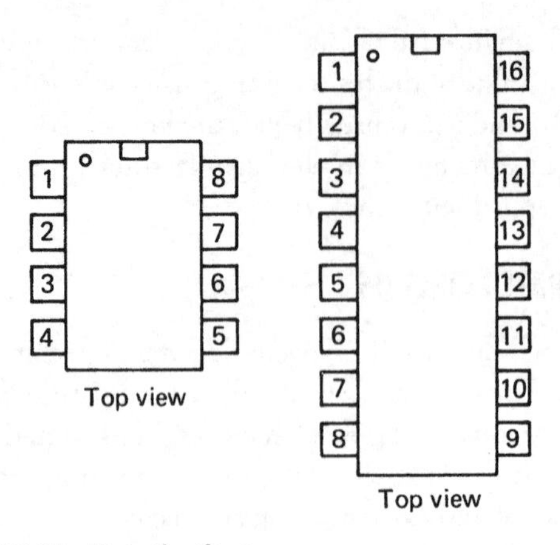

Fig. 10.31 IC pin identification.

When building circuits, it is necessary to be able to identify the IC pin connection by number. The number 1 pin of any IC is indicated by a dot pressed into the encapsulation; it is also the pin to the left of the cutout (Fig. 10.31). Since the packaging of ICs has two rows of pins they are called DIL (dual in-line) packaged integrated circuits and their appearance is shown in Fig. 10.32.

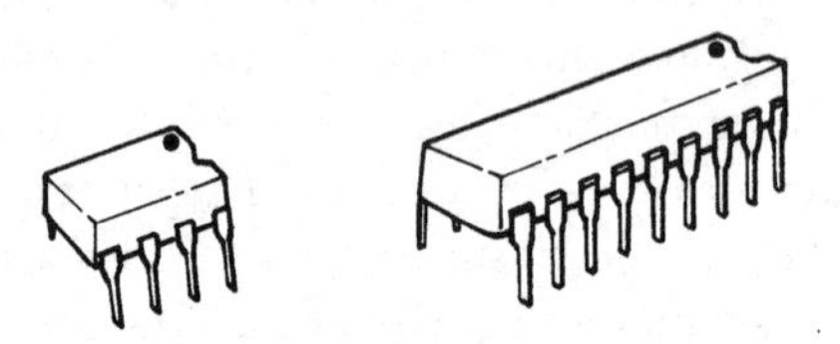

Fig. 10.32 DIL packaged integrated circuits.

Integrated circuits are sometimes connected into DIL sockets and at other times are soldered directly into the circuit. The testing of ICs is beyond the scope of a practising electrician, and when they are suspected of being faulty an identical or equivalent replacement should be connected into the circuit, ensuring that it is inserted the correct way round, which is indicated by the position of pin number 1 as described earlier.

THE THYRISTOR

The *thyristor* was previously known as a 'silicon controlled rectifier' since it is a rectifier which controls the power to a load. It consists of four pieces of semiconductor material sandwiched together and connected to three terminals, as shown in Fig. 10.33.

The word thyristor is derived from the Greek word *thyra* meaning door, because the thyristor behaves like a door. It can be open or shut, allowing or preventing current flow through the device. The door is opened – we say the thyristor is triggered – to a conducting state by applying a pulse voltage to the gate connection. Once the thyristor is in the conducting state, the gate loses all control over the devices. The only way to bring the thyristor back to a non-conducting state is to reduce the voltage across the anode and cathode to zero or apply reverse voltage across the anode and cathode.

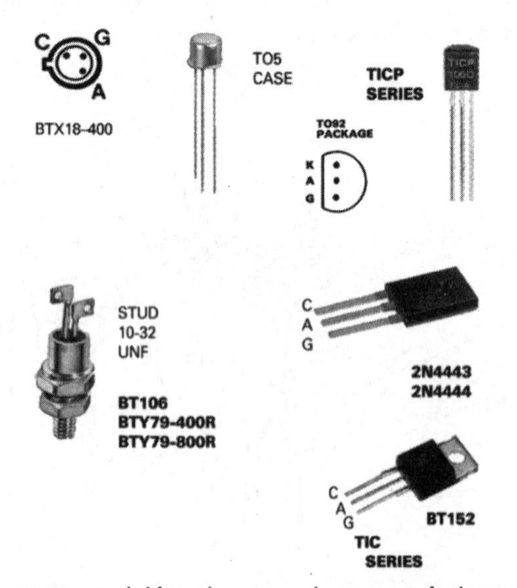

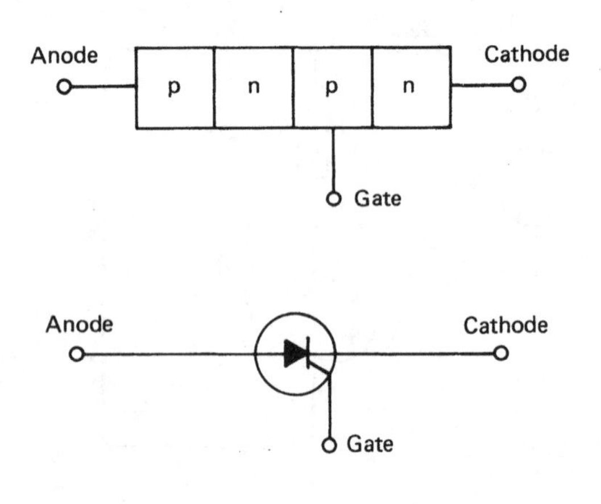

Fig. 10.33 Symbol for and structure and appearance of a thyristor.

We can understand the operation of a thyristor by considering the circuit shown in Fig. 10.34. This circuit can also be used to test suspected faulty components.

When SWB only is closed the lamp will not light, but when SWA is also closed, the lamp lights to full brilliance. The lamp will remain illuminated even when SWA is opened. This shows that the thyristor is operating correctly. Once a voltage has been applied to the gate the thyristor becomes forward conducting, like a diode, and the gate loses control.

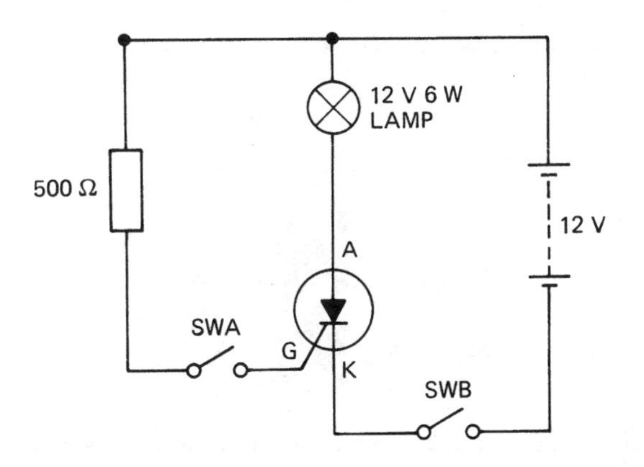

Fig. 10.34 Thyristor test circuit.

A thyristor may also be tested using an ohmmeter as described in Table 10.6, assuming that the red lead of the ohmmeter is positive as described in Chapter 12.

The thyristor has no moving parts and operates without arcing. It can operate at extremely high speeds, and the currents used to operate the gate are very small. The most common application for the thyristor is to control the power supply to a load, for example, lighting dimmers and motor speed control.

Table 10.6 Thyristor testing using an ohmmeter

A 'good' thyristor will give the following readings:
Black to cathode and red on gate = low resistance Red to cathode and black on gate = a higher resistance value
The value of the second reading will depend upon the thyristor, and may vary from only slightly greater to very much greater.
Connecting the test instrument leads from cathode to anode will result in a very high resistance reading, whatever polarity is used.

The power available to an a.c. load can be controlled by allowing current to be supplied to the load during only a part of each cycle. This can be achieved by supplying a gate pulse automatically at a chosen point in each cycle, as shown by Fig. 10.35. Power is reduced by triggering the gate later in the cycle.

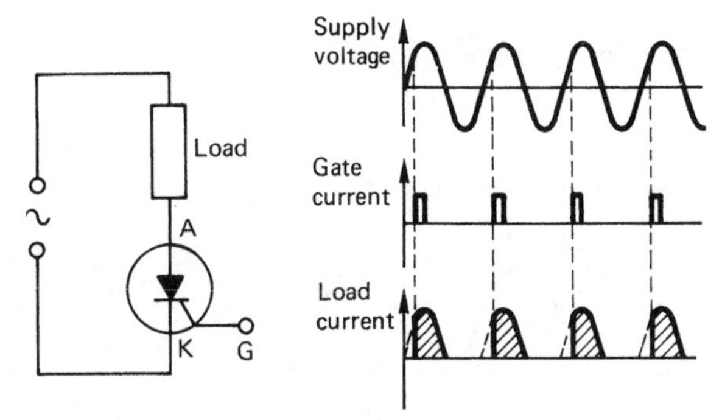

Fig. 10.35 Waveforms to show the control effect of a thyristor.

The thyristor is only a half-wave device (like a diode) allowing control of only half the available power in an a.c. circuit. This is very uneconomical, and a further development of this device has been the triac which is considered next.

THE TRIAC

The triac was developed following the practical problems experienced in connecting two thyristors in parallel, to obtain full wave control, and in providing two separate gate pulses to trigger the two devices.

The triac is a single device containing a back-to-back, two-directional thyristor which is triggered on both halves of each cycle of the a.c. supply by the same gate signal. The power available to the load can, therefore, be varied between zero and full load.

Its symbol and general appearance are shown in Fig. 10.36. Power to the load is reduced by triggering the gate later in the cycle, as shown by the waveforms of Fig. 10.37.

The triac is a three-terminal device, just like the thyristor, but the terms anode and cathode have no meaning for a triac. Instead, they are called main terminal one (MT_1) and main terminal two (MT_2). The device is triggered by applying a small pulse to the gate (G). A gate current of 50 mA is sufficient to

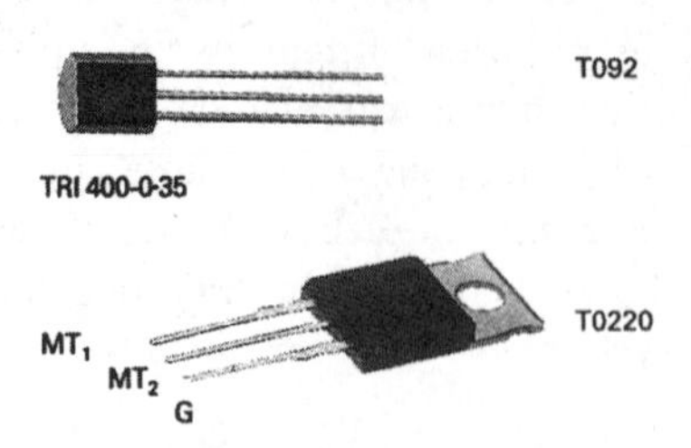

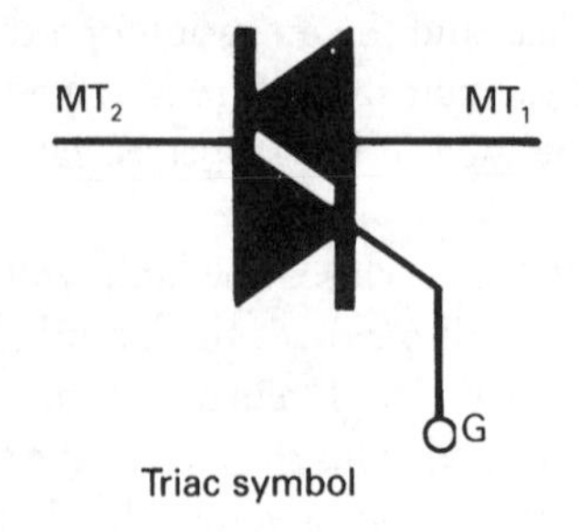

Fig. 10.36 Appearance of a triac.

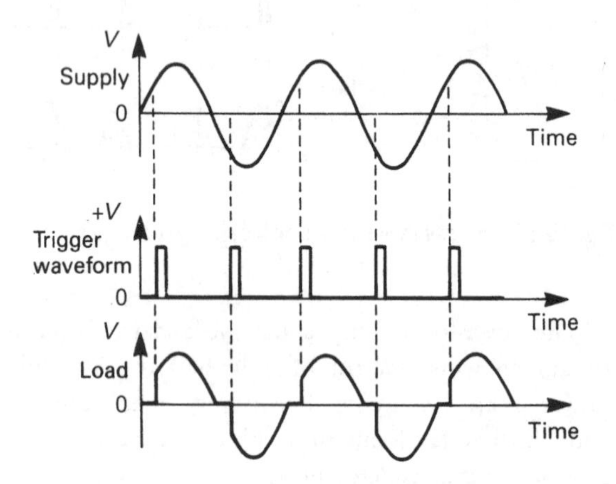

Fig. 10.37 Waveforms to show the control effect of a triac.

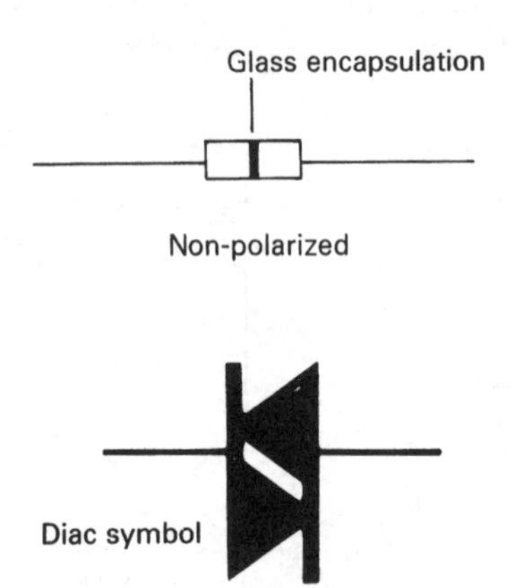

Fig. 10.38 Symbol for and appearance of a diac used in triac firing circuits.

trigger a triac switching up to 100 A. They are used for many commercial applications where control of a.c. power is required, for example, motor speed control and lamp dimming.

THE DIAC

The diac is a two-terminal device containing a two-directional Zener diode. It is used mainly as a trigger device for the thyristor and triac. The symbol is shown in Fig. 10.38.

The device turns on when some predetermined voltage level is reached, say 30 V, and, therefore, it can be used to trigger the gate of a triac or thyristor each time the input waveform reaches this predetermined value. Since the device contains back-to-back Zener diodes it triggers on both the positive and negative half cycles.

Voltage divider

In Chapter 8 we considered the distribution of voltage across resistors connected in series. We found that the supply voltage was divided between the series resistors in proportion to the size of the resistor. If two identical resistors were connected in series across a 12 V supply, as shown in Fig. 10.39(a), both common sense and a simple calculation would confirm that 6 V would be measured across the output. In the circuit shown in Fig. 10.39(b), the 1 kΩ and 2 kΩ resistors divide the input voltage into three equal parts. One part, 4 V, will appear across the 1 kΩ resistor and two parts, 8 V, will appear across the 2 kΩ resistor. In Fig. 10.39(c) the situation is reversed and, therefore, the voltmeter will read 4 V. The division of the voltage

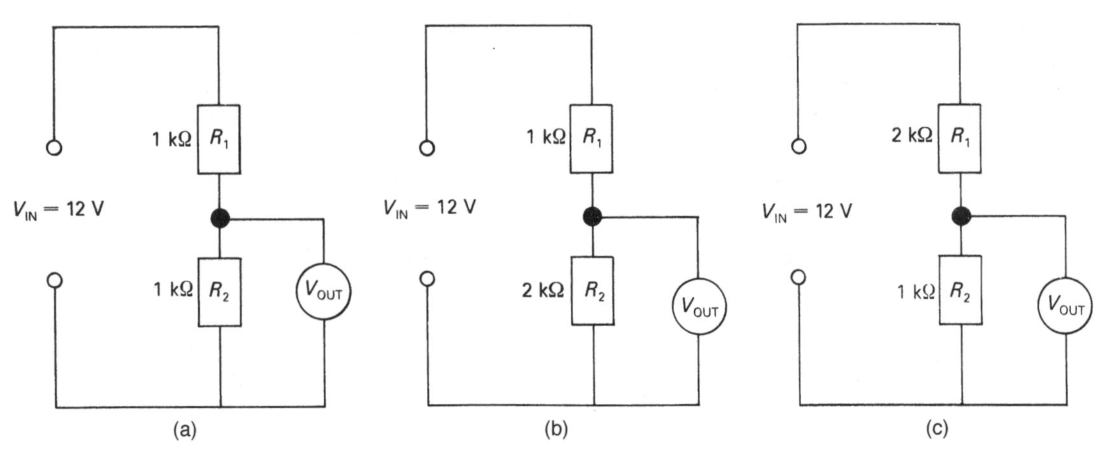

Fig. 10.39 Voltage divider circuit.

is proportional to the ratio of the two resistors and, therefore, we call this simple circuit a *voltage divider* or *potential divider*. The values of the resistors R_1 and R_2 determine the output voltage as follows:

$$V_{OUT} = V_{IN} \times \frac{R_2}{R_1 + R_2} \text{ (V)}$$

For the circuit shown in Fig. 10.39(b),

$$V_{OUT} = 12\text{ V} \times \frac{2\text{ k}\Omega}{1\text{ k}\Omega + 2\text{ k}\Omega} = 8\text{ V}$$

For the circuit shown in Fig. 10.39(c),

$$V_{OUT} = 12\text{ V} \times \frac{1\text{ k}\Omega}{2\text{ k}\Omega + 1\text{ k}\Omega} = 4\text{ V}$$

EXAMPLE 1

For the circuit shown in Fig. 10.40, calculate the output voltage.

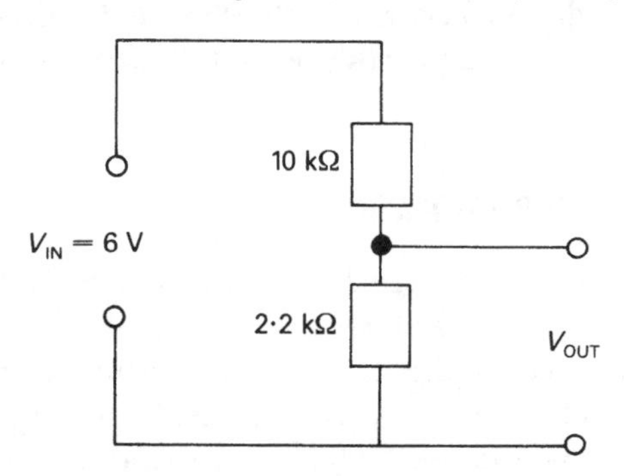

Fig. 10.40 Voltage divider circuit for Example 1.

$$V_{OUT} = 6\text{ V} \times \frac{2.2\text{ k}\Omega}{10\text{ k}\Omega + 2.2\text{ k}\Omega} = 1.08\text{ V}$$

EXAMPLE 2

For the circuit shown in Fig. 10.41(a), calculate the output voltage.

We must first calculate the equivalent resistance of the parallel branch:

$$\frac{1}{R_T} = \frac{1}{R_1} + \frac{1}{R_2}$$

$$\frac{1}{R_T} = \frac{1}{10\text{ k}\Omega} + \frac{1}{10\text{ k}\Omega} = \frac{1+1}{10\text{ k}\Omega} = \frac{2}{10\text{ k}\Omega}$$

$$R_T = \frac{10\text{ k}\Omega}{2} = 5\text{ k}\Omega$$

The circuit may now be considered as shown in Fig. 10.41(b):

$$V_{OUT} = 6\text{ V} \times \frac{10\text{ k}\Omega}{5\text{ k}\Omega + 10\text{ k}\Omega} = 4\text{ V}$$

Voltage dividers are used in electronic circuits to produce a reference voltage which is suitable for operating transistors and integrated circuits. The volume control in a radio or the brightness control of a cathode-ray oscilloscope requires a continuously variable voltage divider and this can be achieved by connecting a variable resistor or potentiometer, as shown in Fig. 10.42. With the wiper

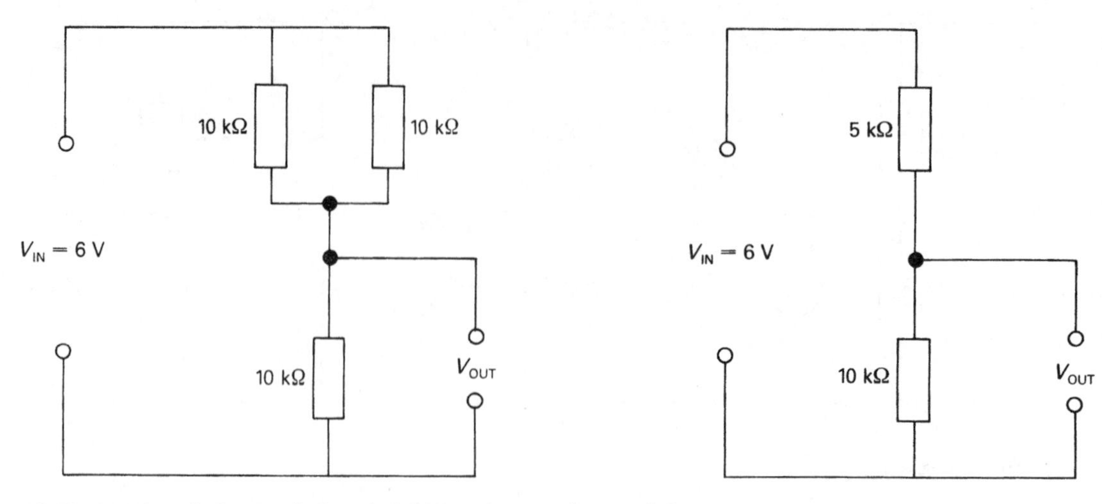

Fig. 10.41 (a) Voltage divider circuit for Example 2; (b) Equivalent circuit for example 2.

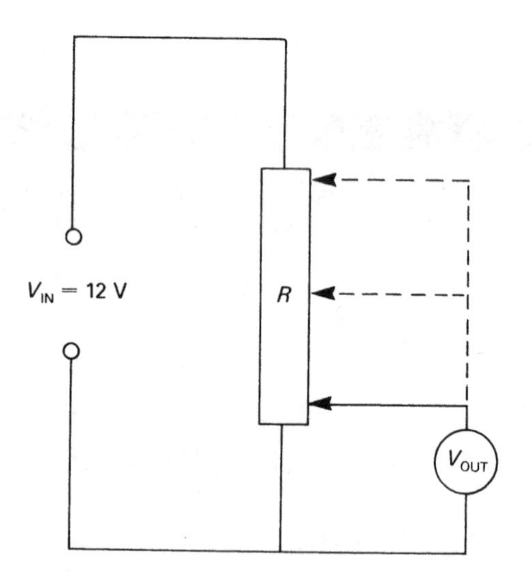

Fig. 10.42 Constantly variable voltage divider circuit.

arm making a connection at the bottom of the resistor, the output would be zero. When connection is made at the centre, the voltage would be 6 V, and at the top of the resistor the voltage would be 12 V. The voltage is continuously variable between 0 V and 12 V simply by moving the wiper arm of a suitable variable resistor such as those shown in Fig. 10.3.

When a load is connected to a voltage divider it 'loads' the circuit, causing the output voltage to fall below the calculated value. To avoid this, the resistance of the load should be at least ten times as great as the value of the resistor across which it is connected. For example, the load connected across the voltage divider shown in Fig. 10.39(b) must be greater than 20 kΩ and across 10.39(c) greater than 10 kΩ. This problem of loading the circuit also occurs when taking voltage readings, as discussed in Chapter 12.

Rectification of a.c.

When a d.c. supply is required, batteries or a rectified a.c. supply can be provided. Batteries have the advantage of portability, but a battery supply is more expensive than using the a.c. mains supply suitably rectified. Rectification is the conversion of an a.c. supply into a unidirectional or d.c. supply. This is one of the many applications for a diode which will conduct in one direction only, that is when the anode is positive with respect to the cathode.

HALF-WAVE RECTIFICATION

The circuit is connected as shown in Fig. 10.43. During the first half cycle the anode is positive with respect to the cathode and, therefore, the diode will conduct. When the supply goes negative during the second half cycle, the anode is negative with respect to the cathode and, there-

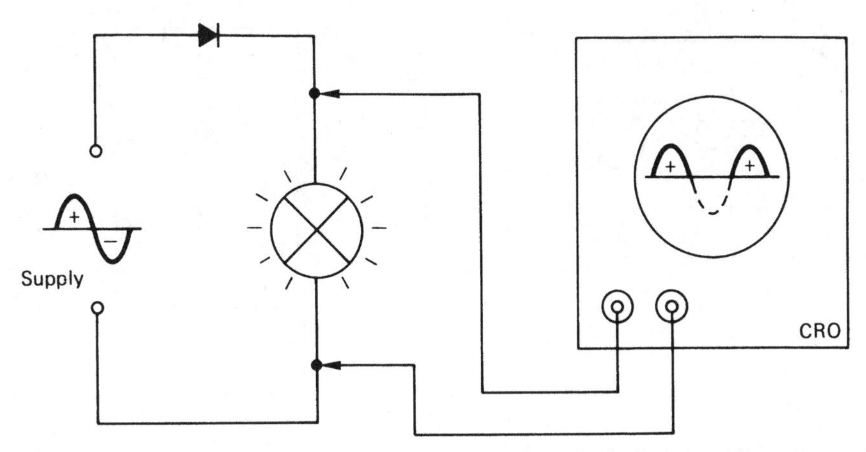

Fig. 10.43 Half-wave rectification.

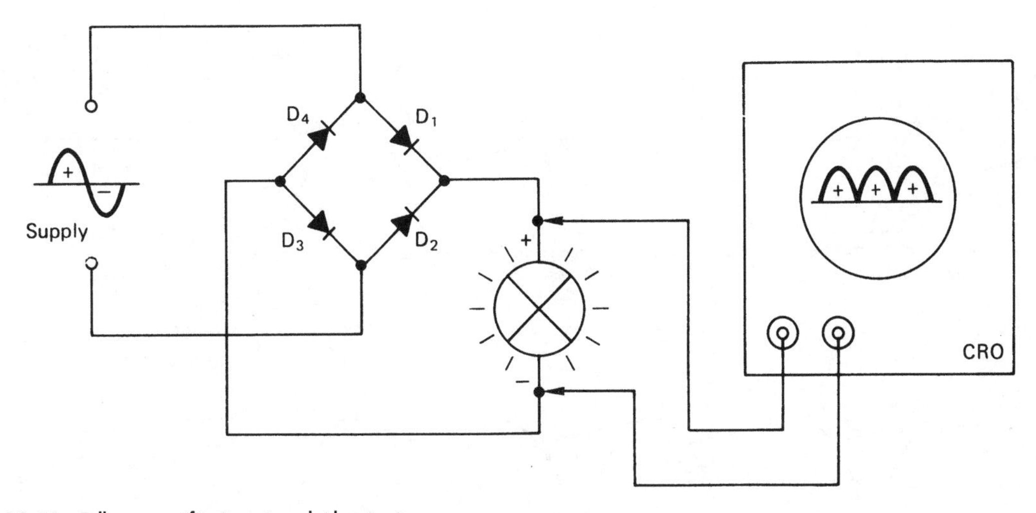

Fig. 10.44 Full-wave rectification using a bridge circuit.

fore, the diode will not allow current to flow. Only the positive half of the waveform will be available at the load and the lamp will light at reduced brightness.

FULL-WAVE RECTIFICATION

Figure 10.44 shows an improved rectifier circuit which makes use of the whole a.c. waveform and is, therefore, known as a full-wave rectifier. When the four diodes are assembled in this diamond-shaped configuration, the circuit is also known as a *bridge rectifier*. During the first half cycle diodes D_1 and D_3 conduct, and diodes D_2 and D_4 conduct during the second half cycle. The lamp will light to full brightness.

Full-wave and half-wave rectification can be displayed on the screen of a CRO and will appear as shown in Figs 10.43 and 10.44.

Smoothing

The circuits of Figs 10.43 and 10.44 convert an alternating waveform into a waveform which never goes negative, but they cannot be called continuous d.c. because they contain a large alternating component. Such a waveform is too bumpy to be used to supply electronic equipment but may be used for battery charging. To be useful in electronic circuits the output must be smoothed. The simplest way to smooth

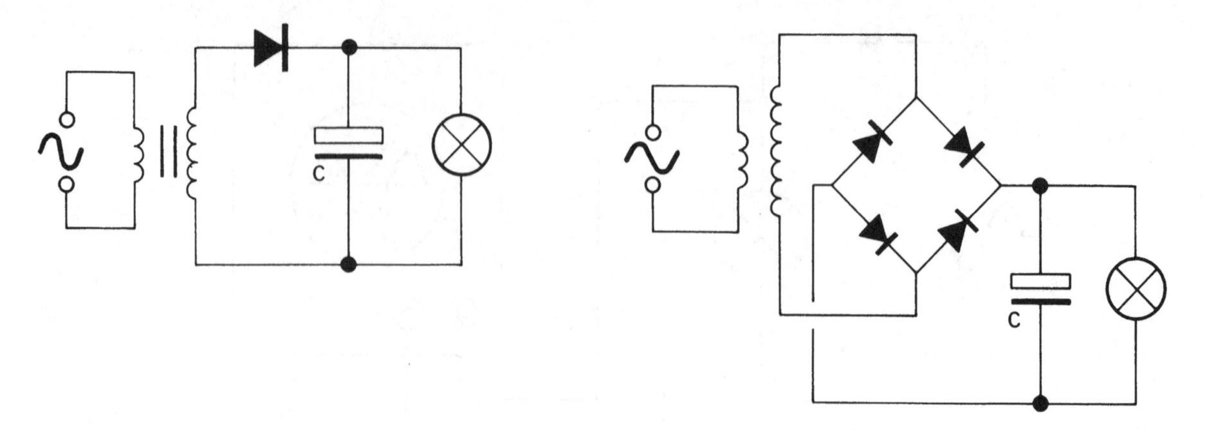

Fig. 10.45 Rectified a.c. with smoothing capacitor connected.

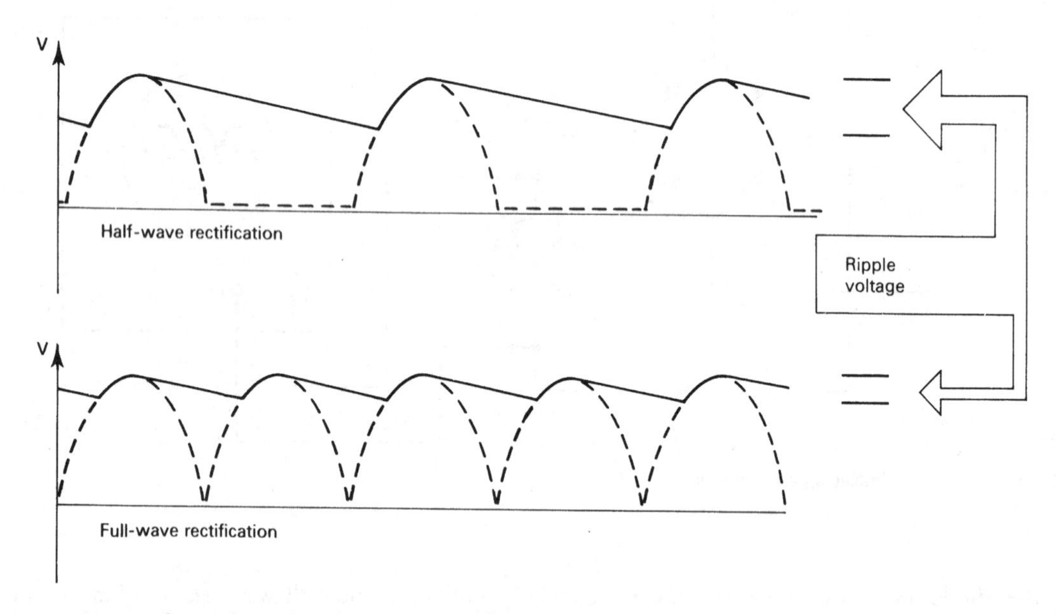

Fig. 10.46 Output waveforms with smoothing showing reduced ripple with full wave.

an output is to connect a large-value capacitor across the output terminals as shown in Fig. 10.45.

When the output from the rectifier is increasing, as shown by the dotted lines of Fig. 10.46, the capacitor charges up. During the second quarter of the cycle, when the output from the rectifier is falling to zero, the capacitor discharges into the load. The output voltage falls until the output from the rectifier once again charges the capacitor. The capacitor connected to the full-wave rectifier circuit is charged up twice as often as the capacitor connected to the half-wave circuit and, therefore, the output ripple on the full-wave circuit is smaller, giving better smoothing. Increasing the current drawn from the supply increases the size of the ripple. Increasing the size of the capacitor reduces the amount of ripple.

LOW-PASS FILTER

The ripple voltage of the rectified and smoothed circuit shown in Fig. 10.45 can be further reduced by adding a low-pass filter, as shown in Fig. 10.47. A low-pass filter allows low frequencies to pass while

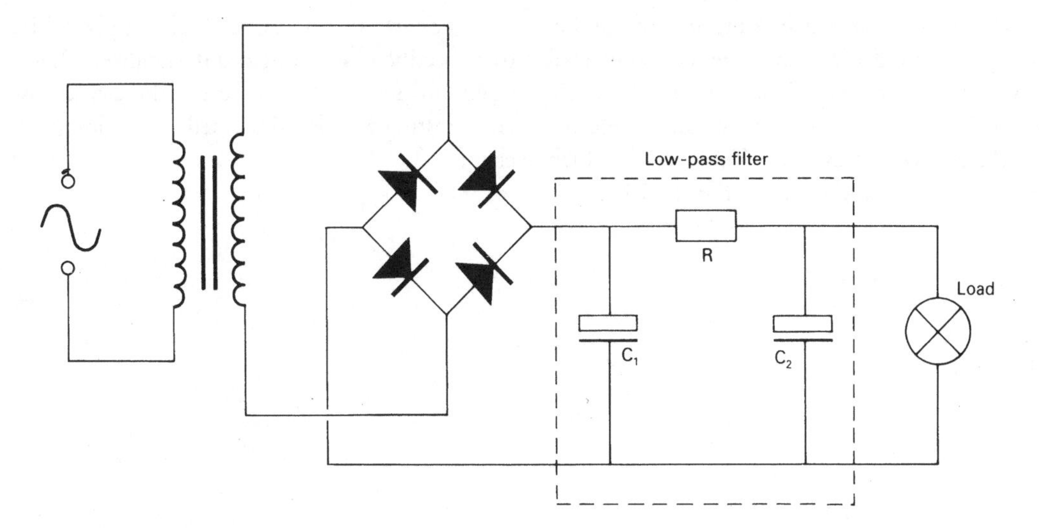

Fig. 10.47 Rectified a.c. with low-pass filter connected

blocking higher frequencies. Direct current has a frequency of zero hertz, while the ripple voltage of a full-wave rectifier has a frequency of 100 Hz. Connecting the low-pass filter will allow the d.c. to pass while blocking the ripple voltage, resulting in a smoother output voltage.

The low-pass filter shown in Fig. 10.47 does, however, increase the output resistance, which encourages the voltage to fall as the load current increases. This can be reduced if the resistor is replaced by a choke, which has a high impedance to the ripple voltage but a low resistance, which reduces the output ripple without increasing the output resistance.

Stabilized power supplies

The power supplies required for electronic circuits must be ripple-free, stabilized and have good regulation, that is the voltage must not change in

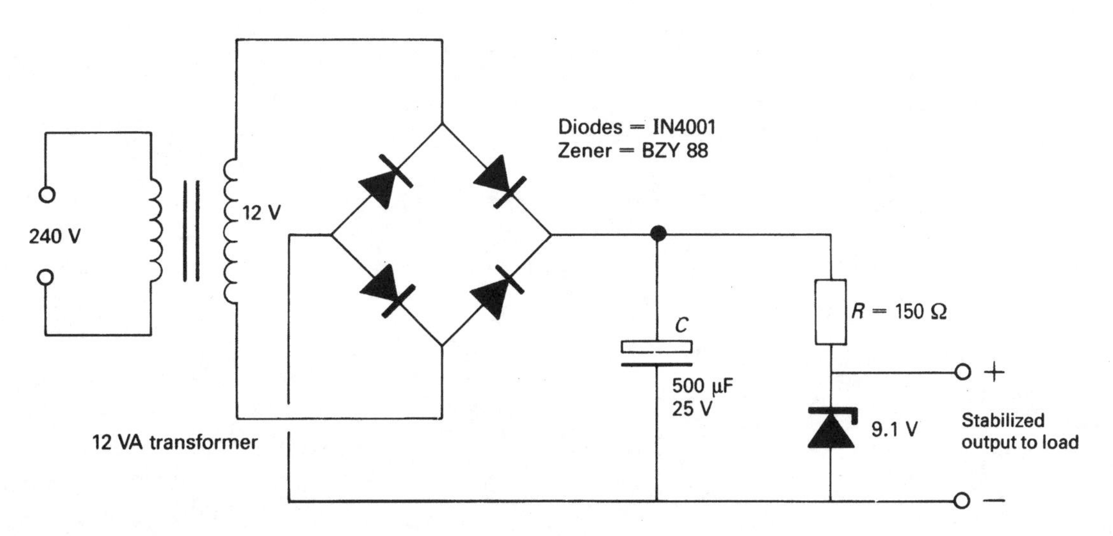

Fig. 10.48 Stabilized d.c. supply.

value over the whole load range. A number of stabilizing circuits are available which, when connected across the output of the circuit shown in Fig. 10.45, give a constant or stabilized voltage output. These circuits use the characteristics of the Zener diode which was described by the experiment in Fig. 10.19.

Figure 10.48 shows an a.c. supply which has been rectified, smoothed and stabilized. You could build and test this circuit using the circuit assembly and testing skills described in the next two chapters.

11

ELECTRONIC CIRCUIT ASSEMBLY

To get a 'feel' for electronics you should take the opportunity to build some of the simple circuits described in Chapter 10 using the constructional methods described in this chapter. Practical electronics can be carried out with very few tools and limited resources. A kitchen table, suitably protected, or a small corner of the electrical workshop is all that is required. The place chosen should be well lit, have a flat and dry area of about 1 m × 1 m and have access to a three-pin socket.

Working with others can also be a valuable source of inspiration and encouragement. Many technical colleges and evening institutes offer basic electronics courses which give someone new to electronics an opportunity to use the tools and equipment under guidance and at little cost. The City and Guilds of London offer many electronics examinations which are particularly suitable for electricians and service engineers who require a formal qualification in electronics.

Fig. 11.1 A plug-in RCD for safe electrical assembly.

Safety precautions

Electricity can be dangerous. It can give a serious shock and it does cause fires. For maximum safety, the sockets being used for the electronic test and assembly should be supplied by a residual current circuit breaker. These sense fault currents as low as 30 mA so that a faulty circuit or piece of equipment can be isolated before the lethal limit to human beings of about 50 mA is reached. Plug-in RCDs of the type shown in Fig. 11.1 can now be bought very cheaply from any good electrical supplier or DIY outlet. All equipment should be earthed and fitted with a 2 A or 5 A fuse, which is adequate for most electronics equipment. Larger fuses reduce the level of protection.

Another source of danger in electronic assembly is the hot soldering iron, which may cause burns or even start a fire. The soldering iron should always be placed in a soldering iron stand when not being used. The chances of causing a fire or burning yourself can be reduced by storing the soldering iron in its stand at the back of the workspace so that you do not have to lean over it when working.

So far we have been discussing the sensible safety precautions which everyone working with electricity should take. However, you or someone else in your workplace may receive an electric shock and Chapter 3 offers some guidance under the subheading 'Electric shock'.

Hand-tools

Tools extend the physical capabilities of the human body. Good-quality, sharp tools are important to any craftsman. An electrician or electronic service engineer is no less a craftsman than a wood carver. Each must work with a high degree of skill and expertise and each must have sympathy and respect for the materials which they use. The basic tools required by anyone working with electrical equipment are those used to strip, cut and connect conductors and components. The tools required for successful electronic assembly are wire strippers, diagonal cutters or snips, long-nose pliers and ordinary or combination pliers (Fig. 11.2). Electricians and electronic service engineers have traditionally chosen insulated hand-tools.

Soldering irons and guns

An electric soldering iron with the correct-size bit is essential for making good-quality, permanent connections in electronic circuits. A soldering iron consists of a heat-insulated handle, supporting a heating element of between 15 W and 25 W. The bit is inserted into this element and heats up to a temperature of about 210°C by conduction. Various sizes of bits are available and they are interchangeable.

Copper bits can be filed clean or rubbed with emery cloth until the tip is a bright copper colour. *Ironclad* bits must *not* be cleaned with a file or emery cloth but should be rubbed clean when they are hot, using a damp cloth or wet sponge.

Before the soldering iron can be used to make electrical connections, the bit must be *tinned* as follows:

- First, clean the bit as described above.
- Plug in the soldering iron and allow it to heat up.
- Apply cored solder to the clean hot bit.
- Wipe off the excess solder with a damp cloth or damp sponge.

This will leave the soldering iron brightly 'tinned' and ready to be used. Figure 11.3 shows a 230 V general-purpose soldering iron and stand suitable for use in electronic assembly.

Soldering guns of the type shown in Fig. 11.4 are trigger-operated soldering irons. Within 10 seconds of pressing the trigger the bit is at the working temperature of 315°C. The working temperature can be arrived at even more quickly with constant use. The plastic case holds a 230 V transformer having an isolated low-voltage high-current secondary circuit which is completed by the copper soldering bit. The bits are interchangeable and should be tinned and

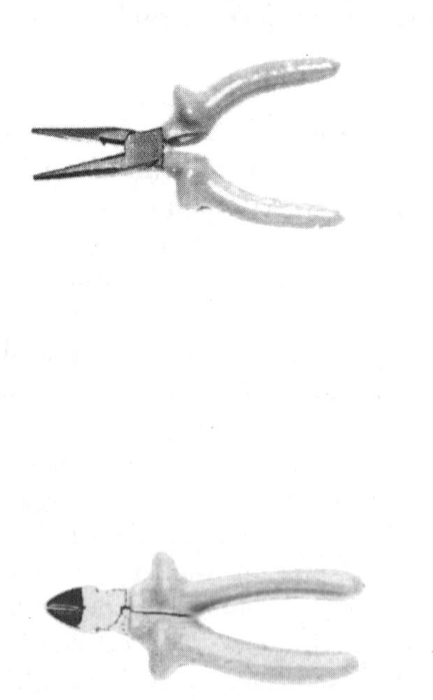

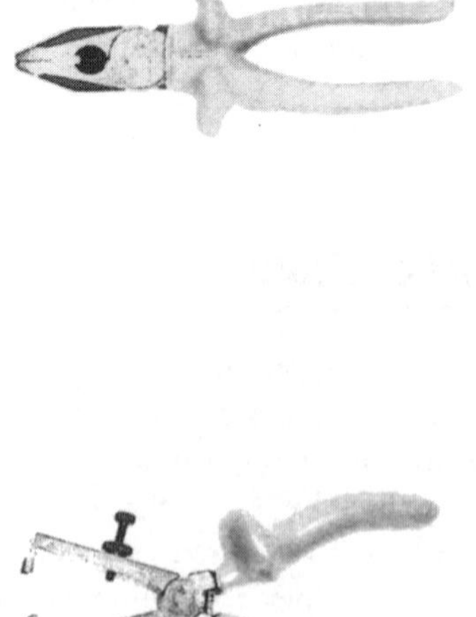

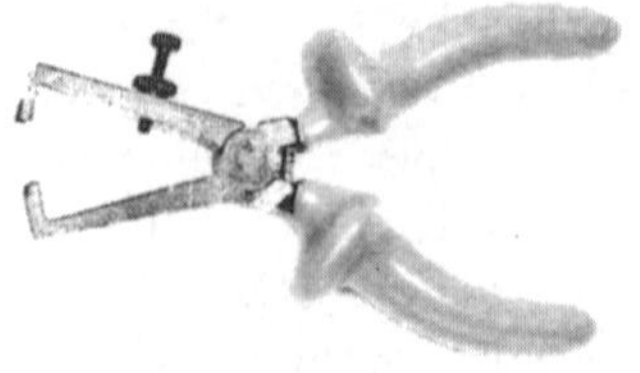

Fig. 11.2 Basic tools required for electronic assembly.

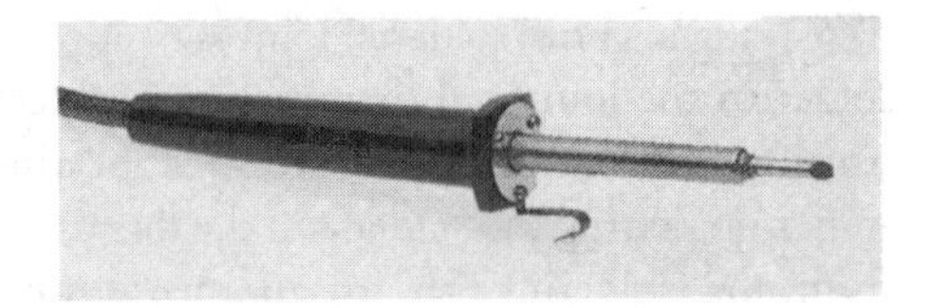

Fig. 11.3 Electronic soldering iron and stand.

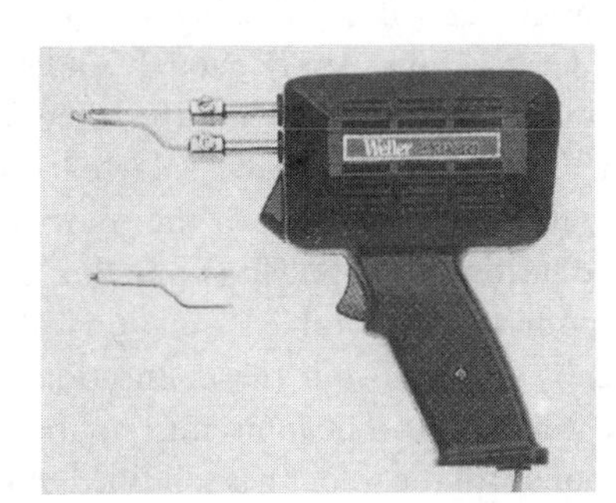

Fig. 11.4 Instant soldering gun.

used in the same way as the general-purpose iron considered above.

Butane gas-powered soldering irons are also available. In appearance they are very similar to the general-purpose soldering iron shown in Fig. 11.3, but without the mains cable. The handle acts as the fuel tank, various sizes of soldering bits are available and a protective cap is supplied to cover the hot end of the tool when not in use. The advantage of a gas soldering iron is that it can be used when a mains supply is not easily available.

The final choice of soldering iron will be influenced by many factors, frequency of use, where used, personal preference and cost. In 1997 the relevant costs were approximately £12 for the general-purpose iron, £30 for the soldering gun, £55 for the gas soldering iron and about £200 for a low-voltage temperature-controlled iron.

Soldering

There are many ways of making suitable electrical connections and in electrical installation work a screwed terminal is the most common method. In electronics, the most common method of making permanent connections is by soldering the components into the circuit. Good soldering can only be achieved by effort and practice and you should, therefore, take the opportunity to practise the technique before committing your skills to the 'real' situation.

Soldering is an alloying process, whereby a small amount of soft metal (the solder) is made to run between the two metals to be joined, therefore mixing or alloying them. Solder can be used to join practically any metals or alloys except those containing large amounts of chromium or aluminium, which must be welded or hard-soldered.

SOFT SOLDERS

Soft solders are so called because they are made up of the rather soft metals tin and lead in the proportion 40 to 60. Solders containing tin will adhere very firmly to most metals, providing that the surfaces of the metals to be joined are clean. Solder will not adhere to a tarnished or oxidized metal surface. This is because solder adheres by forming an alloy with the metal of the connection and this alloy cannot form if there is a film of oxide in the way.

FLUXES

Fluxes are slightly acid materials which dissolve an oxide film, leaving a perfectly clean surface to which the solder can firmly adhere. There are two types of flux in common use – salt flux and rosin flux. Salt fluxes are rather corrosive and are therefore used when joining iron, steel, nickel and stainless steel, which oxidize easily when hot. Rosin fluxes are less corrosive, and, when used in the form of a sticky paste, are the preferred fluxes for soldering tin, copper and brass; they are thus the most suitable for electrical and electronic work. In electronics it is not convenient to apply the flux and solder separately, so they are combined as flux-cored solder wire. This is solder wire with a number of cores of flux running the whole length of the wire. The multicore construction shown in Fig. 11.5 ensures the correct proportions of flux for each soldered joint.

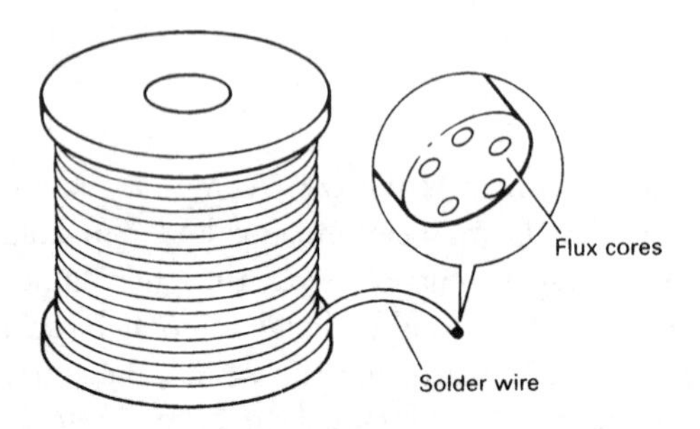

Fig. 11.5 Construction of flux-cored solder wire.

SOLDERING TECHNIQUES

As already mentioned, when soldering with an iron it is important to choose an iron with a suitable bit size. A 1.5 mm or 2.0 mm bit is suitable for most electronic connections, but a 1.0 mm bit is better when soldering dual-in-line IC packages. The bit should be clean and freshly 'tinned'. The materials to be soldered must be free from grease and preferably pretinned. Electronic components should not need more cleaning than a wipe to remove dust or grease. The purpose of the soldering iron is to apply heat to the joint. If solder is first melted on to the bit, which is then used to transfer the solder to the joint, the active components of the flux will evaporate before the solder reaches the joint, and an imperfect or 'dry' joint will result. Also applying the iron directly to the joint oxidizes the component surfaces, making them more difficult to solder effectively. The best method of making a 'good' soldered joint is to apply the cored solder to the joint and then melt the solder with the iron. This is the most efficient way of heating the termination, letting the solder and the flux carry the heat from the soldering bit on to the termination, as shown in Fig. 11.6.

While the termination is heating up, the solder will appear dull, and then quite suddenly the solder will become bright and fluid, flowing around and 'wetting' the termination. Apply enough solder to cover the termination before removing the solder and then the iron. The joint should be soldered quickly. If attempts are made to improve the joint merely by continued heating and applying more flux and solder, the component or the cable insulation will become damaged by the heat and the connection will have excessive solder on it. The joint must not be moved or blown upon until the solder has solidified. A good soldered joint will appear smooth and bright, a bad connection or *dry joint* will appear dull and the solder may be in a 'blob' or appear spiky.

Dry joints

Dry joints may occur because the components or termination are dirty or oxidized, or because the soldering temperature was too low, or too little flux was used. Dry joints do not always make an electrical connection, or the connection will have a high resistance which deteriorates with time and may cause trouble days or weeks later. A suspected dry joint can be tested as shown in Fig. 11.7. If the joint is dry the voltmeter will read 12 V at position B, just to the right of the

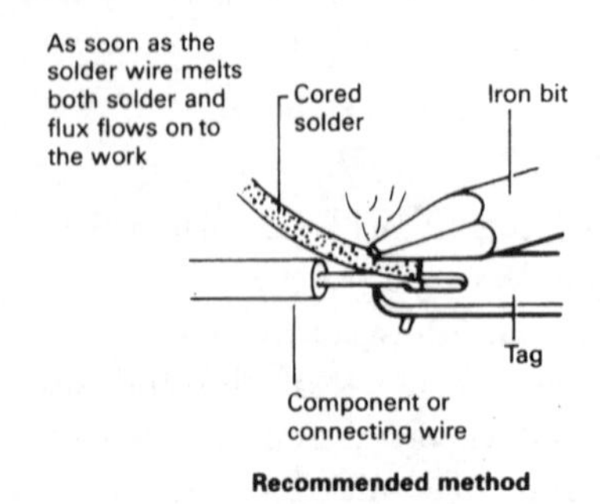

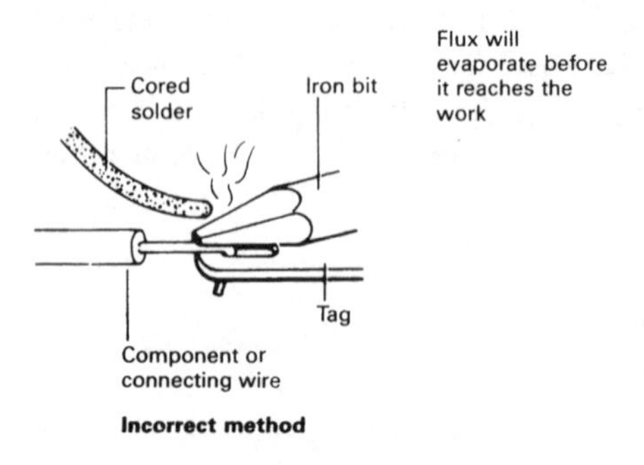

Fig. 11.6 Soldering technique with multicore solders.

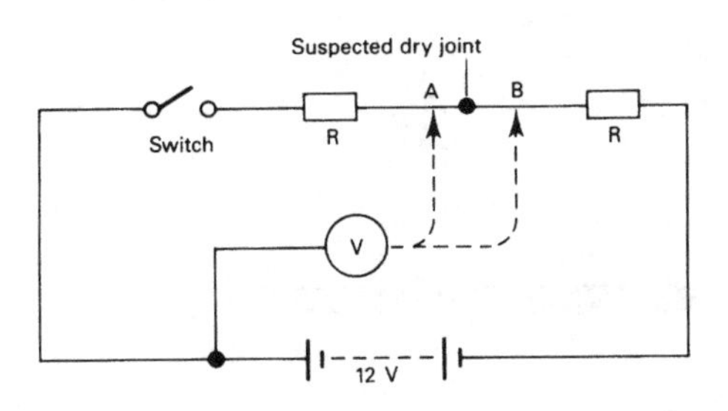

Fig. 11.7 Testing a suspected dry joint.

final pinch-through, identified by a sharp click, may fracture the soldered joint or damage the component.

Most electronic components are very sensitive and are easily damaged by excess heat. Soldered joints must not, therefore, be made close to the body of the component or the heat transferred from the joint may cause some damage. When components are being soldered into a circuit the heat from the soldering iron at the connection must be diverted or 'shunted' away from the body of the component. This can be done by

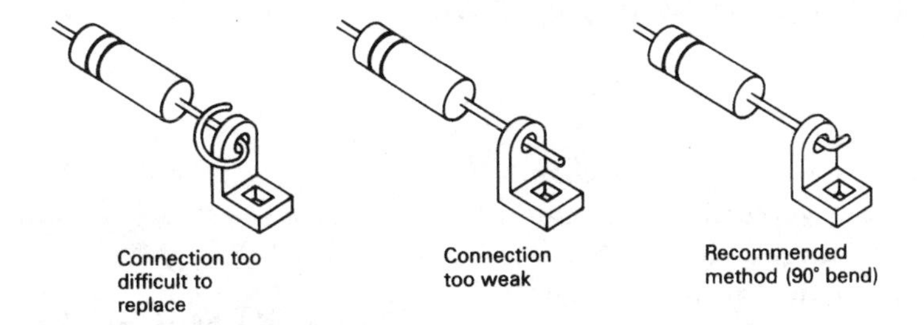

Fig. 11.8 Shaping conductors to give strength to electrical connections to tag terminals.

joint, and 0 V at position A, to the left. If the joint is found to be dry, the connection must be remade.

COMPONENT ASSEMBLY AND SOLDERING

Soft solder is not as strong as other metals and, therefore, the electronic components must be shaped at the connection site to give extra strength. This can be done by bending the connecting wires so that they hook together or by making the joint area large. Special *lead-forming* or *wire-shaping* tools are available which both cut and shape the components' connecting wires ready for soldering. Figure 11.8 shows a suitable tag terminal connection, Fig. 11.9 a suitable pin terminal connection and Fig. 11.10 a suitable stripboard connection.

All wires must be cut to length before assembly because it is often difficult to trim them after soldering. Also the strain of cutting after soldering may weaken the joint and encourage dry joints. If the wires must be cut after soldering, cutters with a shearing action should be used, as shown in Fig. 11.11. Side cutters have a pinching action and the shock of the

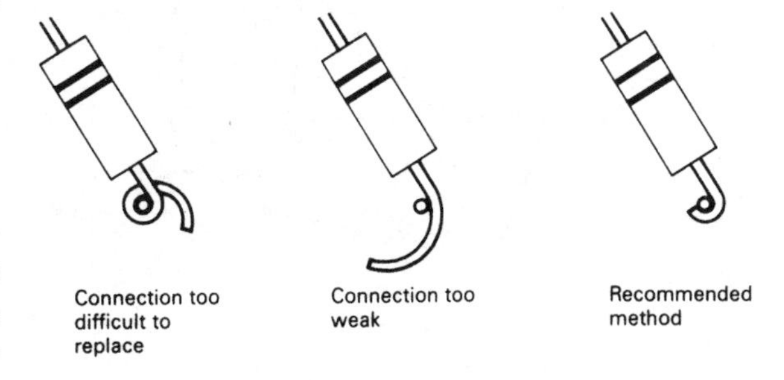

Fig. 11.9 Shaping conductors to give strength to electrical connections to pin terminals (plan view).

placing a pair of long-nosed pliers or a crocodile clip between the soldered joint and the body of the component, as shown in Fig. 11.12.

Components such as resistors, capacitors and transistors are usually cylindrical, rectangular or disc-shaped, with round wire terminations. They should be shaped, mounted and soldered into the circuit as previously described and shown in Fig. 11.13. A small clearance should be left between the body of the com-

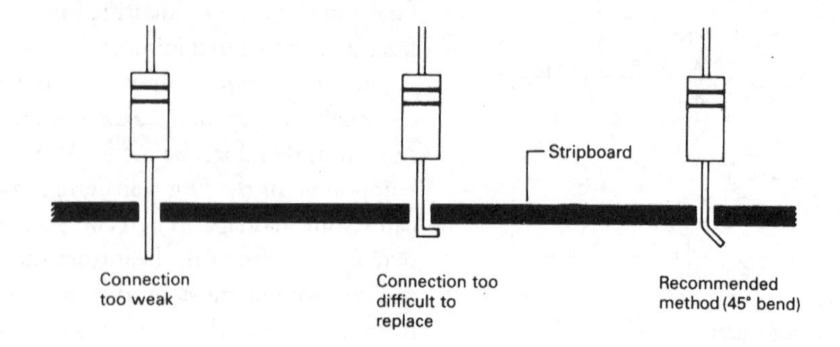

Fig. 11.10 Shaping conductors to give strength to electrical connections to stripboard.

ponent and the circuit board, to allow convection currents to circulate, which encourages cooling. The vertical mounting method permits many more components to be mounted on the circuit board but the horizontal method gives better mechanical support to the component.

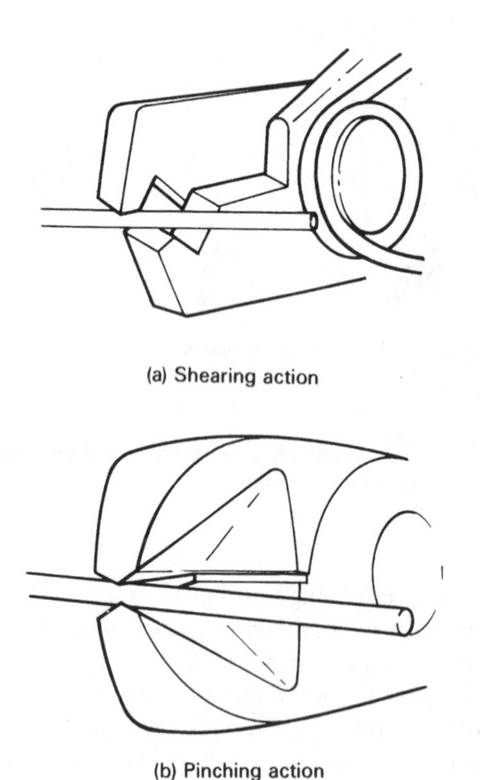

Fig. 11.11 Wire cutting.

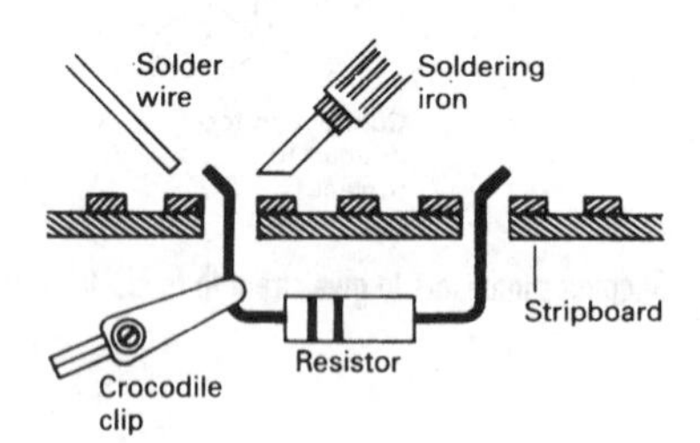

Fig. 11.12 Using a crocodile clip as a heat shunt.

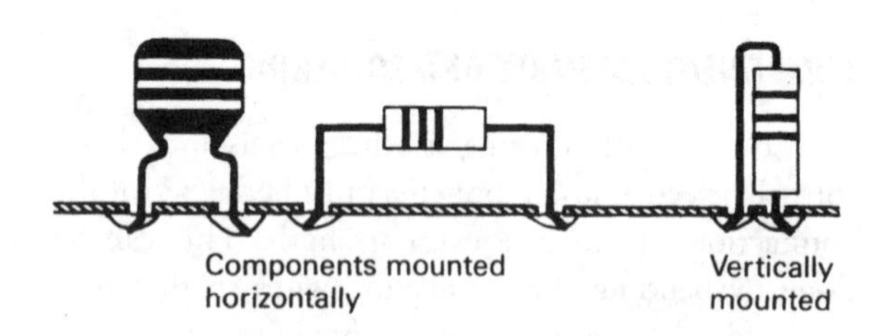

Fig. 11.13 Vertical and horizontal mounting of components.

Desoldering

If it is necessary to replace an electronic component, the old, faulty component must first be removed from the circuit board. To do this the solder of the old joint is first liquefied by applying a hot iron to the joint. The molten solder is then removed from the joint with a desoldering tool. The desoldering tool works like a bicycle pump in reverse and is shown in Fig. 11.14. The tool is made ready by compressing the

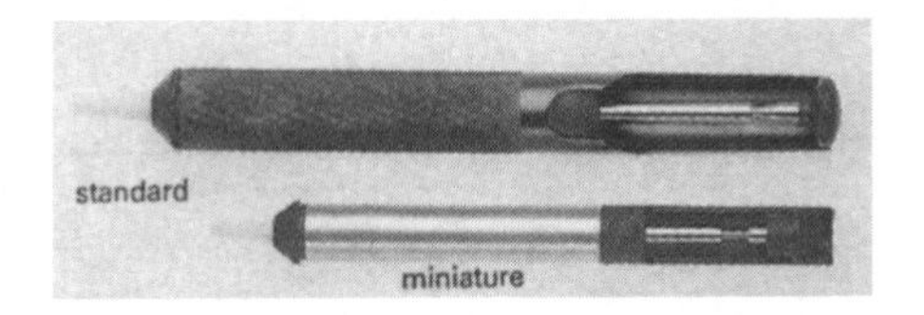

Fig. 11.14 A desoldering tool.

piston down on to a latch position which holds it closed. The nozzle is then placed into the pool of molten solder and the latch release button pressed. This releases the plunger which shoots out, sucking the molten solder away from the joint and into the body of the desoldering tool.

REMOVING FAULTY TRANSISTORS

First identify the base, collector and emitter connections so that the new component can be correctly connected into the circuit. Remove the solder from each leg with the soldering iron and desoldering tool before removing the faulty transistor. Then, with the aid of a pair of long-nosed pliers, pull the legs of the transistor out of the circuit board. An alternative method is to cut the three legs with a pair of side cutters before desoldering and then remove the individual legs with a pair of long-nosed pliers.

REMOVING FAULTY INTEGRATED CIRCUITS (CHIPS)

Remove the solder from each leg of the IC with the soldering iron and desoldering tool and then pull the IC clear of the circuit board. If it has been firmly established that the IC is faulty, it may be removed from the circuit board by cutting the body from the connecting pins before desoldering and removing the individual pins with a pair of long-nosed pliers.

Circuit boards

Permanent circuits require that various discrete components be soldered together on some type of insulated board. Three types of board can be used – matrix, strip and printed circuit board – the base material being synthetic resin bonded paper (SRBP).

MATRIX BOARDS

Matrix boards have a matrix of holes on 0.1 inch centres as shown in Fig. 11.15. Boards are available in various sizes: the 149 × 114 mm board is pierced with 58 × 42 holes and the 104 × 65 mm board has 39 × 25 holes. Matrix pins press into any of the holes in the board and provide a terminal post to which components and connecting wires can be soldered. Single-sided or double-sided matrix pins are available. Double-sided pins have the advantage that connections can be made on the underside of the board as well as on the top. The hole spacing of 0.1 inch makes the board compatible with many electronic components. Plug-in relays, DIL integrated circuits and many sockets and connectors all use 0.1 inch spacing at their connections.

Matrix board is probably the easiest and cheapest way to build simple electronic circuits. It is recommended that inexperienced circuit builders construct the circuit on the matrix board using a layout which is very similar to the circuit diagram to reduce the possibility of mistakes.

Suppose, for example, that we intend to build the very simple circuit shown in Fig. 11.16. First we would insert four pins into the matrix board as shown. The diode would then be connected between

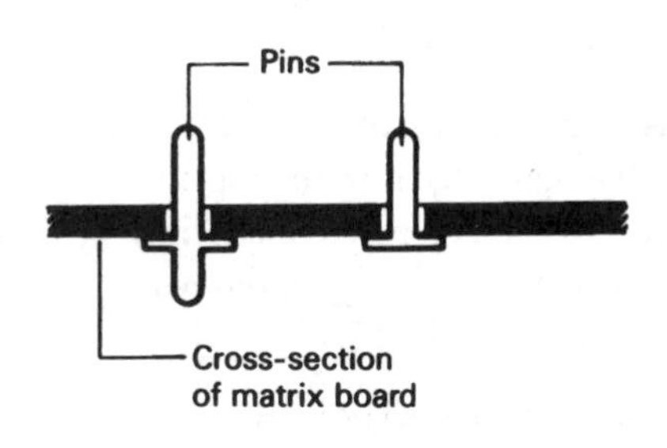

Fig. 11.15 Matrix board and double-sided and single-sided pin inserts.

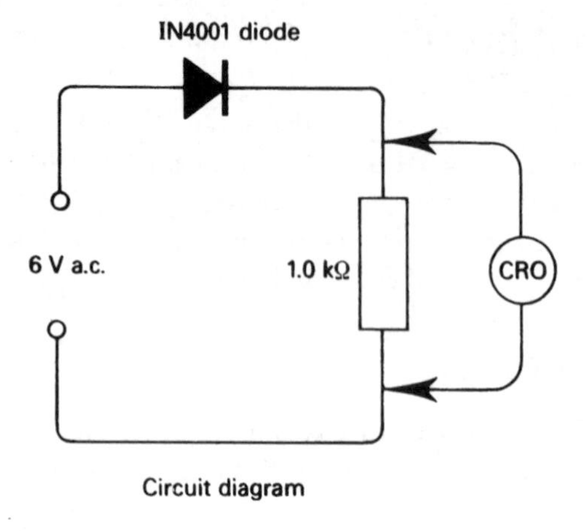

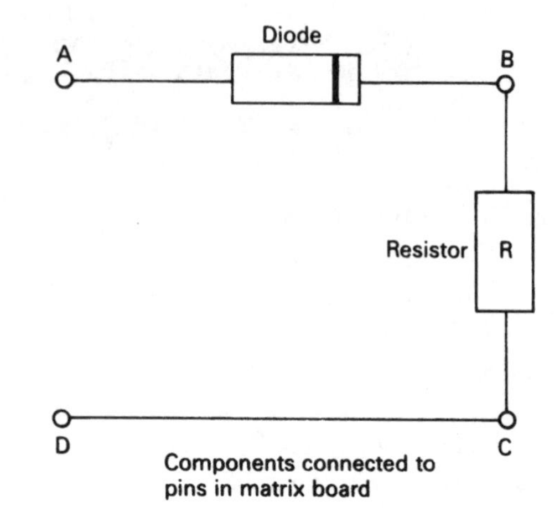

Fig. 11.16 Circuit diagram converted to a component layout.

pins A and B, taking care that the anode was connected to pin A. The resistor would be connected between pins B and C and a wire linked between pins C and D. The a.c. supply from a signal generator would be connected to pins A and D by 'flying' leads and the oscilloscope leads to pins B and C. This circuit would show half-wave rectification. When planning the conversion of circuit diagrams into a matrix board layout it helps to have a positional reference system so that we know where to push the pins in the matrix board.

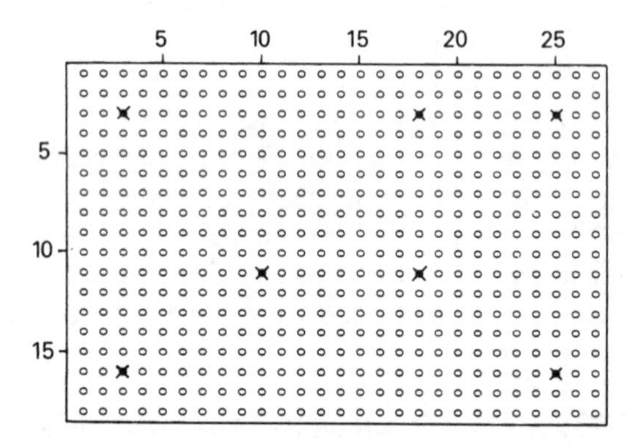

Fig. 11.17 Matrix board pin reference system.

The positional reference system

The positional reference system used with matrix boards uses a simple grid reference system to identify holes on the board. This is achieved by counting along the columns at the top of the board, starting from the left and then counting down the rows. For example, the position reference point 4 : 3 would be 4 holes from the left and 3 holes down. The board should be prepared as follows:

- Turn the matrix board so that a manufactured straight edge is to the top and left-hand side.
- Use a felt tip pen to mark the holes in groups of five along the top edge and down the left-hand edge as shown in Fig. 11.17.

The pins can then be inserted as required. Figure 11.17 shows a number of pin reference points. Counting from the left-hand side of the board there are 3 : 3, 3 : 16, 10 : 11, 18 : 3, 18 : 11, 25 : 3 and 25 : 16.

STRIPBOARD OR VEROBOARD

Stripboard or Veroboard is a matrix board with a continuous copper strip attached to one side by adhesive. The copper strip links together rows of holes so that connections can be made between components inserted into holes on a particular row, as shown in Fig. 11.18. The components are assembled on the plain board side with the component leads inserted through the holes and soldered in place to the copper strip.

The copper strips are continuous but they can be broken using a strip cutter or small drill. The drill or cutter is placed on the hole where the break is to be made and then rotated a few turns between the fingers until the very thin copper strip is removed leaving a circle, as shown in Fig. 11.18.

Stripboard is very useful because the copper strips take the place of the wire links required with plain

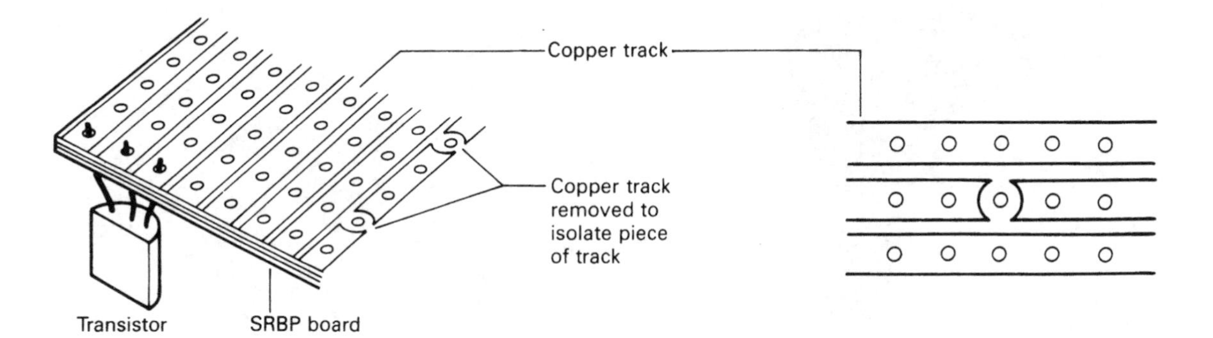

Fig. 11.18 Stripboard or Veroboard.

matrix boards. Components can easily be mounted vertically on stripboard, which leads to high-density small-area circuits being assembled. It is, however, more expensive than matrix board because of the additional cost of the copper, most of which is not used in the circuit. Also some translation of the circuit diagram is required before it can be assembled on the stripboard. Excessive heat from the soldering iron can melt the adhesive and cause the copper to peel from the insulating board. Heat should only be applied long enough to melt the solder and secure the component.

PRINTED CIRCUIT BOARDS

Printed circuit boards (PCBs) are produced by chemically etching a copper-clad epoxy glass board so that a copper pattern is engraved on one side of the board. The pattern provides the copper conducting paths making connections to the various components of the circuit. After etching, small holes are drilled for the components which are inserted from the plain side of the board and soldered on to the copper conductor. The copper pattern replaces the lengths of wire used to connect components on to the matrix board.

The copper foil is very thin and is attached to the board with an adhesive. Excessive heat from the soldering iron can melt the adhesive and cause the track to peel from the insulating board. The board should not be flexed, otherwise hairline cracks may appear but go unnoticed until intermittent faults occur in the circuit later.

The process of designing and making a PCB is quite simple but does require specialized equipment, and for that reason we will not consider it further here.

Wire wrapping

As the electronic industry increasingly uses advanced technology, the demand for a faster, more reliable and inexpensive method of making electrical connections has increased. Electronic equipment today has lots of components with many terminals in a very small space. To make electrical connections to these high-density electronic circuits, the electronic industry has developed a new solderless wire-wrapping technique.

A wire-wrapping connection is made by winding a special insulated wire of 30 AWG (0.25 mm) around the sharp corners of a square pin inserted into the circuit board. The winding tension causes the corners of the pin to cut into the wire, producing a good electrical connection which will not unwind and is as good as or better than a soldered joint. This method of connection was developed by the Bell Telephone Laboratories of the Western Electric Company.

Pneumatic or electric tools are preferred for production work but battery- or hand-operated tools, as shown in Fig. 11.19, are used for service and repair work.

To make the connection, proceed as follows:

- The end of the wire is inserted into the bit of the tool and then bent back at 90° to the tool.
- The bit is placed over the terminal pin.
- The tool is rotated clockwise a few turns without undue pressure.
- The bit is removed from the terminal pin and the connection is made.

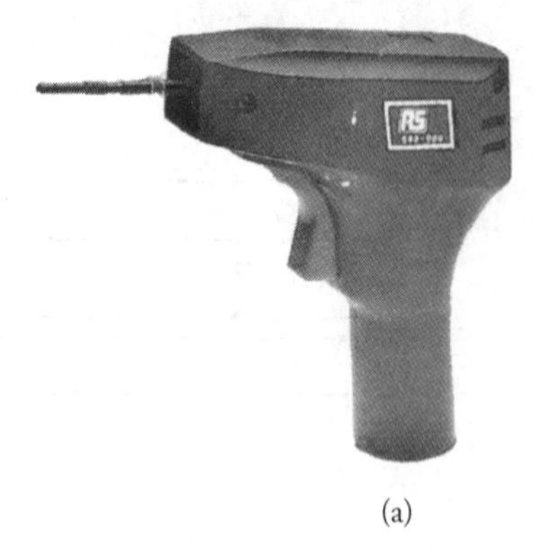

(a)

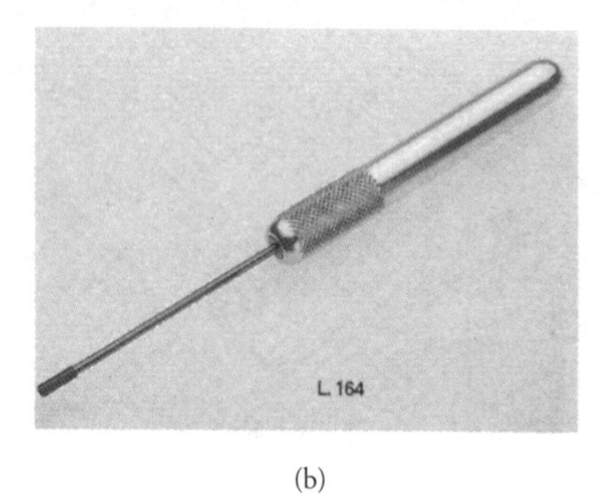

(b)

Fig. 11.19 Wire-wrapping tools: (a) battery-operated; (b) hand-operated.

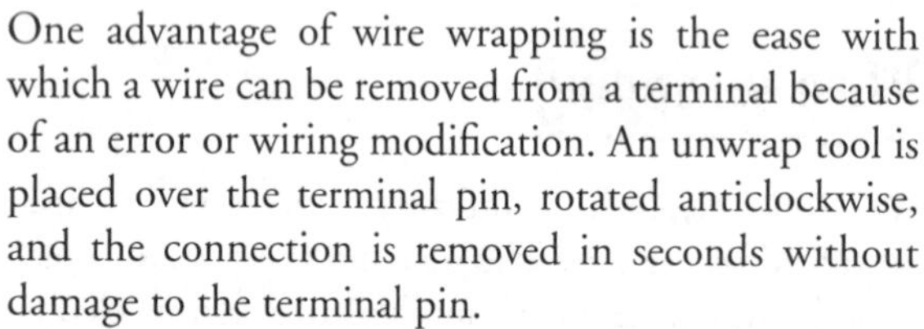

One advantage of wire wrapping is the ease with which a wire can be removed from a terminal because of an error or wiring modification. An unwrap tool is placed over the terminal pin, rotated anticlockwise, and the connection is removed in seconds without damage to the terminal pin.

Wire wrapping is a precision technique and the bit size and wire diameter must be compatible with the terminal pin size if a good electrical connection is to be made.

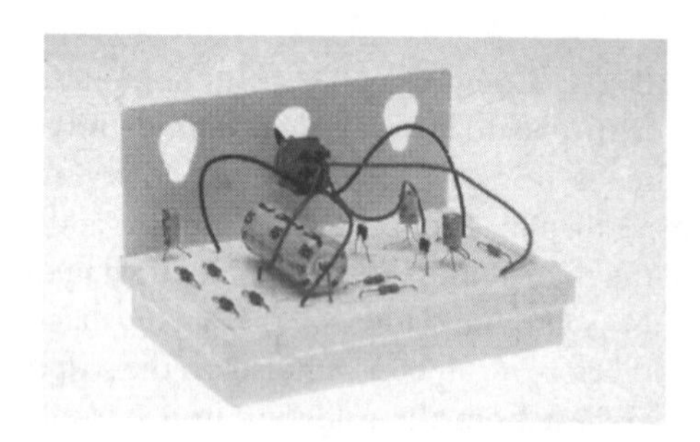

Fig. 11.20 S-DeC prototype board used for temporary circuit building.

Breadboards

Breadboarding is the name given to solderless temporary circuit building by pressing wires and component leads into holes in the prototype board. This method is used for building temporary circuits for testing or investigation.

THE S-DEC

The S-DeC prototype board is designed for interconnecting discrete components. The hole spacing and hole connections do not suit DIL IC packages. Each board has 70 phosphor-bronze contact points arranged in two sections, each of which has seven parallel rows of five connected contact points. The case is formed in high-impact polystyrene and the individual boards may be interlocked to create a larger working area. The S-DeC can be supplied with a vertical bracket for mounting switches or variable resistors, as shown in Fig. 11.20.

THE PROFESSIONAL

The professional prototype board is designed for the interconnection of many different types of component. The hole spacing of 0.1 inch allows DIL IC packages to be plugged directly into the board. Each board has 47 rows of five interconnected nickel-silver contacts each side of a central channel and a continuous row at the top and bottom which may be used as power supply rails. The case is formed in high-impact thermoplastic and the individual boards may be interlocked to create a larger working area. A vertical side bracket is also supplied for mounting switches or variable resistors, as shown in Fig. 11.21.

Interconnection methods

A plug and socket provide an ideal method of connecting or isolating components and equipment

Fig. 11.21 Professional prototype board used for temporary circuit building.

which cannot be permanently connected. In electrical installation work we usually need to make plug and socket connections between three conductors on single-phase circuits and five conductors on three-phase systems. In electronics we often need to make multiple connections between circuit boards or equipment. However, the same principles apply, that is, the plug and socket must be capable of separation, but while connected they must make a good electrical contact. Also the plug and socket must incorporate some method of preventing reverse connection.

PCB EDGE CONNECTORS

A range of connectors are available which make direct contact to printed circuit boards, as shown in Fig. 11.22. Multiple connectors are available with a contact pitch of 0.1 inch so that they can be soldered into circuit boards. The plug and socket can then be used as edge connectors to make board-to-board and cable-to-board connections.

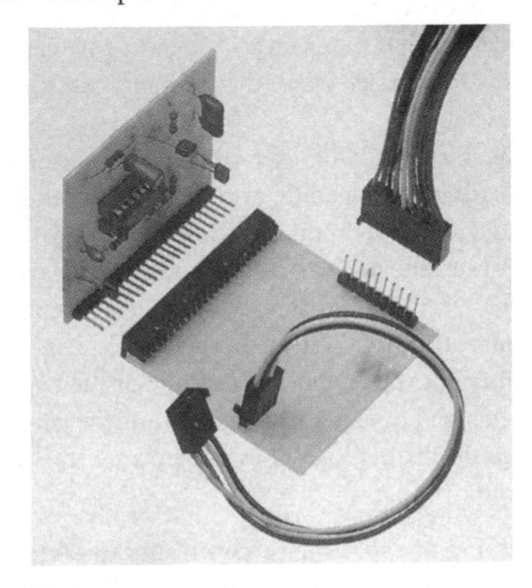

Fig. 11.22 PCB edge connectors.

RIBBON CABLE CONNECTORS

A ribbon cable is a multicore cable laid out as flat strip or ribbon strip. A range of connectors is made to connect ribbon cable, as shown in Fig. 11.23, which is used for making board-to-board interconnections and to connect computer peripherals such as VDUs and printers.

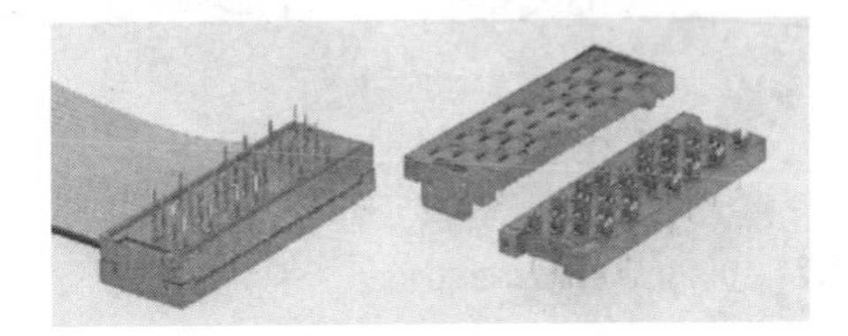

Fig. 11.23 Ribbon cable connector.

DIN CONNECTORS

DIN-style audio connectors are available for making up to eight connections, as shown in Fig. 11.24, and used when frequent connection and disconnection is required between a small number of contacts.

JACK CONNECTORS

Jack connectors are used when frequent connections are to be made between two or three poles on, for example, headphones or microphones. They are available in three sizes: subminiature (2.5 mm), miniature (3.5 mm) and commercial (0.25 inch). Examples are shown in Fig. 11.25.

All the above connectors may be terminated on to the cable end by a soldered, crimped or cable displacement method. When a soldered connection is to be made the cable end must be stripped of its insulation, tinned and then terminated. A crimped connection also requires that the insulation be removed and the prepared cable end inserted into a lug, which is then crimped using an appropriate tool. The insulation displacement method of connection is much quicker to make because the cable ends do not require stripping or preparing. The connection is made by pressing insulation piercing tines or prongs into the cable which displaces the insulation to make an

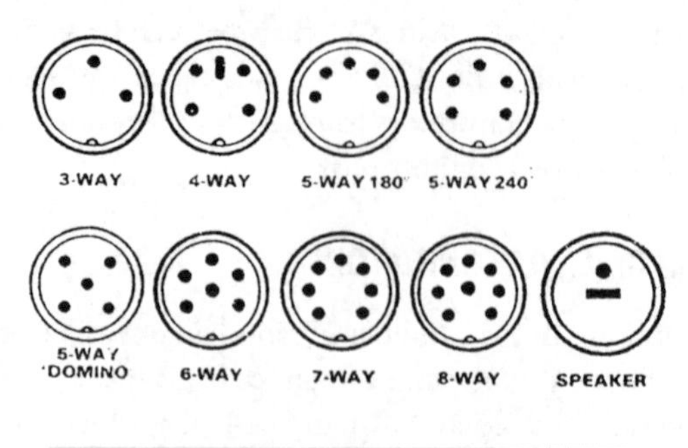

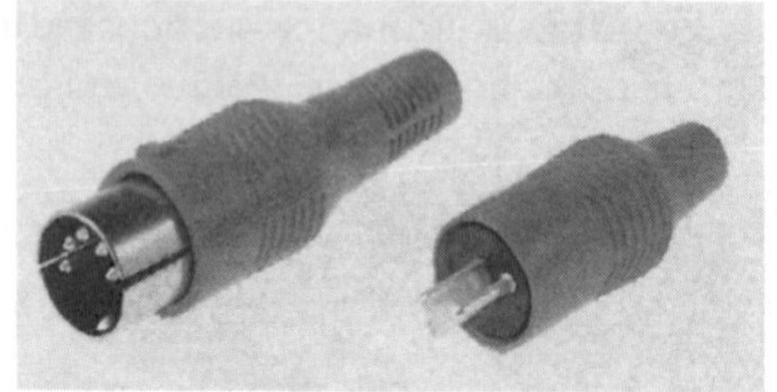

Fig. 11.24 DIN-style audio connectors.

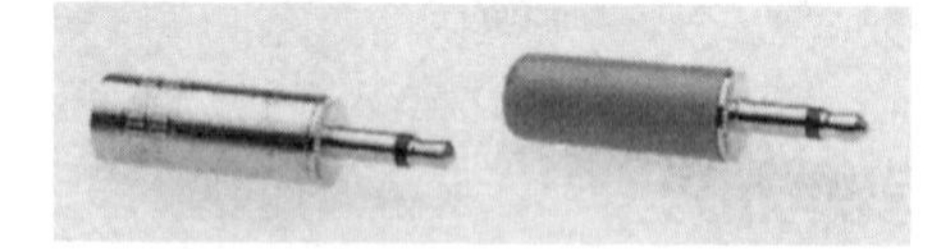

Fig. 11.25 Jack connectors.

electrical connection with the conductor. This method is used extensively when terminating ribbon cable and for making rapid connections to the existing wiring system of motor vehicles.

Fault finding

The best way to avoid problems in electronic circuit assembly is to be always alert while working, to think about what you are doing and to try always to be neat. If, despite your best efforts, the circuit does not work as it should when tested, then follow a logical test procedure which will usually find the faults in the shortest possible time. First carry out a series of visual tests:

1 Is the battery or supply correctly connected?
2 Is the battery flat or the supply switched off?
3 Is the circuit constructed *exactly* as it should be according to the circuit diagram?
4 Are all the components in place?
5 Check the values of all the components.
6 Are all the components such as diodes, capacitors, transistors and ICs connected the correct way round?
7 Have all connections and links been made?
8 Have all the necessary breaks been made in the stripboard?
9 Are all the soldered joints good?
10 Are any of the components hot or burnt?

If the fault has not been identified by the first ten tests, ask someone else to carry them out. You may have missed something which will be obvious to a fresh pair of eyes. If the visual tests have failed to identify the fault, then further meter tests are called for as follows:

11 Check the input voltage and the output voltage. Check the mid-point voltage between components which are connected in series with the supply.
12 Variable resistors may suffer from mechanical wear. Check the voltage at the wiper as well as across the potentiometer.
13 Check the coil voltage on relays; if this is low, the coil contacts may not be making.
14 Is the diode connected correctly? Short circuit the diode momentarily with a wire link to see if the circuit works. If it does the diode is open-circuit.
15 Check capacitor–resistor circuits by momentarily shorting out the capacitor and then observing the charging voltage. If it does not charge, the resistor may be open-circuit. If it charges instantly, the resistor may be short-circuit. Check the polarity of electrolytic capacitors. Check the capacitor leads for breaks where the lead enters the capacitor body.
16 Check the base–emitter voltage of the transistor. A satisfactory reading would be between 0.6 V and 1.0 V. Temporarily connect a 1 kΩ resistor between the positive supply and the base connection. If the transistor works, the base feed is faulty. If it does not work, the transistor is faulty.
17 Short out the anode and cathode of the thyristor. If the load operates, the thyristor or the gate pulse is faulty. If the load does not operate, the load is faulty.

The testing of capacitors, resistors and discrete semiconductor components is dealt with in Chapter 10.

12

ELECTRONIC TEST EQUIPMENT

The use of electronic circuits in all types of electrical equipment has increased considerably over recent years. Electronic circuits and components can now be found in leisure goods, domestic appliances, motor starting and control circuits, discharge lighting, emergency lighting, alarm circuits and special-effects lighting systems. There is, therefore, a need for the installation electrician and service engineer to become familiar with some basic electronic test equipment, which is the aim of this chapter.

Test instruments

Electrical installation circuits usually carry in excess of 1 A and often carry hundreds of amperes. Electronic circuits operate in the milliampere or even microampere range. The test instruments used on electronic circuits must have a *high impedance* so that they do not damage the circuit when connected to take readings. All instruments cause some disturbance when connected into a circuit because they consume some power in order to provide the torque required to move the pointer. In power applications these small disturbances seldom give rise to obvious errors, but in electronic circuits, a small disturbance can completely invalidate any readings taken. We must, therefore, choose our electronic test equipment with great care. Let us consider some of the problems.

Let me first of all define what is meant by the terms 'error' and 'accuracy' used in this chapter. When the term *error* is used it means the *deviation of the meter reading from the true value* and *accuracy* means the *closeness of the meter reading to the true value.*

INSTRUMENT ERRORS

Consider a voltmeter of resistance 100 kΩ connected across the circuit shown in Fig. 12.1(a).

Connection of the meter loads the circuit by effectively connecting a 100 kΩ resistor in parallel with the circuit resistor as shown in Fig. 12.1(b), which changes the circuit to that shown in Fig. 12.1(c).

Common sense tells us that the voltage across each resistor will be 100 V but the meter would read about 67 volts because connection of the meter has changed the circuit. This loading effect can be reduced by choosing instruments which have a very high impedance. Such instruments impose less load on the circuit and give an indication much closer to the true value.

Test instruments used to measure a.c. supplies are also frequency-dependent. The important practical consideration is the *frequency range* of the test instrument. This is the range of frequencies over which the instrument may be considered free from frequency errors and is indicated on the back of the instrument or in the manufacturer's information. Frequency limitations are not a normal consideration for an electrician since electrical installations operate at the fixed mains frequency of 50 Hz.

The scale calibration of an instrument assumes a sinusoidal supply unless otherwise stated. Non-sinusoidal or complex waveforms contain harmonic frequencies which may be outside the instrument frequency range. The chosen instrument must, therefore, be suitable for the test circuit waveform.

The maximum permissible errors for various instruments and their applications are indicated in British Standard 89. When choosing an instrument for electronic testing an electrician or service engineer

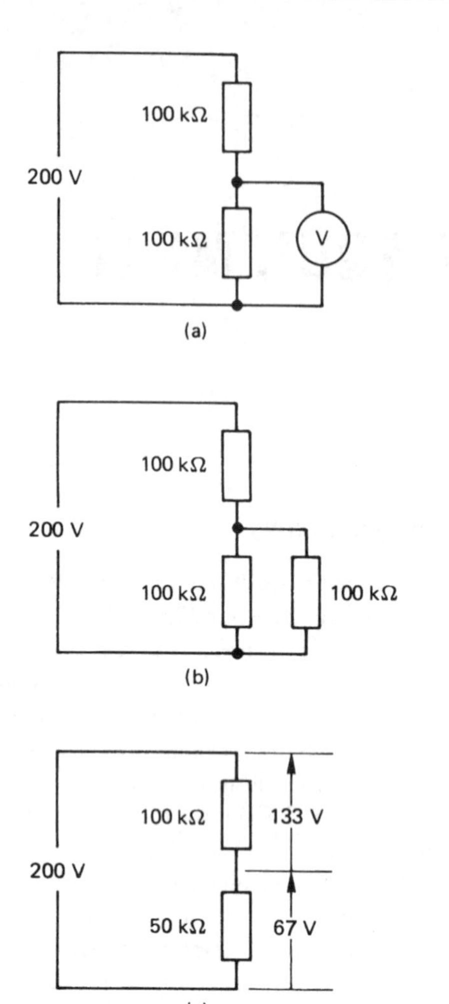

Fig. 12.1 Circuit disturbances caused by the connection of a voltmeter.

will probably be looking for an instrument with about a 2% maximum error, that is, 98% accurate. Instrument manufacturers will provide detailed information for their products.

OPERATOR ERRORS

Errors are not restricted to the instrument being used; operators can cause errors, too. Operator errors are errors such as misreading the scale, recording the measurement incorrectly or reading the wrong scale on a multirange instrument. The test instrument must be used on the most appropriate scale: do not try to read 12 V on a 250 V scale, the reading will be much more accurate if the 25 V scale is used.

Analogue and digital displays

The type of instrument to be purchased for general use is a difficult choice because there are so many different types on the market and every manufacturer's representative is convinced that his company's product is the best. However, most instruments can be broadly grouped under two general headings: those having *analogue* and those with *digital* displays.

ANALOGUE METERS

These meters have a pointer moving across a calibrated scale. They are the only choice when a general trend or variation in value is to be observed. Hi-fi equipment often uses analogue displays to indicate how power levels vary with time, which is more informative than a specific value. Red or danger zones can be indicated on industrial instruments. The fuel gauge on a motor car often indicates full, half full or danger on an analogue display which is much more informative than an indication of the exact number of litres of petrol remaining in the tank.

These meters are only accurate when used in the calibrated position – usually horizontally.

Most meters using an analogue scale incorporate a mirror to eliminate parallax error. The user must look straight at the pointer on the scale when taking readings and the correct position is indicated when the pointer image in the mirror is hidden behind the actual pointer. A good-quality analogue multimeter suitable for electronic testing is shown in Fig. 12.4.

The input impedance of this type of instrument is typically 1000 Ω per volt or 20 000 Ω per volt, depending upon the scale chosen.

DIGITAL METERS

These provide the same functions as analogue meters but they display the indicated value using a seven-segment LED (see Fig. 10.22) to give a numerical value of the measurement. Modern digital meters use semiconductor technology to give the instrument a very high input impedance, typically about 10 MΩ and, therefore, they are ideal for testing most electronic circuits.

The choice between an analogue and a digital display is a difficult one and must be dictated by specific circumstances. However, if you are an electrician or service engineer intending to purchase a new instru-

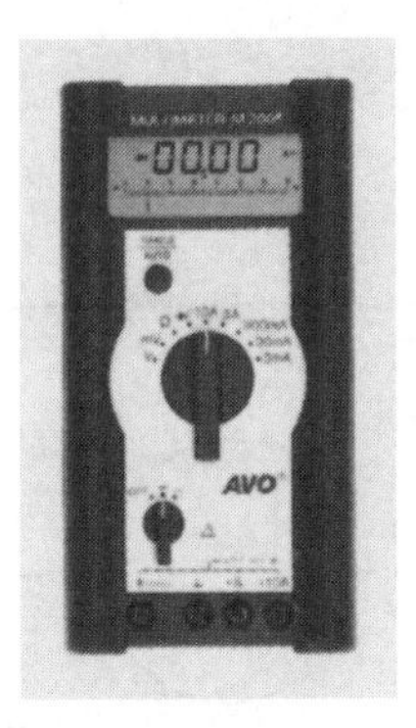

Fig. 12.2 Digital multimeter suitable for testing electronic circuits.

ment which would be suitable for electronic testing, I think on balance that a good-quality digital multimeter such as that shown in Fig. 12.2 would be best. Having no moving parts, digital meters tend to be more rugged and, having a very high input impedance, they are ideally suited to testing electronic circuits.

The multimeter

Multimeters are designed to measure voltage, current or resistance. Before taking measurements the appropriate volt, ampere or ohm scale should be selected. To avoid damaging the instrument it is good practice first to switch to the highest value on a particular scale range. For example, if the 10 A scale is first selected and a reading of 2.5 A is displayed, we then know that a more appropriate scale would be the 3 A or 5 A range. This will give a more accurate reading which might be, say, 2.49 A. When the multimeter is used as an ammeter to measure current it must be connected in series with the test circuit, as shown in Fig. 12.3(a). When used as a voltmeter the multimeter must be connected in parallel with the component, as shown in Fig. 12.3(b).

The ohmmeter

When using a commercial multirange meter as an ohmmeter for testing electronic components, care must be exercised in identifying the positive terminal. The red terminal of the meter, identifying the positive input for testing voltage and current, usually becomes the negative terminal when the meter is used as an ohmmeter because of the way the internal battery is connected to the meter movement. To reduce confusion when using a multirange meter as an ohmmeter it is advisable to connect the red lead to the black terminal and the black lead to the red terminal so that

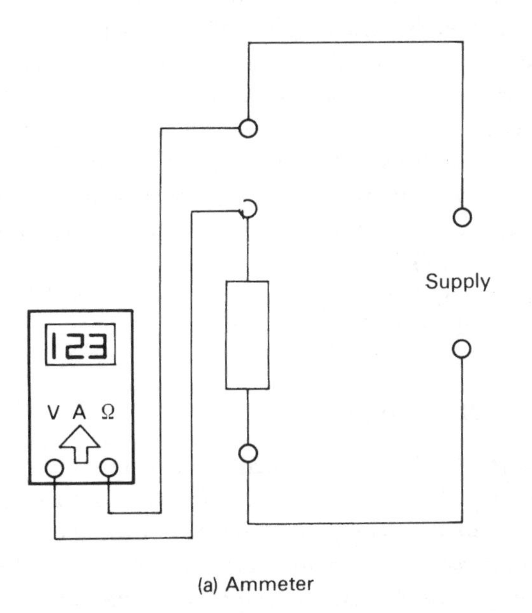

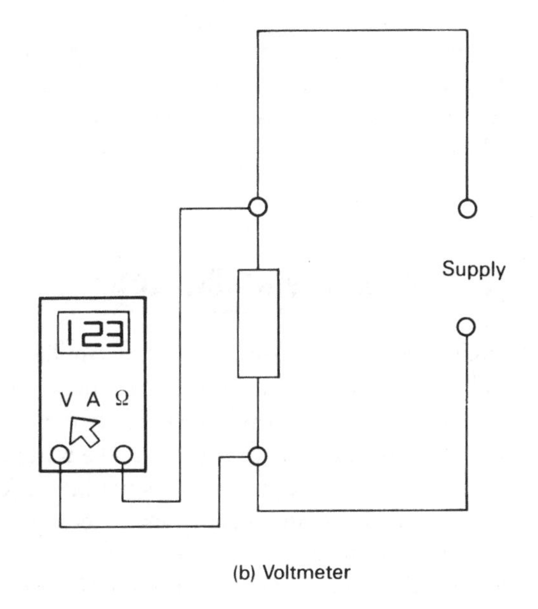

Fig. 12.3 Using a multimeter (a) as an ammeter and (b) as a voltmeter.

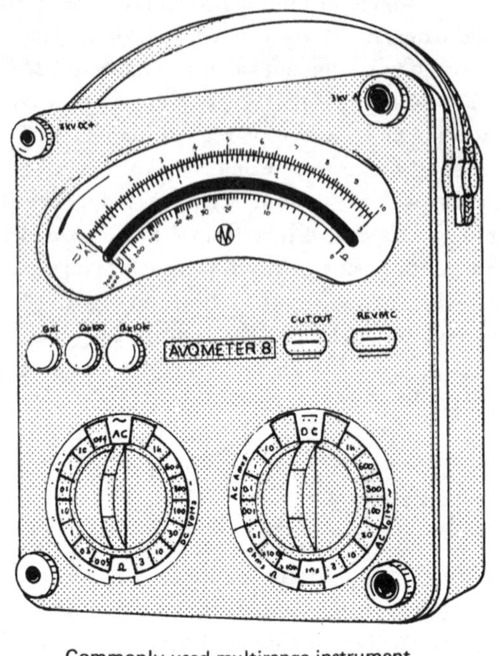

Commonly used multirange instrument

Meter movement
Internal battery for ohm ranges
Meter terminals
Red
Black
Black lead
Red lead

Fig. 12.4 Multirange meter used as an ohmmeter.

the red lead indicates positive and the black lead negative, as shown in Fig. 12.4. The ohmmeter can then be successfully used to test diodes, transistors and thyristors as described in Chapter 10.

Commercial multirange instruments reading volts, amperes and ohms are usually the most convenient test instrument for an electrician or service engineer, although a cathode ray oscilloscope (CRO) can be invaluable for bench work.

The cathode ray oscilloscope

The CRO is probably one of the most familiar and useful instruments to be found in an electronic repair service workshop or college laboratory. It is a more useful instrument for two reasons: it is a high impedance voltmeter and, therefore, takes very little current from the test circuit; and it allows us to 'look into' a circuit and 'see' the waveforms present. 'Cathode ray' is the name given to a high-speed beam of electrons generated in the cathode ray tube and was first used during the Second World War as part of the *radar* system. The beam of electrons is deflected horizontally across the screen at a constant rate by the *time-base circuit* and vertically by the test voltage. The many controls on the front of the CRO are designed so that the operator can stabilize and control these signals. Figure 12.5 shows the front panel of a simple CRO. Electricians and service engineers who are unfamiliar with the CRO should not be baffled by the formidable array of knobs and switches – take them one at a time, and give yourself time to become familiar with these controls.

The single most important component in the CRO is the cathode ray tube.

CATHODE RAY TUBE

Figure 12.6 shows a simplified diagram of the cathode ray tube. This is an evacuated glass tube containing the *electron gun* components on the left and the fluorescent screen, which the operator looks at, on the right. On the far left of the diagram is the wire filament through which a current is passed. This heats a

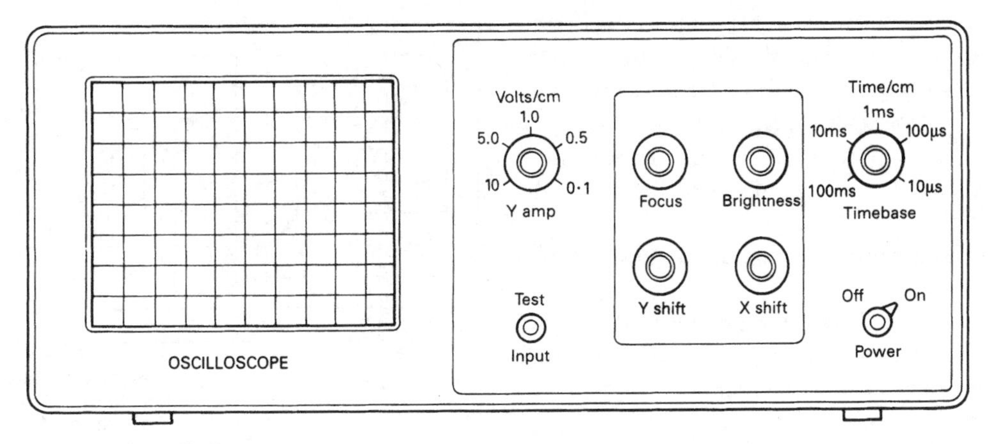

Fig. 12.5 Front panel of a simple CRO.

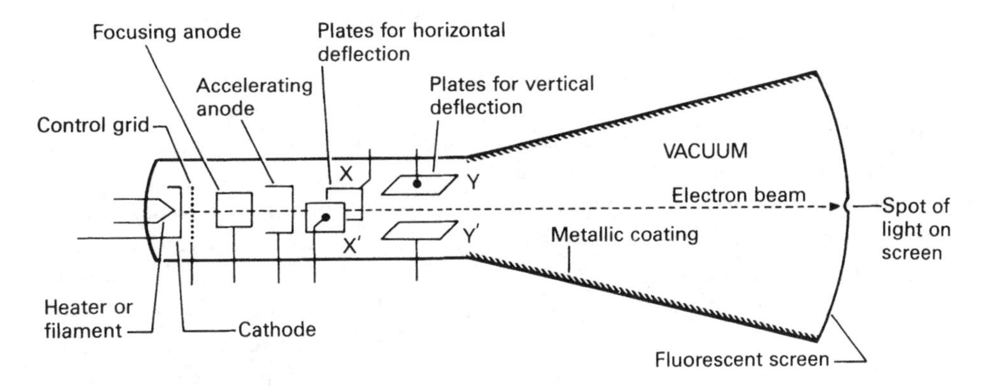

Fig. 12.6 Simplified diagram of the cathode ray tube.

metal plate called a cathode, which emits the electrons to be accelerated. The rate at which the electrons are accepted for acceleration could be modified by making changes to the temperature of the cathode, but in practice it is more convenient to have a metal control grid with a hole in it. By varying the voltage of the control grid it is possible to influence the number of electrons passing through the hole in the grid. The electrons which pass through the grid tend to be moving in various directions and the purpose of the next component therefore is to focus the beam. The electrons are then further accelerated by the accelerating anode to give them sufficient final velocity to produce a bright spot on the screen.

The electrons, on emerging through a hole in this anode, pass through two pairs of parallel plates X–X′ and Y–Y′, each pair being at right angles to the other. If an electric field is established between X and X′ the beam can be deviated horizontally, the direction and magnitude of the deflection depending upon the polarity of the plates. The negative beam of electrons is attracted towards the more positive plate. Likewise, an electric field between plates Y and Y′ produces a vertical deviation. Therefore, a suitable combination of electric fields across X–X′ and Y–Y′ directs the beam to any desired point on the screen.

Upon reaching the screen, the electrons bombard the fluorescent coating on the inside of the screen and emit visible light. The brightness of the spot depends upon the speed of the electrons and the number of electrons arriving at that point.

USE OF THE CRO

The function of the various controls is as follows:

1 Power on switch. Switch on and wait a few seconds for the instrument to warm up. An LED usually indicates a satisfactory main supply.

2 Brightness or intensity. This controls the brightness of the trace. This should be adjusted until bright, but not too brilliant, otherwise the fluorescent powder may be damaged.
3 Scale illumination. This illuminates and highlights the 1 cm square grid lines on the screen.
4 Focus. The spot or trace should be adjusted for a sharp image.
5 Gain controls. 'Adjust' for 'calibrate'.
6 X-shift. The spot or trace can be moved to the left or right and should be centralized.
7 Y-shift. The spot or trace can be moved up or down.
8 TRIG control. This allows the time base to be synchronized to the applied signal to enable a steady trace to be obtained. Set the switch to either Auto or to the Y-input which is connected to the test voltage.
9 AC/GND/DC. It is quite common for a signal to be made up of a mixture of a.c. and d.c. Select DC for all signals and AC to block out the d.c. components of a.c. signals. The GND position disconnects the signal from the Y-amplifier and connects the Y-plates to ground or earth.
10 Chop/Alt. When a double-beam oscilloscope is used, it is common practice to obtain the two X-traces from one beam by either sweeping the electron beams alternately or by sweeping a very small segment of each beam as the trace moves across the screen, leaving each trace chopped up. Use *chop* for slow time-base ranges and *alt* for fast time-base ranges.
11 Connect the test voltage to the CRO leads and adjust the calibrated Y-shift (volts/cm) and time base (time/cm) controls until a steady trace fills the screen.

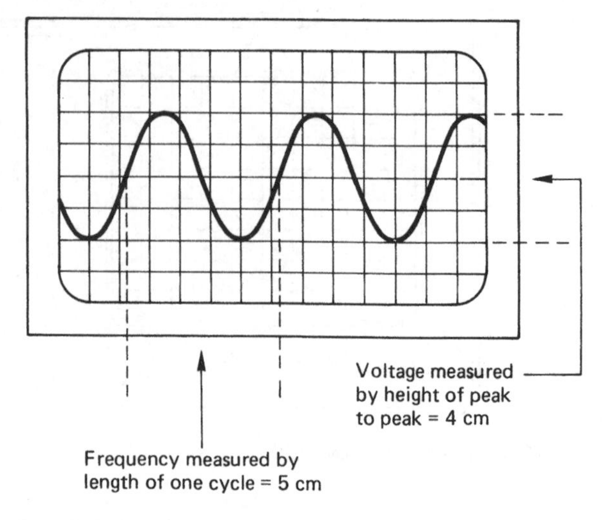

Fig. 12.7 Typical trace on a CRO screen.

USE OF THE CRO TO MEASURE VOLTAGE AND FREQUENCY

The calibrated Y-shift, time base and 1 cm grating on the tube front provide us with a method of measuring the displayed waveform.

With the test voltage connected to the Y-input, adjust all controls to the calibrate position. Adjust the X and Y tuning controls until a steady trace is obtained on the CRO screen, such as that shown in Fig. 12.7.

To measure the voltage of the signal shown in Fig. 12.7 count the number of centimetres from one peak of the waveform to the other using the centimetre grating. This distance is shown as 4 cm in Fig. 12.7. This value is then multiplied by the volts/cm indicated on the Y-amplifier control knob. If the knob was set to, say, 2 V/cm, the peak-to-peak voltage of Fig. 12.7 would be 4 cm × 2 V/cm = 8 V. The peak voltage would be 4 V and the rms voltage 0.7071 × 4 = 2.828 V.

To measure the frequency of the waveform shown in Fig. 12.7 count the number of centimetres for one complete cycle using the 1 cm grating. The distance is shown as 5 cm in Fig. 12.7. This value is then multiplied by the time/cm on the X-amplifier or time-base amplifier control knob. If this knob was set to 4 ms/cm the time taken to complete one cycle would be 5 cm × 4 ms/cm = 20 ms. Frequency can be found from:

$$f = \frac{1}{T} \text{ (Hz)}$$

$$\therefore f = \frac{1}{20 \times 10^{-3}} = \frac{1000}{20} = 50 \text{ Hz.}$$

The waveform shown in Fig. 12.7 therefore has an rms voltage of 2.828 V at a frequency of 50 Hz. The voltage and frequency of any waveform can be found in this way. The relevant a.c. theory is covered in Chapter 8.

EXAMPLE 1

A sinusoidal waveform is displayed on the screen of a CRO as shown in Fig. 12.7. The controls on the *Y*-axis are set to 10 V/cm and the measurement from peak to peak is measured as 4 cm. Calculate the rms value of the waveform.

The peak-to-peak voltage is 4 cm $\times$ 10 V/cm = 40 V
The peak voltage is 20 V
The rms voltage is 20 V $\times$ 0.7071 = 14.14 V

EXAMPLE 2

A sinusoidal waveform is displayed on the screen of a CRO as shown in Fig. 12.7. The controls on the X-axis are set to 2 ms/cm and the measurement for one period is calculated to be 5 cm. Calculate the frequency of the waveform.

The time taken to complete one cycle (T) is 5 cm $\times$ 2 ms/cm = 10 ms.

$$f = \frac{1}{T} \text{ (Hz)}$$

$$\therefore f = \frac{1}{10 \times 10^{-3}} = \frac{1000}{10} = 100 \text{ Hz.}$$

As you can see, the CRO can be used to calculate the values of voltage and frequency. It is not a *direct reading* instrument as were the analogue and digital instruments considered previously. It does, however, allow us to observe the quantity being measured unlike any other instrument and, therefore, makes a most important contribution to our understanding of electronic circuits.

Signal generators

A signal generator is an oscillator which produces an a.c. voltage of continuously variable frequency. It is used for serious electronic testing, fault finding and experimental work. A signal generator is shown in Fig. 12.8.

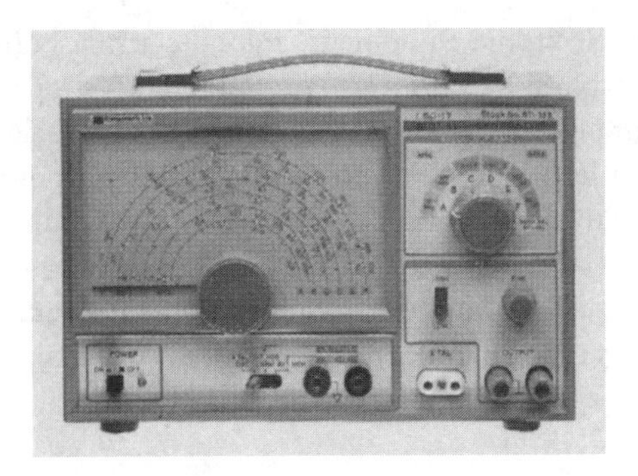

Fig. 12.8 A signal generator.

Power supply unit

A bench power supply unit is a very convenient way of obtaining a variable d.c. voltage from the a.c. mains. The output is very pure, a straight line when observed on a CRO, and continuously variable from zero to usually 30 V. It provides a convenient power source for bench testing or building electronic circuits. A bench power supply unit is shown in Fig. 12.9.

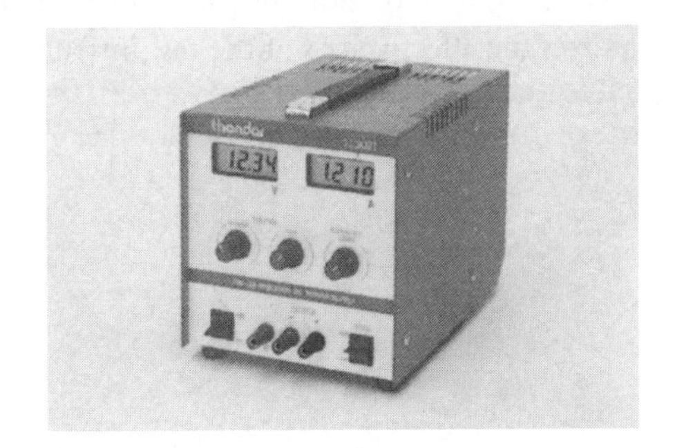

Fig. 12.9 A bench power supply unit.

Mains electricity supply

The mains electricity supply can be lethal, as all electricians and service engineers will know. It is, therefore, a sensible precaution to connect any electronic equipment being tested or repaired to a socket protected by a residual current device. Electronic equipment is protected by in-line fuses and circuit breakers and when testing suspected faulty electronic equipment, a good starting-point is to establish the presence of the mains supply. A multirange meter with the 250 V range selected would be a suitable instrument for this purpose or, alternatively, a voltage indicator as shown in Fig. 12.10 could be used.

When isolating electronic equipment from the mains supply, in order to carry out tests or repairs, the following procedure should be followed:

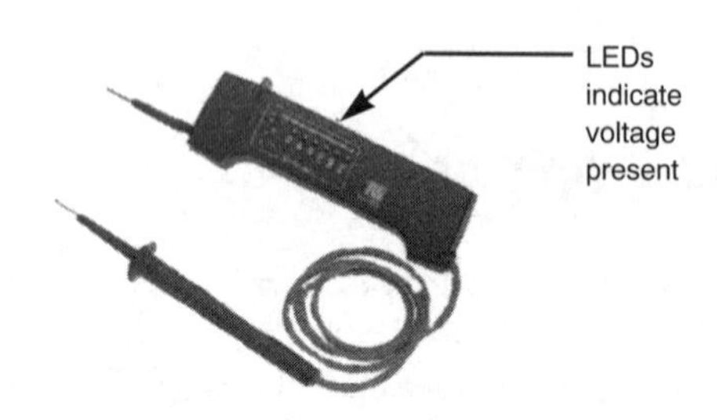

Fig. 12.10 A voltage indicator.

1 Connect the voltage indicator or voltmeter to the incoming supply of the piece of equipment to be isolated. This should indicate the mains voltage and proves the effectiveness of the test instrument.
2 Isolate the supply.
3 Again test the supply to the equipment. If 0 V is indicated the equipment is disconnected from the mains supply.
4 Finally, test the tester again on a known supply or 'proving' unit to be sure the voltage indicator is still working. If the voltage indicator is 'proved' to be working, the equipment is safe to work on.

Insulation tester

The use of an insulation resistance test as described by the IEE Regulations must be avoided with any electronic equipment. The working voltage of this instrument can cause total devastation to modern electronic equipment. When carrying out an insulation resistance test as part of the prescribed series of tests for an electrical installation, all electronic equipment must first be disconnected or damage will result.

Any resistance measurements made on electronic circuits must be achieved with a battery-operated ohmmeter as described previously to avoid damaging the electronic components.

13

LOGIC GATES AND DIGITAL ELECTRONICS

Introduction

Digital electronics embraces all of today's computer-based systems. These are decision-making circuits which use simple little circuits called 'gates' in applications such as industrial robots, industrial hydraulic and pneumatic systems such as programmable logic controllers (PLCs), telephone exchanges, motor vehicle and domestic appliance control systems, children's toys and their parents' personal computers and audio equipment. Digital electronics is concerned with straightforward two-state switching circuits. The simplicity and reliability of this semiconductor transistor switching has encouraged designers to look for new digital markets. Traditional applications which have analogue inputs, such as audio recordings, are now using digital techniques, with the development of analogue-to-digital converters. These convert the analogue voltage signals into digital numbers. A digital and an analogue waveform are shown in Fig. 13.1.

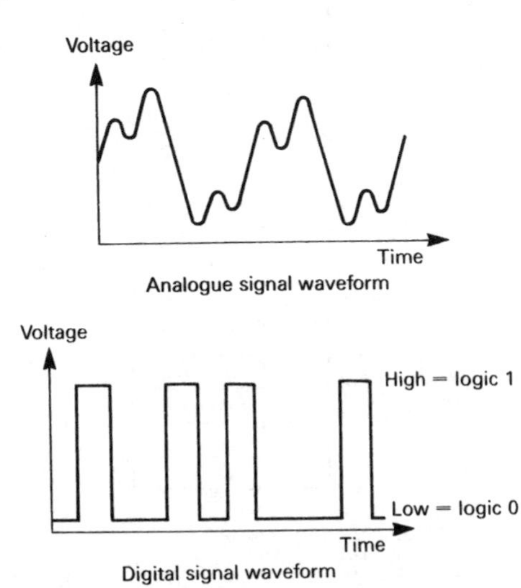

Fig. 13.1 Comparison of analogue and digital waveforms.

The digital waveform has two quite definite states, either on or off, and changes between these two states very rapidly. An analogue waveform changes its value smoothly and progressively between two extremes.

In an analogue system, changes in component values due to ageing and temperature can affect the circuit's performance. Digital systems are very much less susceptible to individual component changes. Another significant advantage of digital circuits is their immunity to noise and interference signals. With analogue circuitry this is a nuisance, particularly when signal levels are very small and, therefore, easily contaminated by noise. Digital signals, however, have a very large amplitude and can, therefore, be made relatively free of noise, which helps manufacturers to achieve a very high quality of sound reproduction, as anyone who listens to a compact disc recording can testify. Logic circuits have been developed to deal with these digital, two-state switching circuits. Information is expressed as *binary numbers*, that is, numbers which consist of ones and zeros. These two binary states are represented by low and high voltages, where low voltage is 0 V and high voltage is, say, +5 V. The low level is called logic 0 and the high level logic 1. When the voltage level of a digital signal is not rapidly changing it remains steady at one of these two levels. Information is processed according to rules built into circuits made up of single units called *logic gates*. These units can allow information to pass through or stop it, and behave according to rules which can be described by logical or predictable statements. A logic gate may have a number of inputs but has only one output which can only be either logic 1 or logic 0; no other value exists. The basic various types of logic gate are known by the names AND, OR, NOT, NOR and NAND.

The AND logic gate

The operation of this gate can probably best be understood by drawing a simple switch-equivalent circuit, as shown in Fig. 13.2. The logic symbol is also shown. The signal lamp will only illuminate if both switch A *and* switch B are closed, or we could say the output F of the gate will only be at logic 1 if both input A and input B are both at logic 1.

If the AND gate was operating a car handbrake warning lamp, it would only illuminate when both the handbrake and the ignition were on. The *truth table* shows the output state for all possible combinations of inputs.

The OR gate

The OR gate can be represented by parallel connected switches, as shown in Fig. 13.3 which also shows the logic symbol. In this case the signal lamp will only illuminate if switch A *or* switch B *or* both switches are closed. Alternatively, we could say that the output F will only be at logic 1 if input A or input B or both inputs are at logic 1.

If the OR gate was operating an interior light in a motor car, it would illuminate when the nearside door was opened *or* the offside door was opened *or* when both doors were opened. The truth table shows the output state for all possible combinations of inputs.

The exclusive-OR gate

The exclusive-OR gate is an OR gate with only two inputs which will give a logic 1 output only if input A *or* input B is at logic 1, but *not* when both A and B are at logic 1. The symbol and truth table are given in Fig. 13.4.

The NOT gate

The NOT gate is a single input gate which gives an output that is the opposite of the input. For this reason it is sometimes called an *inverter* or a *negator* or simply a *sign changer*. If the input is A, the output is *not* A, which is written as $\bar{A}$ (A bar). The small circle on the output of the gate always indicates a change of sign.

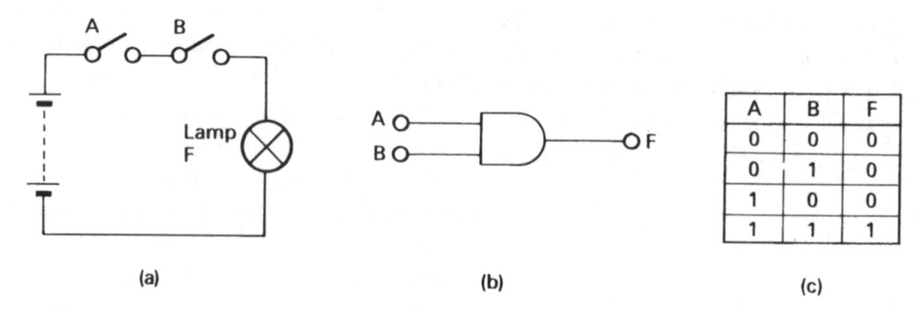

A	B	F
0	0	0
0	1	0
1	0	0
1	1	1

Fig. 13.2 The AND gate: (a) simple switching circuit; (b) logic symbol; (c) truth table.

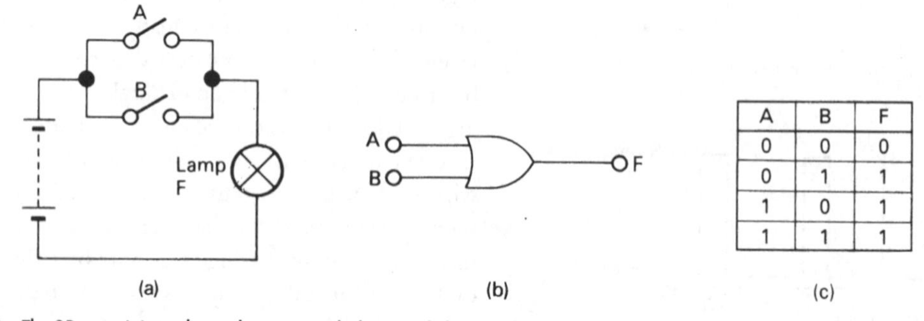

A	B	F
0	0	0
0	1	1
1	0	1
1	1	1

Fig. 13.3 The OR gate: (a) simple switching circuit; (b) logic symbol; (c) truth table.

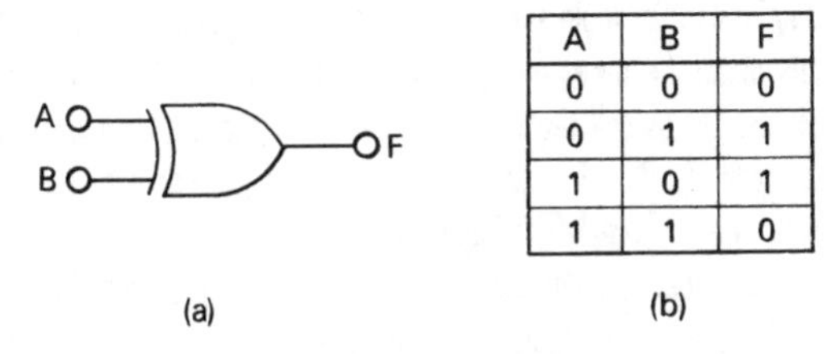

A	B	F
0	0	0
0	1	1
1	0	1
1	1	0

Fig. 13.4 The exclusive-OR gate: (a) logic symbol; (b) truth table.

If the NOT gate was operating a spin dryer motor it would only allow the motor to run when the lid was *not* open. The truth table shows the output state for all possible inputs in Fig. 13.5.

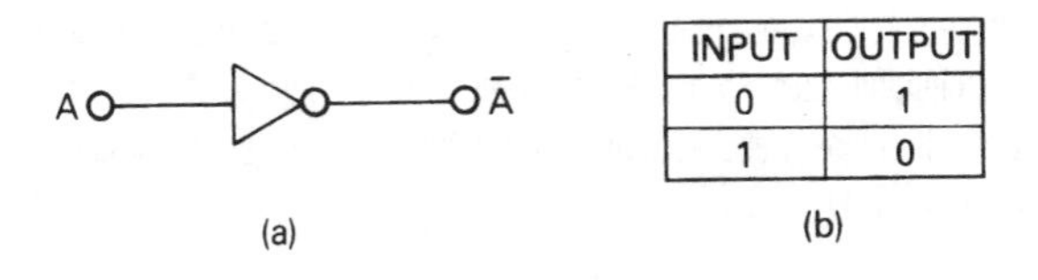

INPUT	OUTPUT
0	1
1	0

Fig. 13.5 The NOT gate: (a) logic symbol; (b) truth table.

The NOR gate

The NOR gate is a NOT gate and an OR gate combined to form a NOT-OR gate. The output of the NOR gate is the opposite of the OR gate, as can be seen by comparing the truth table for the NOR gate in Fig. 13.6 with that of the OR gate.

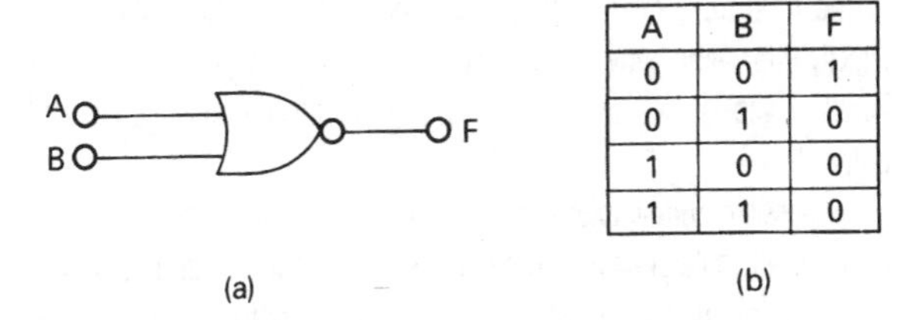

A	B	F
0	0	1
0	1	0
1	0	0
1	1	0

Fig. 13.6 The NOR gate: (a) logic symbol; (b) truth table.

The NAND gate

The NAND gate is a NOT gate and an AND gate combined to form a NOT-AND gate. The output of the NAND gate is the opposite of the AND gate, as can be seen by comparing the truth table for the NAND gate in Fig. 13.7 with that of the AND gate.

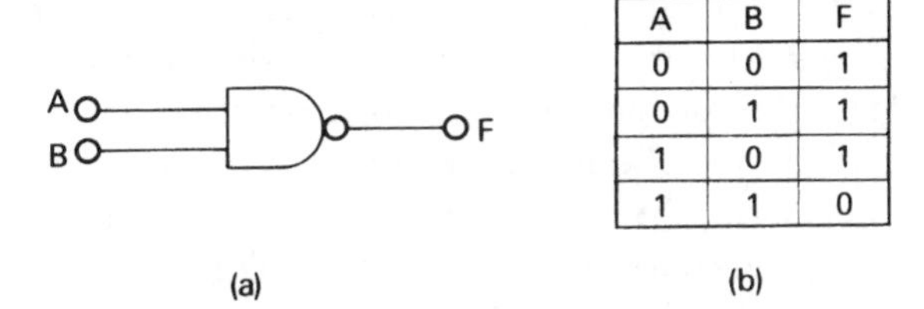

A	B	F
0	0	1
0	1	1
1	0	1
1	1	0

Fig. 13.7 The NAND gate: (a) logic symbol; (b) truth table.

Buffers

The simplest of all logic devices is the buffer. This device has only one input and one output, and its logical output is exactly the same as its logical input. Given that this device has no effect upon the logic levels within a circuit, you may be wondering what the purpose of such an apparently redundant device might be! Well, although the input and output voltage levels of the buffer are identical, the *currents* present at the input and output can be *very* different. The output current can be much greater than the input current and, therefore, buffers can be said to exhibit *current gain*. In this way, buffers can be used to interface logic circuits to other circuits which demand more current than could be supplied by an unbuffered logic circuit. The symbol used to represent a buffer is shown in Fig. 13.8.

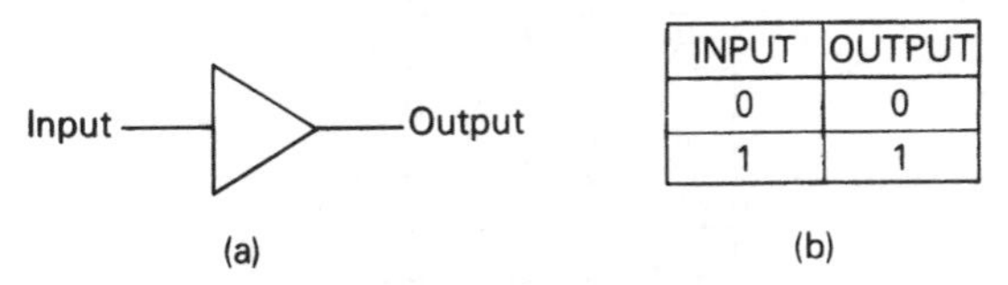

INPUT	OUTPUT
0	0
1	1

Fig. 13.8 The buffer: (a) logic symbol; (b) truth table.

Logic networks

Individual logic gates may be interconnected to provide any desired output. The results of any combination can be found by working through each individual gate in the combination or logic system in turn, and producing the truth table for the particular network. It can also be very instructive to build up logic gate combinations on a logic simulator and to confirm the theoretical results. This facility will undoubtedly be available if the course of study is being undertaken at a technical college, training centre or evening institute.

EXAMPLE 1

Two logic gates are connected together as shown in Fig. 13.9. Complete the truth table for this particular logic network.

In considering Fig. 13.9 and working as always from left to right, we can see that an AND gate feeds a NOT gate. The whole network has two inputs, A and B, and one output F. The first step in constructing the truth table for the combined logic gates is to label the outputs of *all* the gates and prepare a blank truth table as shown in Fig. 13.3. Let us call the output of the AND gate C (it could be any letter except A, B or F) and work our way progressively through the individual gates from left to right. For any two-input logic gate, there are four possible combinations, 00, 01, 10 and 11. When these are included on the truth table it will appear as shown in Table 13.1. The next step is to complete column C. Now, C is the output of an AND gate and can, therefore, only be at logic 1 when both A *and* B are logic 1. The truth table can, therefore, be completed as shown in Table 13.2. The final step is to complete column F, the output of a NOT gate whose input combinations are given by column C. A NOT gate is a single-input gate whose output is the opposite of the input and, therefore, the output column F must be the opposite of column C, as shown by Table 13.3. The truth table tells us that this particular combination of gates will give a logic 1 output with any input combination *except* when A and B are both at logic 1. This combination, therefore, behaves like a NAND gate, as can be confirmed by referring to Fig. 13.7.

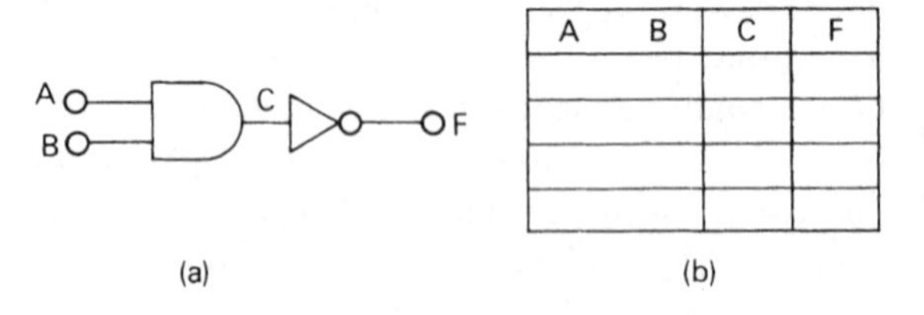

A	B	C	F

Fig. 13.9 (a) Logic network and (b) blank truth table for Example 1.

A	B	C	F
0	0		
0	1		
1	0		
1	1		

Table 13.1 Truth table for Example 1.

A	B	C	F
0	0	0	
0	1	0	
1	0	0	
1	1	1	

Table 13.2 Truth table for Example 1.

A	B	C	D
0	0	0	1
0	1	0	1
1	0	0	1
1	1	1	0

Table 13.3 The completed truth table for Example 1.

EXAMPLE 2

A NAND and NOT gate are connected together as shown in Fig. 13.10. Complete a truth table for this particular network.

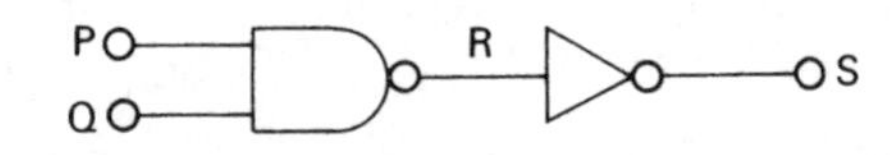

Fig. 13.10 Logic network for Example 2.

The truth table for this particular combination can be constructed in exactly the same way as for Example 1. The NAND gate has two inputs P and Q and an output R. The NOT gate has an input R and output S.

All possible combinations of inputs are shown in columns P and Q of the truth table shown in Table 13.4. A NAND gate will give a logic 1 output for *all* combinations of inputs *except* when input A *and* B are at logic 1 as shown by column R of Table 13.4. The second and

P	Q	R	S
0	0	1	0
0	1	1	0
1	0	1	0
1	1	0	1

Table 13.4 Truth table for Example 2.

final gate in this network is a NOT gate which provides an output which is the reverse of the input. The output, given by column S of the truth table, will therefore be the reverse of column R, as shown in Table 13.4.

This particular combination will, therefore, give a logic 1 output only when input P *and* input Q are at logic 1. Therefore, it can be seen that the combination of a NAND and a NOT gate produces the equivalent of an AND gate. This can be checked by referring back to Fig. 13.2.

EXAMPLE 3

A NAND gate has a NOT gate on each of its inputs as shown in Fig. 13.11. Construct a truth table for this particular network.

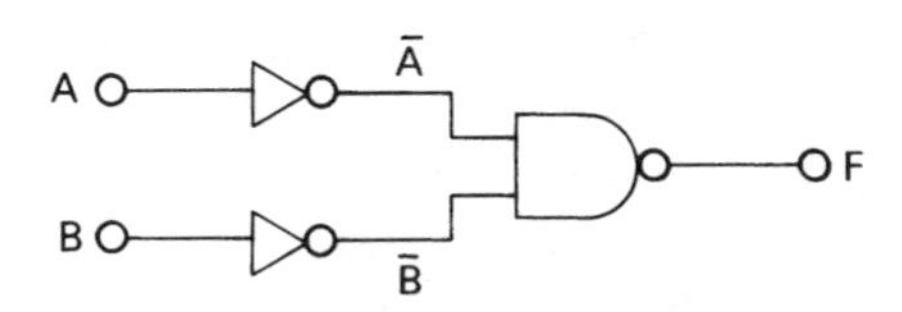

Fig. 13.11 Logic network for Example 3.

The NOT gates will invert or reverse the input. We can, therefore, call the output of these NOT gates not A and not B, written as $\overline{A}$ and $\overline{B}$. This then provides the input to the NAND gate. A NAND gate will provide a logic 1 output for any input combination *except* when both inputs are at logic 1. The truth table can, therefore, be developed as shown in Table 13.5. It can be seen by referring back to Fig. 13.3, and comparing the inputs A and B and output F, that this combination gives the network equivalent of an OR gate. That is, the output is at logic 1 if the input A *or* input B *or* both are at logic 1.

A	B	$\overline{A}$	$\overline{B}$	F
0	0	1	1	0
0	1	1	0	1
1	0	0	1	1
1	1	0	0	1

Table 13.5 Truth table for Example 3.

EXAMPLE 4

A logic network is assembled as shown in Fig. 13.12. Develop a truth table and describe in a sentence the relationship between the input and output.

The truth table for this particular combination can be drawn up as

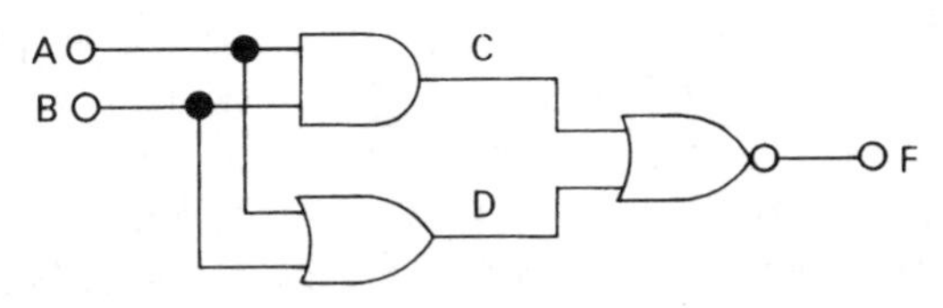

Fig. 13.12 Logic network for Example 4.

shown in Table 13.6. There are only two inputs A and B. The output C of the AND gate and the output D of the OR gate provide the input to a NOR gate, which provides the output F.

A	B	C	D	F
0	0	0	0	1
0	1	0	1	0
1	0	0	1	0
1	1	1	1	0

Table 13.6 Truth table for Example 4.

The output of an AND gate is high, that is at logic 1, only when input A *and* input B are at logic 1. Column C of the truth table shows the output of the AND gate for all combinations of input. The output of an OR gate is high when input A *or* input B *or* both are high. This is shown by column D of the truth table. The input to the final NOR gate is provided by the logic levels indicated in columns C and D and the output F is, therefore, as shown in column F. The output of this combination of logic gates is high only when input A *and* input B are low. This is equivalent to a single NOR gate.

In the examples considered until now, the inputs have been restricted to only two variables. In practice, logic gates may be constructed with many inputs and the truth tables developed as shown above. However, when there are more than three inputs the truth table becomes very cumbersome because the number of lines required for the truth table follows the law of 2^n where n is equal to the number of inputs. Therefore, a two-input gate requires 2^2 (4) lines, as can be seen in the previous examples, a three-input gate 2^3 (8) lines, a four-input gate 2^4 (16) lines, etc.

EXAMPLE 5

A logic system having three inputs is assembled as shown in Fig. 13.13. Develop a truth table and describe in a sentence the relationship between the input and output.

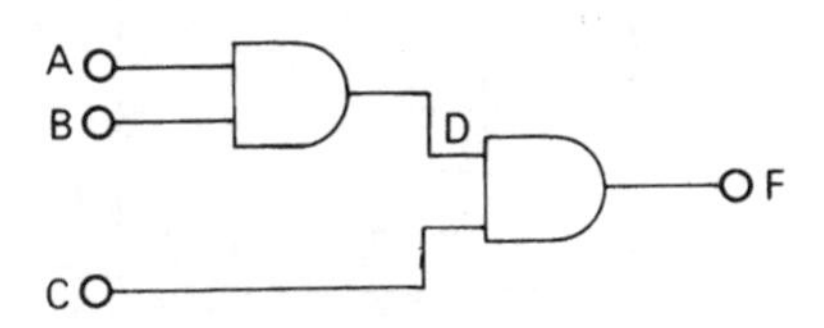

Fig. 13.13 Logic network for Example 5.

The truth table for this combination of logic gates can be drawn up as shown in Table 13.7. Three inputs mean that the truth table must have $2^3 = 8$ rows. All possible combinations of input are shown in columns A, B and C. The first AND gate will give a logic 1 output only when input A and B are both logic 1. There are two such occasions, as shown by column D. The second AND gate will give a logic 1 output only when input C and D are both logic 1. This occurs on only one occasion. That is, the output is at logic 1 only when all three inputs are at logic 1.

A	B	C	D	F
0	0	0	0	0
0	0	1	0	0
0	1	0	0	0
0	1	1	0	0
1	0	0	0	0
1	0	1	0	0
1	1	0	1	0
1	1	1	1	1

Table 13.7 Truth table for Example 5.

EXAMPLE 6

A three-input logic network is assembled as shown in Fig. 13.14. Develop a suitable truth table and use this to describe the relationship between the three inputs and the output Z.

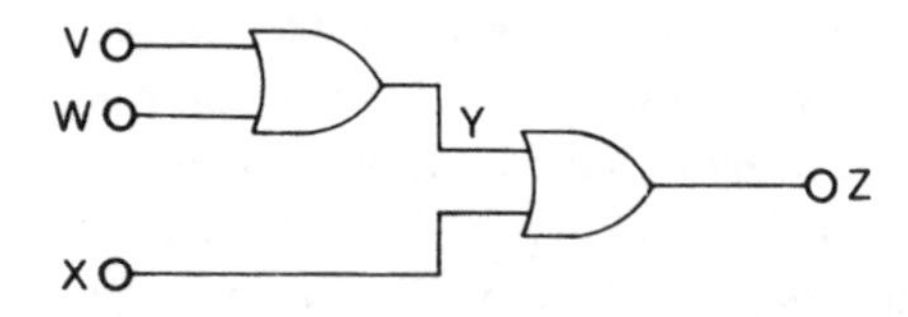

Fig. 13.14 Logic network for Example 6.

The truth table for this network can be constructed as shown in Table 13.8. All possible combinations of the input are shown in columns V, W and X.

V	W	X	Y	Z
0	0	0	0	0
0	0	1	0	1
0	1	0	1	1
0	1	1	1	1
1	0	0	1	1
1	0	1	1	1
1	1	0	1	1
1	1	1	1	1

Table 13.8 Truth table for Example 6.

The first OR gate will give a logic 1 output when V *or* W *or* both are at logic 1. This occurs on all but two occasions as can be seen by considering column Y of the truth table. The second OR gate will give a logic 1 output when X *or* Y *or* both are at logic 1. This occurs on all but one occasion. Therefore, we can say that the output Z is at logic 0 *only* when all three inputs are at logic 0. If any input is at logic 1, the output Z is also at logic 1.

Logic families

The simplicity of digital electronics, with its straightforward on–off switching, means that many logic elements can be packed together in a single integrated circuit and packaged as a standard dual-in-line IC, as shown in Figs 10.30 and 10.32. Different types of semiconductor circuitry can be used to construct the logic gates. Each type is called a logic family because all members of that integrated circuit family will happily work together in a circuit.

Two main families of digital logic have emerged as the most popular with designers of general-purpose digital circuits in recent years. These are the TTL and CMOS families. The older of these is the TTL (transistor–transistor logic) family which was introduced in 1964 by Texas Instruments Ltd. The standard TTL family is designated the 7400 series. Figure 13.15 shows the internal circuitry of a TTL 7400 IC. This contains a quad 2-input NAND gate, that is, it contains four NAND gates each with two inputs and one output. Thus, with two power supply connections, the 7400 IC has 14 connections and is manufactured as the familiar 14 pin dual-in-line package. Many other combinations are available and each has its own unique number which, in this family, always begins with 74 and is followed by two other numbers. The final two numbers indicate the type of logic gate: for example, a 7432 is a quad 2-input OR gate, and a 7411 a triple 3-input AND gate. Details are given in manufacturers' data sheets.

The CMOS family, pronounced 'see-mos', is the complementary metal oxide semiconductor family of logic ICs which was introduced in 1968. The best-known CMOS family is designated the 4000 series and, like its TTL equivalent, is housed in a 14 pin dual-in-line package. The 4011B is a quad 2-input NAND gate, as shown in Fig. 13.16. This is *similar* to the TTL 7400 shown in Fig. 13.15,

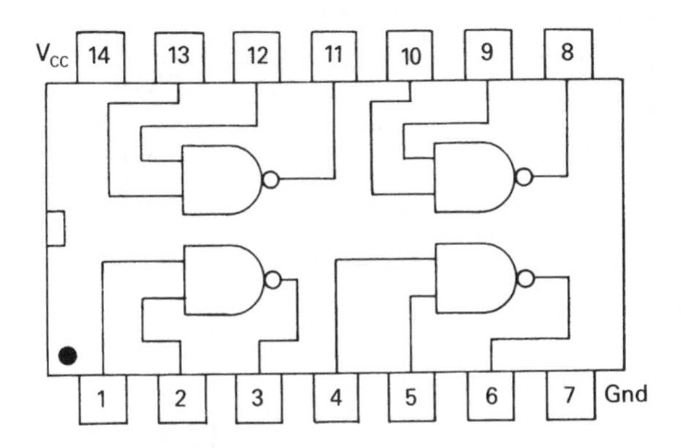

Fig. 13.15 The 7400 TTL logic family.

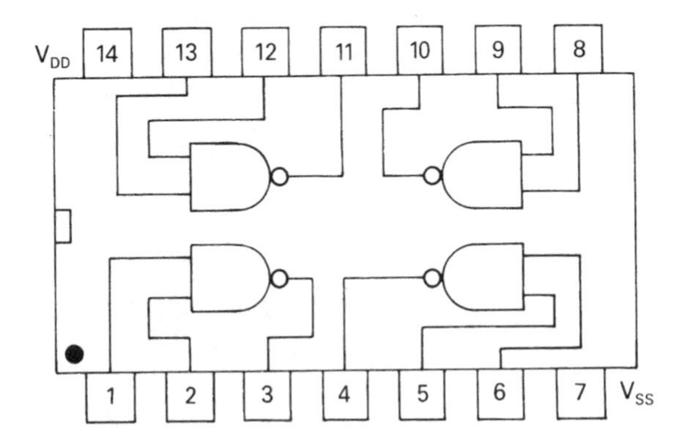

Fig. 13.16 The 4011B CMOS logic family.

but it is not identical because the pin connections differ and, therefore, a TTL package cannot replace a CMOS package.

The theory of digital logic is the same for all logic families. The differences between the families are confined to the practical aspects of the circuit design. Each logic family has its own special characteristics which make it appropriate for particular applications.

COMPARISON OF TTL AND CMOS

A CMOS device dissipates about 1 mW per logic gate, compared with about 20 mW for a standard TTL logic gate. Therefore, CMOS has a much lower power consumption than TTL, which is particularly important when the circuitry is to be battery-powered.

The output of a logic gate may be connected to the input of many other logic gates. The drive capability of a gate to hold its input at logic 0 or logic 1 while delivering current to the other gates in the circuit is called the *fan-out* capability. The fan-out for TTL is 10, which means that ten other TTL logic gates can take their input from one TTL output and still switch reliably before overloading occurs. A fan-out of 50 is typical for CMOS because they have a very high input impedance and low power consumption.

The power supply for TTL must be 5 V ± 0.25 V with a ripple of less than 5% peak to peak. A TTL device will be damaged if voltages in excess of these limits are applied. CMOS devices can tolerate a much wider variation of supply voltages, typically +3 V to +15 V.

Another advantage of CMOS logic circuits is that they require only about one-fiftieth of the 'floor space' on a silicon chip compared with TTL. CMOS is, therefore, ideal for complex silicon chips, such as those required by microprocessors and memories.

Table 13.9 Properties of logic families

Property	TTL	CMOS
Power consumption	high: 20 mW	low: 1 mW
Operating current	high: mA range	low: μA range
Power supply	5 V ± 0.25 V d.c.	3 V to 15 V d.c.
Switching speeds	fast: 10 ns	slow: 100 ns
Input impedance	low	high
Fan-out	10	50

The switching times of any logic network are infinitesimal when compared with an electro-mechanical relay. However, the switching times for TTL logic are very much faster than for CMOS, although both are measured in nanoseconds. The properties of each family are summarized in Table 13.9.

Working with logic

ICs of the same number will always have the same function regardless of the manufacturer and any suffix or prefix which may accompany the basic gate number. Therefore, an IC package must be

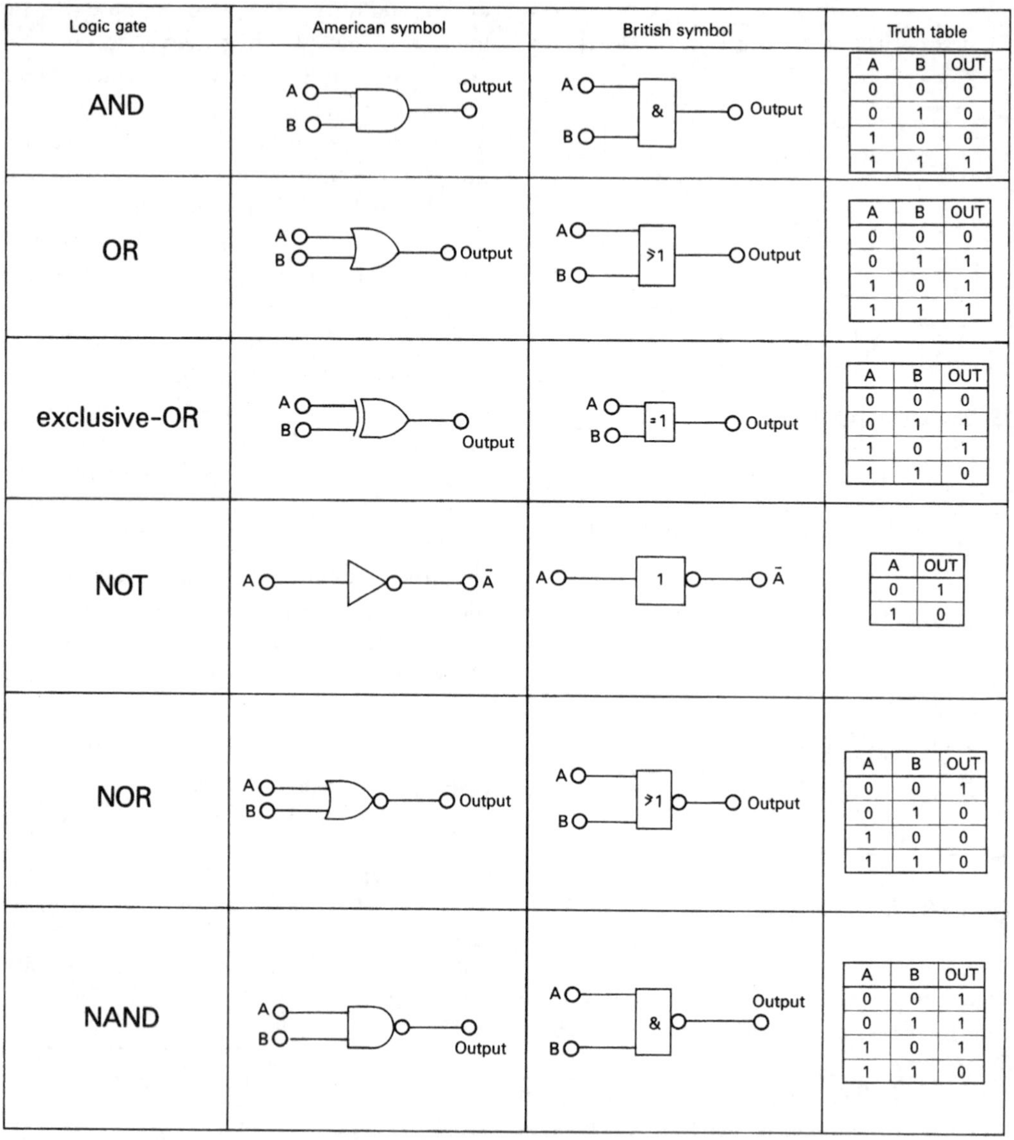

Logic gate	American symbol	British symbol	Truth table
AND	A, B, Output	A, B, &, Output	A B OUT: 0 0 0 / 0 1 0 / 1 0 0 / 1 1 1
OR	A, B, Output	A, B, ≥1, Output	A B OUT: 0 0 0 / 0 1 1 / 1 0 1 / 1 1 1
exclusive-OR	A, B, Output	A, B, =1, Output	A B OUT: 0 0 0 / 0 1 1 / 1 0 1 / 1 1 0
NOT	A, $\bar{A}$	A, 1, $\bar{A}$	A OUT: 0 1 / 1 0
NOR	A, B, Output	A, B, ≥1, Output	A B OUT: 0 0 1 / 0 1 0 / 1 0 0 / 1 1 0
NAND	A, B, Output	A, B, &, Output	A B OUT: 0 0 1 / 0 1 1 / 1 0 1 / 1 1 0

Fig. 13.17 Comparison of British and American logic gate symbols.

replaced with another of the same number. The very high input impedance of CMOS accounts for its low power consumption but it does mean that static electricity can build up on the input pins if they come into contact with plastic, nylon or the man-made fibres of workers' clothing during circuit assembly or repair. This does not happen with TTL because the low input impedance ensures that any static charges leak harmlessly away through the junctions in the IC. Static voltages on CMOS devices can destroy them; they are supplied with anti-static carriers and these should not be removed until wiring is completed. Internal protection is also provided by buffered inputs, but these cannot become effective until the supply is connected. Inputs must, therefore, be disconnected before the mains connections when disconnecting CMOS devices. Alternatively, the power supplies must be connected before the inputs when assembling CMOS chips. Input signals must not be applied until the power supply is connected and switched on.

When operating CMOS devices with normal positive logic signals V_{SS} is the common line (0 V) and V_{DD} is the positive connection, 3–15 V (inputs 7 and 14 in Fig. 13.16). Unused inputs must not be left *floating*. They must always be connected in parallel with similar used inputs, or connected to the supply rail.

Working with CMOS has created many new problems for electronic technicians. These can be overcome by

- working on a copper plate working surface which is connected to earth
- ensuring that all equipment is properly earthed and
- wearing a conductive wristband which is connected to the earth of the working surface.

When these precautions are observed the problems of handling CMOS ICs can be overcome without too much difficulty.

British Standard symbols

Although the British Standards recommend symbols for logic gates, much of the manufacturers' information uses the American 'MilSpec' Standard symbols. For this reason I have reluctantly used the American standard symbols in this chapter. However, there is some pressure in the UK to adopt the BS symbols, and for this reason the British Standard and American Standard symbols are cross-referenced in Figure 13.17.

Exercises

1 A voltage signal which changes smoothly and progressively between two extremes is called:
(a) a logical waveform
(b) an analogue waveform
(c) an interference signal
(d) a digital waveform.

2 A voltage signal which has two quite definite states, either on or off, is called
(a) a logical waveform
(b) an analogue waveform
(c) an interference signal
(d) a digital waveform.

3 A single logic gate has two inputs X and Y and one output Z. The output Z will be at logic 1 only when input A and input B are at logic 1 if the gate is
(a) a NOT gate
(b) an AND gate
(c) an OR gate
(d) a NOR gate.

4 Develop the truth table for an exclusive-OR gate.

5 Develop the truth table for a NOR gate.

6 For the circuit shown in Fig. 13.18 develop the truth table.

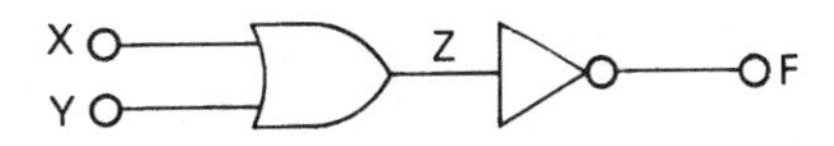

Fig. 13.18 Logic network for Exercise 6.

7 Develop the truth table for the network shown in Fig. 13.19 and describe the relationship between the inputs and output.

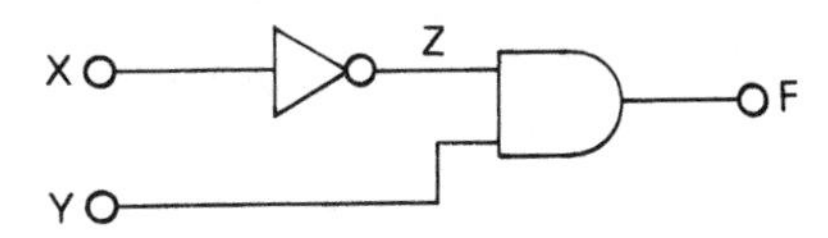

Fig. 13.19 Logic network for Exercise 7.

8 Develop the truth table for the logic system shown in Fig. 13.20 and describe the relationship between the output and inputs.

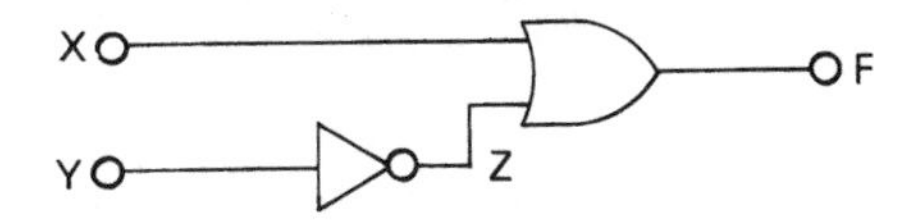

Fig. 13.20 Logic network for Exercise 8.

9 Work out the truth table for the circuit shown in Fig. 13.21. Describe in a sentence the behaviour of this circuit.

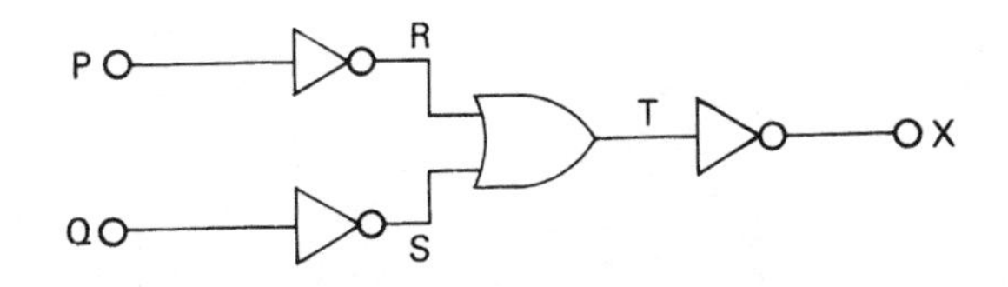

Fig. 13.21 Logic network for Exercise 9.

10 Complete the truth table for the circuit shown in Fig. 13.22 and describe the circuit behaviour.

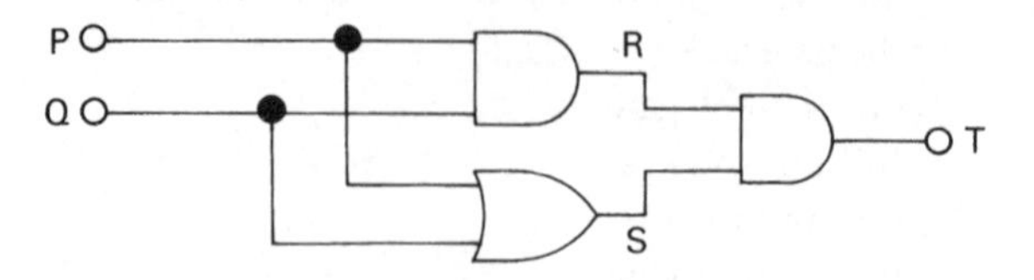

Fig. 13.22 Logic network for Exercise 10.

11 Using a truth table describe the output of the logic system shown in Fig. 13.23.

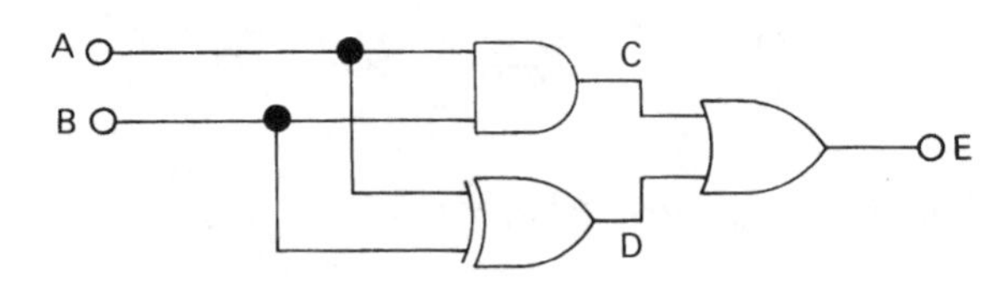

Fig. 13.23 Logic network for Exercise 11.

12 Use a truth table to describe the output of the logic network shown in Fig. 13.24.

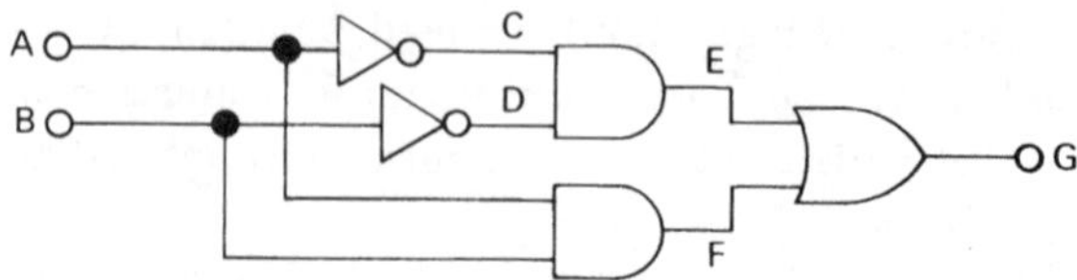

Fig. 13.24 Logic network for Exercise 12.

13 Inputs A, B and C of Fig. 13.25 are controlled by three separate key switches. Determine the sequence of key switch positions which will give an output at F.

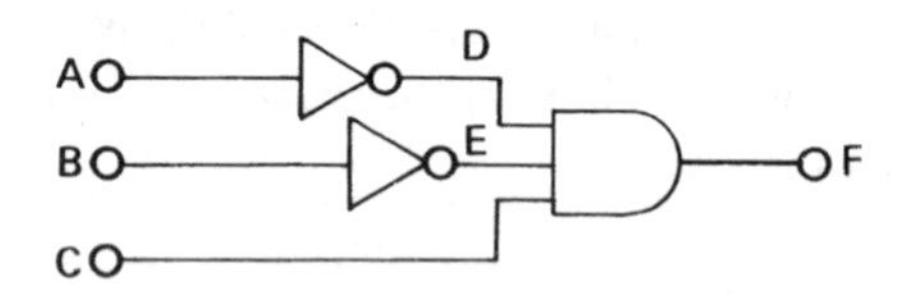

Fig. 13.25 Logic network for Exercise 13.

SOLUTIONS TO EXERCISES

CHAPTER 1

1: a and d 2: d 3: c 4: c 5: d 6: c 7: d 8: b 9: a 10: b 11: a 12: a 13: b 14: c 15–20: Answers in text

CHAPTER 2

1: c 2: d 3: a 4: b 5: b 6: c 7–12: Answers in text

CHAPTER 3

1: d 2: c 3: b 4: a 5–10: Answers in text

CHAPTER 4

1: b 2: c 3: d 4: a 5: b 6: c 7: d 8: a 9: d 10: c 11: b

CHAPTER 5

1: b 2: c 3: d 4: c 5: b 6: a 7: b 8: c 9: a 10: b 11: d 12, 13: Answers in text

CHAPTER 6

1: d 2: a 3: b 4: a 5: b 6: b 7: d 8: b 9: d 10: c 11: b 12: d 13: b 14: b 15: b 16: c 17: c 18: c 19: d 20: d 21–31: Answers in text 32: c 33: d 34: a 35: d 36: b 37: b 38: 25 mm 39: 100 × 25 40: 17 41: 5 42: 25 mm 43: 155 44: 75 × 25

CHAPTER 7

1: b 2: d 3: d 4: b 5: c 6: c 7: c 8: 1.27 Ω 9: 190 mΩ 10–13: Answers in text

CHAPTER 8

1: c 2: d 3: d 4: a 5: c 6: d 7, 8: Answers in text 9: c 10: c 11: d 12: c 13: b 14: c 15: d 16: b 17: a 18: b 19: c 20: a 21: d 22: Answer in text 23: d 24: a 25: d 26: a 27: c 28: c 29: a 30: b 31: b 32: b 33: b 34: a 35: a 36: d 37: Answer in text 38: d 39: c 40: b 41: b 42: b 43: a 44: b 45: Answer in text

CHAPTER 9

1: b 2: d 3: c 4: a 5: c 6: d 7: d 8: c 9: d 10: c 11: c 12: b 13: Answer in text 14: d 15: b 16: Answer in text

CHAPTER 13

1: b 2: d 3: b 4: See Fig. 13.4 5: See Fig. 13.6 6: See Table S.1

X	Y	Z	F
0	0	0	1
0	1	1	0
1	0	1	0
1	1	1	0

Table S.1 Truth table for Exercise 6 of Chapter 13.

7: See Table S.2. The output is high only when the input X is low and the input Y is high. For all other input combinations the output is low.

X	Y	Z	F
0	0	1	0
0	1	1	1
1	0	0	0
1	1	0	0

Table S.2 Truth table for Exercise 7 of Chapter 13.

8: See Table S.3. The output is high for all input combinations except when input X is low and input Y is high.

X	Y	Z	F
0	0	1	1
0	1	0	0
1	0	1	1
1	1	0	1

Table S.3 Truth table for Exercise 8 of Chapter 13.

9: See Table S.4. The output X is at logic 1 only when input P and input Q are at logic 1. For all other input combinations the output is logic 0.

P	Q	R	S	T	X
0	0	1	1	1	0
0	1	1	0	1	0
1	0	0	1	1	0
1	1	0	0	0	1

Table S.4 Truth table for Exercise 9 of Chapter 13.

10: See Table S.5. The output T is logic 1 only when both inputs are at logic 1. for all other input combinations the output is logic 0.

P	Q	R	S	T
0	0	0	0	0
0	1	0	1	0
1	0	0	1	0
1	1	1	1	1

Table S.5 Truth table for Exercise 10 of Chapter 13.

11: See Table S.6. The output E is logic 1 for all input combinations except when input A and B are both logic 0.

A	B	C	D	E
0	0	0	0	0
0	1	0	1	1
1	0	0	1	1
1	1	1	0	1

Table S.6 Truth table for Exercise 11 of Chapter 13.

12: See Table S.7. The output is high when both inputs are the same.

A	B	C	D	E	F	G
0	0	1	1	1	0	1
0	1	1	0	0	0	0
1	0	0	1	0	0	0
1	1	0	0	0	1	1

Table S.7 Truth table for Exercise 12 of Chapter 13.

13: See Table S.8. An output is only available at F when keys A and B are off (both at logic 0) and key C is on.

A	B	C	D	E	F
0	0	0	1	1	0
0	0	1	1	1	1
0	1	0	1	0	0
0	1	1	1	0	0
1	0	0	0	1	0
1	0	1	0	1	0
1	1	0	0	0	0
1	1	1	0	0	0

Table S.8 Truth table for Exercise 13 of Chapter 13.

APPENDICES

Appendix A: Obtaining information and components

For local suppliers, you should consult your local telephone directory. However, the following companies distribute electrical and electronic components throughout the UK. In most cases, telephone orders received before 5 pm can be dispatched the same day.

Electromail (RS mail order business)
PO Box 33, Corby, Northants NN17 9EL
Tel: (01536) 204555

Farnell Electronic Components
Canal Road, Leeds LS12 2TU
Tel: (0113) 263 6311

Maplin Electronics
PO Box 777, Rayleigh, Essex SS6 8LV
Tel: (01702) 552961

Rapid Electronics Ltd
Heckworth Close, Severalls Industrial Estate, Colchester, Essex CO4 4TB
Tel: (01206) 751166

RS Components Ltd
PO Box 99, Corby, Northants NN17 9RS
Tel: (01536) 201234

Verospeed Electronic Components
Boyatt Wood, Eastleigh, Hants SO5 4ZY
Tel: (01703) 644555

Appendix B: Abbreviations, symbols and codes

Abbreviations used in electronics for multiples and submultiples

T	tera	10^{12}
G	giga	10^{9}
M	mega or meg	10^{6}
k	kilo	10^{3}
d	deci	10^{-1}
c	centi	10^{-2}
m	milli	10^{-3}
μ	micro	10^{-6}
n	nano	10^{-9}
p	pico	10^{-12}

Terms and symbols used in electronics

Term	Symbol
Approximately equal to	$\simeq$
Proportional to	$\propto$
Infinity	∞
Sum of	$\sum$
Greater than	$>$
Less than	$<$
Much greater than	$\gg$
Much less than	$\ll$
Base of natural logarithms	e
Common logarithms of x	$\log x$
Temperature	θ
Time constant	T
Efficiency	η
Per unit	p.u.

Electrical quantities and units

Quantity	Quantity symbol	Unit	Unit symbol
Angular velocity	ω	radian per second	rad/s
Capacitance	C	farad	F
		microfarad	μF
		picofarad	pF
Charge or quantity of electricity	Q	coulomb	C
Current	I	ampere	A
		milliampere	mA
		microampere	μA
Electromotive force	E	volt	V
Frequency	f	hertz	Hz
		kilohertz	kHz
		megahertz	MHz
Impedance	Z	ohm	Ω
Inductance, self	L	henry (plural, henrys)	H
Inductance, mutual	M	henry (plural, henrys)	H
Magnetic field strength	H	ampere per metre	A/m
Magnetic flux	θ	weber	Wb
Magnetic flux density	B	tesla	T
Potential difference	V	volt	V
		millivolt	mV
		kilovolt	kV
Power	P	watt	W
		kilowatt	kW
		megawatt	MW
Reactance	X	ohm	Ω
Resistance	R	ohm	Ω
		microhm	μΩ
		megohm	MΩ
Resistivity	ρ	ohm meter	Ω m
Wavelength	λ	metre	m
		micrometre	μm

Capacitor values – conversion table

Capacitance (picofarad, pF)	Capacitance (nanofarad, nF)	Capacitance (microfarad, μF)	Capacitance code*
10	0.01		100
15	0.015		150
47	0.047		470
82	0.082		820
100	0.1		101
330	0.33		331
470	0.47	0.00047	471
1 000	1.0	0.001	102
1 500	1.5	0.0015	152
2 200	2.2	0.0022	222
4 700	4.7	0.0047	472
6 800	6.8	0.0068	682
10 000	10	0.01	103
22 000	22	0.022	223
47 000	47	0.047	473
100 000	100	0.1	104
220 000	220	0.22	224
470 000	470	0.47	474

*First two digits are significant figures; third is number of zeros. Value given in pF.

Resistor and capacitor letter and digit code (BS 1852)

Resistor values are indicated as follows:

0.47 Ω	marked	R47	100 Ω	marked	100R
1 Ω		1R0	1 kΩ		1K0
4.7 Ω		4R7	10 kΩ		10K
47 Ω		47R	10 MΩ		10M

A letter following the value shows the tolerance.
F = ± 1%; G = ± 2%; J = ± 5%; K = ± 10%; M = ± 20%;
R33M = 0.33 Ω ± 20%; 6K8F = 6.8 kΩ ± 1%.

Capacitor values are indicated as:

0.68 pF	marked	p68	6.8 nf	marked	6n8
6.8 pf		6p8	1000 nF	1 μ0	
1000 pF		1n0	6.8 μF		6 μ8

Tolerance is indicated by letters as for resistors. Values up to 999 pF are marked in pF, from 1000 pf to 999 000 pF (= 999 nF) as nF (1000 pF = 1 nF) and from 1000 nF (= 1 μF) upwards as μF.
Some capacitors are marked with a code denoting the value in pF (first two figures) followed by a multiplier as a power of ten ($3 = 10^3$). Letters denote tolerance as for resistors but C = ± 0.25 pf. For example, 123J = 12 pF $\times 10^3 \pm 5\%$ = 12 000 pF (or 0.12 μF).

Appendix C: Greek symbols

Greek letters used as symbols in electronics

Greek letter	Capital (used for)	Small (used for)	
Alpha	–	α	(angle, temperature coefficient of resistance, current amplification factor for common-base transistor)
Beta	–	β	(current amplification factor for common-emitter transistor)
Delta	Δ (increment, mesh connection)	δ	(small increment)
Epsilon	–	ε	(permittivity)
Eta	–	η	(efficiency)
Theta	–	θ	(angle, temperature)
Lambda	–	λ	(wavelength)
Mu	–	μ	(permeability, amplification factor)
Pi	–	π	(circumference/diameter)
Rho	–	ρ	(resistivity)
Sigma	Σ (sum of)	σ	(conductivity)
Phi	Φ (magnetic flux)	ϕ	(angle, phase difference)
Psi	Ψ (electric flux)	–	
Omega	Ω (ohm)	ω	(solid angle, angular velocity, angular frequency)

Appendix D: Types of battery

ALKALINE PRIMARY CELLS

These cells and batteries offer very long service life compared with Leclanché types in equipment having high current drains. In addition, these cells have very low self-discharge currents and are completely sealed.

Available in sizes AAA, AA, C, D and PP3.

SILVER/MERCURIC OXIDE PRIMARY CELLS

These button cells are suitable for use in calculators, small tools, cameras, clocks, watches, etc. They may often be used as replacements for previously fitted alkaline manganese button cell types. Supplied in boxes of individual blister packs.

Available in six of the most popular sizes.

NI-CAD SINTERED CELLS

Applications where extreme ruggedness and/or high peak currents are required. In addition, these cells offer very long service life and can be electrically misused without damage.

Available in sizes N, AAA, AA, C, D and PP9.

NI-CAD HIGH-TEMPERATURE SINTERED CELLS

Primarily for use in emergency lighting installations, these cells and batteries are particularly suitable for charging and discharging at elevated temperatures. Other applications include alarm control panels and emergency and standby areas where higher ambient temperatures are experienced.

Available as single D cells and 3 × D cell battery packs.

NOTE: SINTERED CELLS

These have fairly high self-discharge currents and are therefore not suitable for equipment which has to be operational without recharging, after being left unattended for long periods of time.

NI-CAD MASS PLATE CELLS

Applications where small size and ruggedness are required. These cells have low self-discharge currents and are ideal in small portable equipment.

A range of sizes, including PP3, are available.

Appendix E: Epoxy-potted bridge rectifiers

These are bridge rectifiers packaged as a single block.

Voltage	Current (A)	Device no.
200	2	KBPC 102
400	2	KBPC 104
600	2	KBPC 106
800	2	KBPC 108
200	4	KBU 4D
800	4	KBU 4K
200	6	KBPC 802
800	6	KBPC 808
200	12	SKB 25/02
800	12	SKB 25/08
1200	12	SKB 25/12
50	25	KBPC 25005
200	25	KBPC 2502
600	25	KBPC 2506
200	35	KBPC 3502
600	35	KBPC 3506

Notes:

1. The bridge assembly should be mounted on a heat sink.
2. Current ratings are for resistive loads. When the rectifier is used on a battery or capacitive load the current rating should be multiplied by 0.8.

Appendix F: Transistors

Transistor pin connections

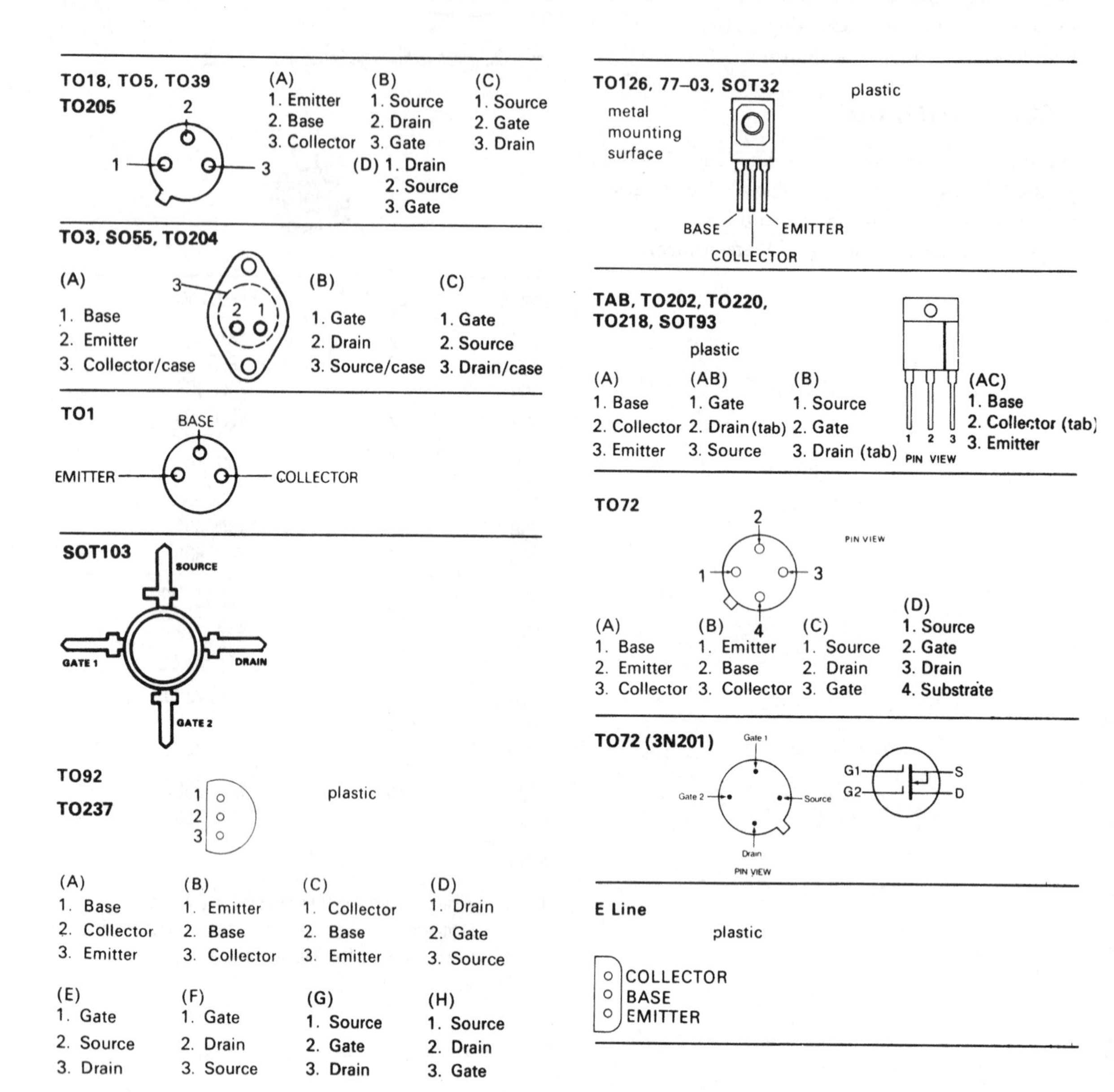

INDEX